ACKNOWLEDGMENTS

The material presented in this publication has been collected and developed within the framework of the European Concerted Action Programme on Daylighting. All those who have contributed to the project are gratefully acknowledged.

Coordinator:	A. FANCHIOTTI	(CIAM, Italy)

Main authors:

	E. BECCHI	(CIAM, Italy)
	E. BRESCIANI	(ICIE, Italy)
	P. CHAUVEL	(CSTB, France)
	H. COCH	(ETSAB, Spain)
	P. DE PASCALI	(ICIE, Italy)
	M. DE WIT	(TUE, Netherlands)
	W. DÖRING	(RWTH, Germany)
	M. FONTOYNONT	(ENTPE, France)
	M. GRANT	(ABACUS, United Kingdom)
	S. LOS	(IUAV, Italy)
	S. MATTEOLI	(TEP, Italy)
	T. MAVER	(ABACUS, United Kingdom)
	A. MONZANI	(CIAM, Italy)
	B. PAULE	(ENTPE, France)
	M. PERRAUDEAU	(CSTB, France)
	N. PULITZER	(Synergia, Italy)
	R. SERRA	(ETSAB, Spain)
	G. STOER	(Philips, Netherlands)
	A. TENNER	(Philips, Netherlands)
	M. VIO	(IUAV, Italy)
	G. WILLBOLD-LOHR	(RWTH, Germany)
	L. ZONNEVELDT	(TUE, Netherlands)

Experts of the
Commission:
A. LOHR (Germany)
J. PAGE (United Kingdom)

Programme Manager
at the Commission:
T. C. STEEMERS

Assistant editor:
L. CROSSBY

PARTICIPATING INSTITUTIONS

The following groups have participated in the book's research and contents:

ABACUS, University of Strathclyde, Glasgow, United Kingdom.
CAR, Cambridge Architectural Research, Cambridge, United Kingdom.
CIAM (Engineering Company), Modena, Italy.
CSTB, Centre Scientifique et Technique du Bâtiment, Nantes, France.
ENTPE-LASH, Laboratoire de Sciences de l'Habitat de l'Ecole Nationale des Travaux Publics de l'Etat, Vaulx-en-Velin, France.
ETSAB, Escuela Tecnica Superior de Arquitectura de Barcelona, Universitat Politécnica de Catalunya, Spain.
ICIE, Istituto Cooperativo per l'Innovazione, Rome, Italy.
IUAV, Istituto Universitario di Architettura, Venice, Italy.
MC, The Martin Centre, University of Cambridge, United Kingdom.
Philips International BV, Eindhoven, The Netherlands.
RWTH, Rheinisch-Westfälische Technische Hochschule Aachen, Lehrstuhlär Baukonstruktion II, Germany.
SIV, Societa Italiana Vetro, San Salvo, Italy
Synergia Progetti, Bassano del Grappa, Italy.
TEP (Software Company), Rome, Italy.
TUE, Technical University Eindhoven, The Netherlands.
University of Munich, Germany
University of Rome, Italy
University of Sevilla, Spain

The editors gratefully acknowledge the use of photographs and diagrams from the following sources:

The Courtauld Institute - figure 1.3; The Dulwich Picture Gallery - figure 1.16
D.Hawkes - figures 1.9, 1.11, 1.24, 1.39; William Heinemann Ltd. - figure 1.33
Judges Postcards Ltd - figure 1.2; Leicester University Press - figure 1.22
Pilkington Bros. PLC - figure 1.34; Sir John Soane's Museum - figures 1.30, 1.31;
A.Tombazis - figures 1.1, 1.38, 1.44. The Architectural Review - figures on pages 11.10, 11.11, 11.12 ;
A&U Publishing Ltd - page 11.15; CONPHOEBUS - page 11.13;
St Martins Press - page 11.8; Van Nostrand Reinhold - page11.14.

Commission of the European Communities
Directorate-General XII for Science Research and Development

DAYLIGHTING IN ARCHITECTURE

A European Reference Book

Edited by

N. BAKER   A. FANCHIOTTI   K. STEEMERS

Published for the Commission by: James & James (Science Publishers) Ltd

This publication has been prepared in The Third Solar R&D programme of the Commission of the European Communities, Directorate General XII for Science Research and Development, within the DAYLIGHTING Action co-ordinated by A. Fanchiotti, Università di Roma.

Publication arrangements have been made under the VALUE Programme (programme for the Dissemination and Utilisation of Community research results) within the Research Dissemination: Energy-Efficient Building project of the Commission of the European Communities DG XIII.

Publication No. EUR 15006 EN of the Commission of the European Communities, Scientific and Technical Communication Unit, Directorate General Telecommunications, Information and Innovation, Luxembourg.

© ECSC-EEC-EAEC 1993 Brussels and Luxembourg. First published in 1993. All rights reserved. No part of this publication may be reproduced in any form or by any means, without permission from the publisher.

Publication arranged by Energy Research Group, School of Architecture, University College, Dublin.

Graphic Design and Typesetting by Mike Baker, DesignBase.

Printed in Hong Kong by Shiny Offset Printing Company Ltd

Published by James & James (Science Publishers) Ltd, 5 Castle Road, London NW1 8PR UK. for the COMMISSION OF THE EUROPEAN COMMUNITIES

ISBN 1-873936-21-4

LEGAL NOTICE
Neither the Commission of the European Communities nor any person acting on behalf of the Commission is responsible for the use which might be made of the information contained within.

Edited by Baker N.V., The Martin Centre for Architectural and Urban Studies, University of Cambridge, Fanchiotti A., Università di Roma, 'La Sapienza', Dipartimento di Fiscia Tecnica, Steemers K.A., Cambridge Architectural Research Ltd, UK.

Cover Photo - Stansted Airport Terminal. Architect: Foster Associates, Photo: Martin Charles

PREFACE

Natural light has always played a dominant role in architecture, both to reveal the architecture of the building and to create a particular atmosphere, as well as to provide the occupants with visual comfort and functional illumination. The optimal use of daylight in buildings was, at the time of cheap energy, often seen as a superfluous design constraint. Illuminance deficiencies in the building were corrected with artificial lighting. The oil crisis and subsequent increase in energy prices, and now the even greater awareness of the impact of energy production on the global environment, has given an impetus to energy-conscious design.

With the growing interest in energy-conscious design in general and solar architecture in particular, the importance attached to energy use for artificial lighting in the non-domestic building sector has grown as well. It is estimated that about half of the energy used in non-domestic buildings goes to artificial lighting. Waste heat from luminaires in winter may contribute to heating, but in summer energy is often wasted getting rid of surplus heat from luminaires by means of air-conditioning systems. No wonder that daylighting has become, next to passive solar heating and passive cooling, a major topic in energy conscious design, and therefore, a major issue in the Commission's Solar Energy and Energy Conservation R&D Programmes.

In an emerging design technology such as daylighting, it seems prudent to start by assessing the state-of-the-art. To this end a team of 25 European experts, whose names are listed in the acknowledgments, have worked together to collect, select, evaluate and sometimes further develop the material from which they finally drafted the contents of this book. Their work represents a significant achievement.

Meanwhile, the Commission's effort to make progress with the development of the technology has continued. Research projects are under way in three domains :
- availability of daylighting data
- further development of daylighting design and control technology
- development of daylighting components.

However, the design of new buildings which make increased use of daylighting will not stand still pending the development of new strategies. This book provides designers with an essential tool; one which will be strengthened in the future as the results of current European research become available.

Theo C. Steemers
Commission of the European Communities

CONTENTS

Acknowledgments

Preface

Introduction

Chapter 1	**DAYLIGHTING EVOLUTION AND ANALYSIS**	**1.1**
	The Pre-industrial Period	1.1
	The Industrial Revolution	1.4
	Daylighting in Art Galleries	1.6
	Daylighting in UK Schools	1.9
	The Analytical Approach	1.12
	Design Tools	1.13
	The Post-fluorescent Era	1.15
	European Research and Development	1.16
Chapter 2	**LIGHT AND HUMAN REQUIREMENTS**	**2.1**
	Design Constraints	2.2
	Design Response	2.10
	Visual Comfort Requirements	2.12
Chapter 3	**DAYLIGHT DATA**	**3.1**
	Review of Sky Models	3.1
	Luminous Efficacy of Daylight	3.2
	Results of Measurements	3.3
	Sky Type Probabilities	3.9
	Luminous Distribution Algorithms	3.9
Chapter 4	**PHOTOMETRY OF MATERIALS**	**4.1**
	Surface Photometry Characterisation	4.1
	Selection of Appropriate Materials	4.7
	New Materials	4.15
Chapter 5	**DAYLIGHTING COMPONENTS**	**5.1**
	General Classification System	5.1
	The Basic Component: The Window	5.7
	Description and Performance	5.8
	Applications: Schools and Offices	5.26
	Experimental Analysis of Selected Components	5.33
	Conduction Component: Atrium	5.34
	Control Elements	5.49
	Prismatic Systems	5.57
	Holographic Optical Elements	5.60
	Recommendations for Control Elements	5.63
	General Checklist for Design	5.64

Chapter 6	**ELECTRIC LIGHTING**		**6.1**
	Lamps		6.1
	Control Gear		6.11
	Luminaires		6.11
	Luminaire mounting systems		6.16
Chapter 7	**CONTROL SYSTEMS**		**7.1**
	Controls for Artificial Lighting Systems		7.3
	Management Strategies		7.7
	Examples		7.12
Chapter 8	**LIGHT TRANSFER MODELS**		**8.1**
	Direct Illumination		8.2
	Reflection and Transmission		8.4
	Calculation Models		8.10
Chapter 9	**EVALUATION AND DESIGN TOOLS**		**9.1**
	Scale Models		9.1
	Review of Simplified Design Tools		9.3
	Review of Computer Codes		9.8
	Comparison and Validation		9.8
Chapter 10	**INTEGRATED ENERGY USE ANALYSIS**		**10.1**
	Example 1 : ESP		10.1
	Example 2 : HEATLUX		10.13
	Example 3 : The LT Method		10.19
	Future Directions		10.22
Chapter 11	**CASE STUDY ANALYSIS**		**11.1**
	The Architectural Design Process		11.1
	The Typological Grammar of Architecture		11.2
	Case Studies		11.3
	Methodology and Criteria for Classification		11.4
	The Morphological Box		11.4
	Selection and Classification of Daylit Buildings		11.5
Glossary			**GL.1**
Appendices			
	A	Sky Type Probability	A.1
	B	Daylight Availability	B.1
	C	Survey of Light Measuring Instruments	C.1
	D	Guide to Scale Models	D.1
	E	Survey of Control Systems	E.1
	F	Review of Design Tools	F.1
	G	Review of Computer Codes	G.1
	H	Survey of Artificial Skies	H.1
Index			**IN.1**

INTRODUCTION

The Scope of this Book

This European Reference Book on Daylighting is the first concrete achievement of the Commission of the European Communities R & D programme on daylighting. It attempts to bring together, on a European-wide basis, existing knowledge on systematic daylighting design. It includes discussion on the interrelationships between daylight design and artificial lighting design, and the significance of their controlled interactions for the saving of energy. Few works can claim to be comprehensive and at the same time balanced. This reference book is the result of contributions from researchers across Europe and inevitably represents the particular interests and viewpoints of individual research groups. Thus it does not set out to be a comprehensive design handbook, rather a review of current European research, which has an overall aim of serving the design community.

The energy economics of daylighting are inextricably mixed with the energy economics of the associated artificial lighting systems and their controls. The economics of daylighting have an energy aspect and a productivity aspect. Good daylighting of work spaces helps promote efficient productive working, and simultaneously increases the sense of well being. However, energy conservation should never become the sole concern of daylighting design to the exclusion of perceptual considerations.

Daylight indoors is provided for people. Daylight design therefore has to respect their visual perceptual needs. Designers need to understand what conditions enable people to see well and comfortably. Human visual comfort, considered from the daylighting point of view, is multidimensional. It is not enough simply to provide the appropriate illumination levels. Direct and reflected glare must be controlled. Patterns of contrast must be appropriate. There is therefore a current need to find better ways of integrating perceptual, physical scientific and engineering approaches, taking proper account of the new advances in materials relevant to daylighting design.

Architectural Aspects

While this book is mainly about the science of daylighting design, great importance should be placed on the architectural quality resulting from visually exciting daylighting design. The cultural traditions involving daylighting design in architecture have always been very important in Europe, and we must make sure the situation remains the same. We also have to bear in mind, when considering indoor functional aspects of daylighting design, that buildings are seen both from the outside and from the inside. The daylighting design exerts a big impact on the external appearance of buildings.

Daylighting design also throws up many town planning issues. Architects therefore have a key role in sustaining and developing the cultural aspects of daylighting. Only some of the architectural issues needing to be addressed have been explored. The importance of daylighting morphologies in relation to practical design has been highlighted. Advances in computer science, and daylight modelling (some recent European advances are described in this book) are enabling architects to "see" on the computer screen for themselves, the modelled quantitative results of their daylighting decisions, expressed in perceptual terms. These advances should help span the gap between building science, illuminating engineering and architectural design.

Historical Background

Organised knowledge about good daylighting practice for buildings has a very long history in Europe, and the Romans were pioneers. For example, good daylighting practice is discussed in the classical writings of Vitruvius. Above all he stressed the importance of properly considering window orientation. What was written in Roman times still has a considerable relevance for today, but we now have at our disposal a far wider range of glazing materials with which to tackle daylighting tasks.

While the actual techniques used for summer cooling have altered, the challenge of providing good daylighting without excessive solar gains in the overheated season remains. In surviving Roman structures still in use, like the Pantheon in Rome, and the excavated residential buildings at Pompeii, we can still today perceive the nature of historic daylighting solutions, which differed in the religious and domestic context.

The Romans too provided the first legal structure for safeguarding rights of light in existing properties against unacceptably adverse adjacent developments. Their practices anticipated the complex town planning requirements we need today to safeguard daylighting and sunlighting standards in contemporary urban development.

Before effective artificial lighting became available, it was particularly important to get the daylighting design right. In northern Europe, shortage of daylight, especially in winter, made it necessary, when glass became reasonably affordable, to provide relatively large windows, and to secure good daylighting penetration by use of high ceilings and open plan forms. In the southern countries of Europe, the dominating need to control summer overheating in conjunction with more adequate winter daylight, led to very different window designs, and the use of very different plan forms. The courtyard plan was found to provide very amenable solutions. The sunlight was interreflected into buildings from appropriately placed external surfaces, rather than allowed to penetrate directly.

Vegetation covering the courtyard was often used to regulate and soften summer daylighting, which is otherwise harsh. It also helped to control overheating. In winter, the leaves dropped off, giving more usable daylight and useful passive solar heating gains. An integration of thermal microclimate design and daylight design was achieved. Southern Europe also produced specific daylight inventions, like the Venetian blind, to help in the processes of daylight and thermal control.

Daylighting in the 20th Century

The development of styles of architecture in the 20th century, that were neither very environmentally conscious nor very energy conscious, tended to overlay the earlier European traditions of daylight design, which previously had reflected year round needs in climatically sensitive ways. The availability of cheap fluorescent electric lighting tended to accelerate the neglect of daylighting design. This led on to an architectural impoverishment of our European cultural tradition of daylighting design, embodied in great traditional styles based on the interaction of form and daylight; for example, our Gothic cathedrals, and above all, in the masterly use of daylight in our Baroque buildings. Many buildings of the period between 1945 and 1975 especially must be judged as failures in daylighting design terms, when compared with historic solutions.

Fortunately many of the most successful modern European architects have always resisted the contemporary tendency to ignore the visual richness offered by the creative use of daylight in their buildings, but even the most perceptive still have to assimilate new opportunities, for example new glazing materials like reflective glasses, prismatic glass systems and so on. They have to resolve successfully any inherent difficulties and avoid the pitfalls, like excessive heat gain in summer. This is easier said than done. There are always latent risks in intuitive design, especially in situations involving innovation. A combination of art and science is needed. This book points the way towards these sounder architectural approaches. While it draws on the past, it also points towards the new opportunities presented by scientific advances, for example, the development of new "intelligent" window systems, like electrochromically "smart" windows whose transmission properties can be controlled electrically. It will be necessary to develop new daylighting design, and artificial lighting control procedures to deal with such developments in new materials. Architects will also need to learn how to handle the impacts of these novel glazing materials on the external appearance of their buildings.

Green Issues

Daylight, by displacing electric light use, reduces carbon dioxide emission and, in turn, the greenhouse effect. The European nations are committed to the control of global warming, and impacts on the ozone layer, as well as policies for saving energy *per se*. Thus energy savings achieved by better daylight design are doubly important. We must therefore regard better daylighting design of buildings as playing an ecological role, in addition to its other contributions, like saving energy, improved work performance and increased human well being. Additionally we have to study, in more detail, how improved daylighting design can reduce air conditioning cooling loads, and so reduce the air conditioning plant sizes, and hence the volume of CFCs associated with them, in addition to making substantial savings in lighting and cooling energy consumption.

Systems and Components

While we have reasonable knowledge of the daylighting properties of simple unshaded windows, this book fills the gap in our knowledge of more complex apertures or "daylighting pass through elements" as they are called. The opening is often not a simple hole containing glass in a thin wall. There are the various blinds, overhangs, side fins etc used in practice to control daylighting and overheating. Some of these devices are fixed. Others are moveable and controlled, sometimes by hand, sometimes automatically. These "devices" are often used in combination, and linked in with glazing materials with special transmission characteristics, for example, reflective glazing. Complex components are a very typical feature of window design solutions for the hotter European climates. The courtyard itself can be thought of as a sort of external component.

There is therefore a need to think in terms of "system daylighting performance" of building components, and this book has attempted to systematically classify and characterise building daylighting components. This is of greatest importance in the case of innovative and complex systems.

For example one of the disadvantages of daylighting in rooms with vertical clear windows in one wall alone, is the sharp drop in illumination as one moves back into the room. Advances are now being made by changing the light distribution characteristics of window components and the glazing materials incorporated within them. Such changes can help achieve a better internal distribution of daylight in a side lit room. Redistribution can help reduce the need for artificial light in the deeper parts of the building.

Daylighting devices like light shelves can protect the occupants close to the windows from the direct rays of the sun and, at the same time, make some of that sunlight available at the back of the room for daylighting purposes, by interreflection between the top surface of the light shelf and the ceiling. More advanced optical systems can provide daylighting at greater depths in building, so helping to reduce cooling loads due to day-time electric lighting use. In the past, quantitative design methods for such systems were lacking. This book contains new research based studies to help improve design of such systems.

Knowledge of the lighting properties of materials, especially glazing materials, is very important for daylighting design. New glazing materials, like holographic films and prismatic systems, are emerging. Work is proceeding on various kinds of "smart windows" with controllable transmission properties. We need reasonably precise knowledge of their optical properties to decide on appropriate designs.

This book presents work on simple methods for characterising the lighting properties of materials, important in daylight illumination studies. Additionally basic design methodologies have been enlarged. New studies, based on modelling, are presented for assessing the performance of window light shelves and for considering the daylighting of atria, an increasingly important field in view of the popularity of atria solutions in contemporary European architecture.

Human Factors and Controls

There is a growing awareness in wider fields of building science that it is crucial to take account of the effect of occupants, when considering the performance of buildings. For example the shading of windows interferes with daylighting indoors. Adjustable shading devices are sometimes operated for visual control reasons and sometimes for thermal control reasons, and sometimes for both. When blinds are lowered in hot weather, the daylight is often reduced to such an extent that the electric lights are put on. So, in buildings with shading systems, the energy use of artificial lighting is linked to outdoor lighting levels in a complex way. The problem has a behavioural aspect, as the building occupants are often the controllers of adjustable shading systems.

Comfort aspects of the control problem are certainly very important, and there is considerable evidence that, where automatic shading control has been based on an over-simplistic approach, considering energy factors alone and ignoring human comfort, the results have not been very acceptable. It is no use having an energy efficient building filled with unhappy workers. Only some of these control issues have been considered in this book, and there is still a long way yet to go.

Educating Architects and Engineers

Many architects all over Europe are making decisions without expertise or support, which will crucially effect the daylighting performance of buildings. Planners too are shaping our cities without a sound understanding of the implications of plan, section and site layout for the environmental performance of the urban tissue. Fundamental building characteristics such as these dictate the environmental performance for the building's lifetime.

One of the major obstacles to progress is the present lack of knowledge on the part of lighting engineers of the principles of good daylighting design. This makes it difficult for them to work effectively with architects on the new approaches which integrate, through the use of controls, natural light and electric light. Too easily, in this situation of ignorance, engineering design slips back to the adoption of simple artificial lighting solutions totally unrelated to daylighting needs.

In the final analysis, progress will be determined by the quality of the daylight design education of designers and planners, and by the availability of appropriate design tools, and local climatic information in forms helpful for practical design and town planning control. Simple graphic tools relating to norms and standards are valuable and necessary, especially for the early stages of design. With progress in information technology, wider use is being made of computer based design tools of different levels of complexity. This book has attempted to review current practice in both these areas.

Daylighting in Architecture, A European Reference Book

Daylighting in Architecture is aimed at the architect and engineer who wants to acquire an in-depth understanding of the principles of daylighting design. The contents of this reference book are also a good indication of the current scientific and design support work of the European countries. After an introduction which briefly reviews the historical development of daylighting design, "Light and Human Requirements" sets out the physiological and psychological background to the visual process in relation to design considerations. "Daylight Data" considers the sky as a light source, describing new statistical approaches to describing real sky luminance distribution. This approach reflects the growing use of computer based tools for daylight calculation and simulation, and represents a major step forward from the average and standard (CIE) sky.

"Photometry of Materials" deals with precise photometric descriptions of material surfaces, an important issue in modelling reflected light both quantitatively and for image simulation by extended CAD graphics techniques, as typified by the Genelux model. New transparent materials such as holographic films, optical fibres and aerogels are also described in this chapter.

A systematic taxonomy of window systems is given in "Daylighting Components". Typologies under the two building types, schools and offices, are presented and the role of the atrium in daylighting is dealt with in some detail. The application of materials such as holographic film and components such as light shelves is also described.

The energy saving of daylight use is directly related to the artificial lighting that it displaces. The degree to which artificial light is displaced is not only dependent upon the availability of daylight but also on the switching and control system. These topics are dealt with in the two chapters "Electric Lighting" and "Control Systems".

"Light Transfer Models" presents the background to the mathematical description daylight transfer from the sky to the room. Direct illumination, reflection and transmission are described and calculation models, including ray-tracing and simplified analytical methods, are reviewed.

A major role of this reference book is to provide design guidance. "Evaluation and Design Tools" provides a review of techniques, simplified tools and computer codes, including physical models tested in artificial skies.

The interaction of daylight and its displacement of artificial light, and other energy uses for heating, cooling and ventilation is investigated using the simulation model ESP and is described in "Integrated Energy Use Analysis". A series of case studies in a very concentrated format are described in "Case Study Analysis". This chapter presents a systematic approach to morphological analysis - the morphological box - which enables the daylighting of a building to be described in terms of a few parameters.

Extensive appendices provide data on sky types and illuminances, photometric instruments, surveys of design tools, computer codes, and lighting control systems, and a directory giving the location of artificial sky facilities in Europe. Finally a glossary gives definitions of descriptive words and photometric terms and parameters.

Conclusions

The various discussions, which took place on the way to producing this book have proved extremely fruitful. These have produced some consensus about what we know already, and what we need to know for the future. The discussions have tended to throw up the complexities of the issues, and the dangers of taking too narrow an approach to the daylighting design of window systems. Much has been achieved, but all the contributors realise how much more there is still to be done before we can achieve a fully scientifically-validated European-wide approach to daylighting design based on a combination of climate, human comfort, human visual performance and energy analysis.

Chapter 1
DAYLIGHTING EVOLUTION AND ANALYSIS

DAYLIGHTING AND BUILDINGS

Vision is by far the most developed of all our senses and throughout our cultural development light has been the main prerequisite for sensing our own world, and that of others - through painting, sculpture and literature.

The building of edifices for a whole range of functions, from simple shelter (which has enabled us to colonise almost all of the land surface of the globe) to ritual and ceremonial worship, and as instruments of state and government, has surely been the most significant of human activities and expressions of culture. It is not surprising then that the provision of light in buildings has been a fundamental concern of builders and designers - as fundamental as the building's ability to modify climate or its structural stability.

Although we see daylight as the fundamental light source, it is interesting to note that artificial light sources have been present, in the form of fire, for as long as, if not longer than, the most primitive of shelters. However, up to most recent times man-made light was considered only as an alternative to the darkness of night-time. Now, quite suddenly in the last half century, the majority of the population of the developed world spend most of their working day, in schools, factories and offices, in artificial light. And much of their leisure, at home, in sports halls, shops and restaurants, also no longer relies upon daylight.

Recently, two issues have given us all cause for concern. Firstly the growing realisation that the energy use involved in the provision of artificial lighting contributes significantly to global environmental pollution, and secondly that the deprivation of daylight may have detrimental physiological and psychological effects on the occupants of buildings. These issues, together with architectural and aesthetic questions, make up the case for daylighting.

But has designing for daylight become a forgotten art? Can the skills of daylighting design be identified and disseminated, and what does science offer in support? These are the issues explored in the following chapters of this reference book.

Figure 1 - Modern mastery of daylight. Notre Dame de Haut, Ronchamp, France (Architect: Le Corbusier, 1954).

THE PRE-INDUSTRIAL PERIOD

Pre-industrial man had very different requirements from those of today - most of his time was spent outside tending the crops and animals. The earliest shelters were too primitive for any concern with the admission of light, the priority being protection from the life-threatening elements. Indoor domestic activities were simple and did not demand good lighting. It is in ceremonial and religious buildings that we see the earliest manifestation of conscious design for daylight.

The invention of glass had a crucial influence on the evolution of the window. Glass enabled the light and view-providing function to be maintained while insulating the interior, to some extent, from the external climate. This was of greater significance in the cooler climates of northern Europe than in southern Europe, where the presence of daylight more frequently coincided with a comfortable outside temperature. Nevertheless it was, as with many significant technical achievements, the Romans who first explored the thermal (as distinct from decorative) benefits of glazing their windows.

In medieval domestic architecture, the window pre-dates the use of glass but was equally important as a ventilation device to allow the smoke from the unflued fire to escape. Indeed the word "window" in the English language derives from "wind eye", and it is interesting that some of the earliest rules of orientation related to the sources of healthy and bad air, with particular concern for the plague, and not the sun. These openings were usually shuttered with internal hinged timber panels or surprisingly sophisticated sliding panels. The window opening was formed with vertical mullions which provided structural support, and limited the aperture size for security purposes.

Figure 2 - Late 16th Century farmhouse from Pendean, West Sussex, at the Weald and Downland Open Air Museum, Singleton. Indoor tasks made modest demands on daylighting - the small unglazed windows also provided ventilation.

The unglazed window, and the generally poorly fitting doors and shutters, led to a high infiltration rate. This was probably fortunate before the development of the flued fireplace, keeping pollution levels bearable. Internal temperatures would follow the external temperature quite closely and comfort was attained mainly by the radiant environment provided by the fire. This in turn gave little incentive to incorporate thermal insulation; glazing was needed to provide the crucial containment of internal air.

However, it was in sacred buildings that daylight first played the major role. The admission of light through the massive elements of the fabric carried symbolism which was irresistible to the architects of the great churches and cathedrals.

It was in sacred buildings too that glass made the earliest impact; the use of stained glass in the great cathedrals was widespread by the 12th century. The technique was pioneered in France and Germany and artisans from these countries often travelled Europe-wide to trade their wares and sell their skills. It is difficult to overstate the impact that a sunlit stained glass window, of majestic proportions, would have had on an illiterate peasant. Nowhere else would he

Figure 3 - 16th Century stained glass in King's College Chapel, Cambridge. A Flemish glazier, using French glass, for an English King, - an early European venture.

see such colour, such brilliance and such richness of imagery. For the task of promoting religious beliefs, it was "state-of-the-art" media technology, remaining unchallenged for its visual impact until the cinema nine centuries later.

Figure 4 - Church in Mdina, Malta, illustrates the ultimate in Baroque daylighting design.

Gradually glass manufacture became more established and costs came down. The land-owning classes were the first to adopt glazing, as violence and strife subsided, and the fortification of their dwellings became of lower priority. Windows began to be larger, and by the 16th century we see the secular celebration of the window, as typified by the oriel window in the Cambridge hall.

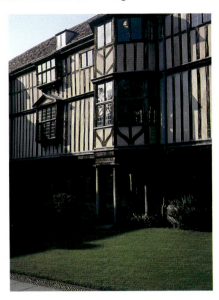

Figure 5 - Oriel window in the Master's Lodge at Queens' College, Cambridge (16th Century). The elaboration of the window indicated the importance of the people who dined adjacent to it.

Even before industrialisation, there were certain indoor activities that made real technical demands on daylighting. Writing, printing and painting all would have needed good light and, with only primitive artificial lighting available, there would have been a heavy reliance on daylight.

Figure 6 - Corpus Christi, Oxford, College Library 1604. The planning provides illumination for the bookstacks and the reading desks.

Figure 7- Detail from Vermeer's "The Music Lesson" (1662) showing the remarkably realistic rendering of the daylighting from the studio window.

It was in spinning and weaving that we see most impact on building design. The need for light would have meant that production was crucially dependent upon prolonging the availability of daylight to a maximum. It must be realised that in the 15th Century the real cost of artificial light (in relation to the cost of living) was about 6000 times (per lumen) more than today.

Figure 8- Clerestory windows at the Guild Hall in Lavenham, UK, (c.1520). The timber-framed structure allowed large window areas, typical of many buildings in this region, famous for spinning and weaving.

Despite a steady increase in the use of glass in the 17th Century, mainly in the form of small leaded panes, total glazing areas remained small by modern standards. This was partly due to structural limitations, especially in masonry buildings, although in grand architecture amazing structural feats where achieved with stone and even moulded

Figure 9 - Hardwick Hall (Architect: Smythson, 1597) "More glass than wall", built in Derbyshire, UK, for the Countess of Shrewsbury. The glass was made using sand and timber from the estate, by glass makers from the continent.

Figure 10.- Casting plate glass in the 18th Century. There then followed the laborious task of grinding and polishing. Although mechanised, this remained in principle the only way to make large sheets of perfect optical quality, until the invention of the Float process, by Pilkington in 1959. The floating of the molten glass on a pool of molten tin ensured that the surface set optically smooth and flat without polishing.

brick mullions. It was also due to method of manufacture, the crown glass technique, which produced panes of small size. Obviously the high cost of glass was also a constraint - the cottage homes of the rural peasant would still have their windows protected by oiled cloth or parchment.

From an early age, institutional interventions such as Regulations and Taxes have influenced the design of buildings. Both the Netherlands and Britain have had taxes which have directly affected window design.

In Britain a window tax, introduced in 1697, paradoxically led to an increase in window size. This was due to the tax being applied to the number of windows rather than the total glass area. The tax had sufficient financial impact to prompt building owners to brick-up windows, and even to erect new buildings with bricked-up, arched recesses, awaiting the time when either the abolition of the tax, or the increased wealth of the owner, would permit glazing to be installed. Some grew tired of waiting almost two centuries for the tax to be dropped, and painted fake windows in the recesses in deference to architectural composition. In the Netherlands a tax on street frontage led to the Flemish narrow gable-fronted house. The relatively narrow span between structural cross walls freed the gables from structural function allowing very large areas of glazing, providing light deep into the building. It is interesting to speculate how this feature of large windows has persisted, even in modern Dutch domestic architecture, even though the technical origins no longer apply.

During this pre-industrial period steady progress had been made in the technology of windows, both in the framing and metal work and in the increasing size and quality of glass panes. At the beginning of this period the sash window was introduced, probably originating in the Netherlands.

THE INDUSTRIAL REVOLUTION

As in many other aspects of building, it was the Industrial Revolution that brought the most rapid changes in both the requirements and the solutions for daylighting. Firstly, the technology made great strides forward. New techniques for glass production were devised; cylindrical blown glass enabled relatively large sheets of moderately good optical quality to be made, without resorting to the expensive process of casting and polishing. Not only did glass become relatively cheaper, and available in larger panes, but major improvements were made in the framing technology. This largely stemmed from the use of iron for glazing bars, together with cast iron trusses and columns. A whole new architecture of light and air was born.

For more than two centuries, horticulture had made its own particular demand on daylighting. Before the days of fast or refrigerated transport, fruits and vegetables had to be grown on site. Social status was enhanced by the presence of table exotica such as the banana, peach and orange. The tender nature of these fruits made them quite unsuitable to the northern European climate unless protected by

Figure 11 - The application of wrought iron technology to glazed structures reduced the thickness of the bars allowing more light to enter. This had obvious benefits in horticultural applications. Iron also offered great architectural opportunities with the ability to form curved surfaces. The Palm House at Kew (Architect: Richard Turner, 1848).

glass. A whole science of glass horticulture developed, and it was here that we see the most significant advances in the understanding of the optical and thermal properties of windows.

This line of development expanded to include different building types - railway stations, libraries and shopping arcades, culminating in the great glass and steel engine sheds of the newly emerging railways, and the vast exhibition halls, representing the ultimate in glass, iron and pre-fabrication technology.

Figure 12 - State-of-the-art glazing technology was applied to 19th Century galleria in many European cities. Galleria St Hubert, Brussels. (Architect: Cluysenaar).

But the Industrial Revolution had another impact. As the rural population flooded into the cities to work in mills, factories and workshops, a much greater demand for daylighting was created, for now a large proportion of the population were working indoors. Considerable advances had been made in artificial light, most notable being the invention of the incandescent gas mantle by Welsbach in 1885. But despite advances in artificial light, it still remained far too costly to use in preference to daylight. The availability of electric light towards the end of the 19th Century reduced the pollution of the indoor environment, and reduced the risk of fire, but did little to reduce the cost.

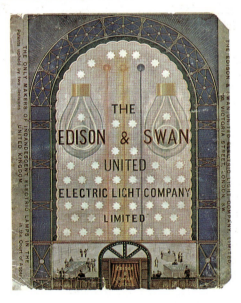

Figure 13 - The invention of the electric lamp seemed to be a quantum leap forward - it provided instant light without the pollution and danger of gas or oil. But it was not cheap, a lumen of electric light in 1880 cost in real terms 600 times more than today.

The requirement for daylighting was a form-giver to both plan and section. Plans of multi-storey buildings had to be narrow (by contemporary standards), with room depth limited to about twice the floor to ceiling height, to allow side lighting. Spaces which demanded a deep plan, such as the manufacturing workshops, mills and printing works, had to be single storey with lighting from above. A whole vocabulary of industrial rooflighting developed.

Where a multi-storey building required a deep floor plan, the section was pierced with lightwells which in some cases became of such proportions that they could be seen as the precursor of the modern atrium form.

1.6 Daylighting in Architecture

Figure 14 - Industrial roofs of the 19th and early 20th Centuries were dominated by the need to provide daylight. In this example the saw-tooth profile provided north-facing glazing, and was integrated with the structural design.

Figure 15 - Large deep buildings needed a new approach to daylighting. Frank Lloyd Wright's Larkin Building in Buffalo, USA, adopted an open top-lit atrium form which provided daylight to the interior.

As with so many advances in the man-made world, we can identify the 19th century as the time when demands were both made and met for daylighting design. The same Industrial Revolution which produced affordable glass and glazing systems also produced large buildings to house both the industry itself and the commerce which grew from it.

It is interesting to trace the development of particular building types, from this period to today, to illustrate the emerging skills and scientific knowledge and, more importantly, their impact on building design. In particular the art gallery combines the need for technical performance in terms of daylighting, with a strong architectural message appropriate for a public cultural building.

DAYLIGHTING IN ART GALLERIES

The history of the art gallery as a specific building type is a fairly recent one, beginning some two hundred years ago with the search for an architectural expression of the highly specialised functions of protecting and displaying art.

The earliest purpose-designed display spaces for art were in the palaces of the Renaissance. Extensions to many European palaces took the form of enfiladed side-lit rooms or galleries. Paintings were usually hung on the walls tier upon tier creating a continuous tapestry of art. Individual paintings would be taken down and studied in more detail in better light. Smaller paintings, or ones with great detail, were hung near eye level, whereas the larger tableaux were positioned high up the wall.

The use of top lighting in rooms designed for the specific purpose of displaying works of art did not become commonplace until later, although this form of lighting offered several advantages. First, light would be more evenly distributed over all the walls; second, more wall space would be available, and third, viewing conditions would be improved by reducing glare and limiting veiling reflections from windows.

The end of the 18th Century saw the transfer of major art collections from private to public hands throughout Europe. One of the first independent public art galleries to be built is Sir John Soane's Dulwich Gallery. The plan is based on the model of the private galleries and consists of five enfiladed rooms. The use of lantern lights is however original and very effective, and the building is considered, up to this day, to provide one of the most satisfying display environments.

Figure 16 - The Dulwich Picture Gallery (Architect: Sir John Soane, 1814) is an early public art gallery where daylight enters through lantern rooflights onto the walls of the enfiladed rooms.

Before electric lighting, the main environmental concern of gallery designers was the effective distribution of natural light. Lighting in art galleries progressed from simple side lighting to clerestory windows, rooflights and lanterns. Soane was a great experimenter with the manipulation of natural light, highlighted by the use of a wide variety of rooflight configurations and mirrors in his own house in London.

Figure 17 - The view over the back of the Sir John Soane Museum (1812-13) reveals a variety of skylights and suggests that the architect's home could be considered as the scientist's laboratory: an experimental, creative research station for the investigation of daylight.

Pollution, dust and moisture brought into the gallery by increasing numbers of visitors were thought to be damaging to paintings. This often resulted in the use of glass to cover art work for protection. Unfortunately this had the repercussion of making reflections more disturbing to the viewer. A series of geometric constructions for the gallery section were proposed, attempting to alleviate the visual deficiencies that existing galleries tended to suffer from.

It was as early as 1850 that the first theories on good gallery daylighting were developed by F. P. Cockerell. His main concern was to light the picture well while reducing glare from the skylight, and reflections of the spectators in the picture glasses. He proposed the use of a 'velarium', or opaque screen, over the central area of a gallery. Weissman in 1895 developed Cockerell's construction further in positioning of skylights, reminiscent at least strategically of Soane's design at Dulwich. Seager in 1912 continued the refinement of Cockerell's principles. Research at the UK National Physical Laboratory led to the configuration used in the Duveen Wing extension of the National Portrait Gallery, London.

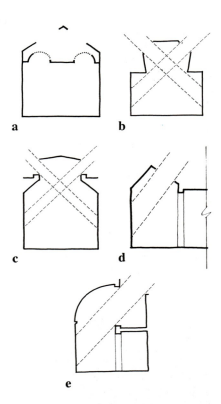

Figure 18 - Investigations of the geometry of gallery sections were carried out in an attempt to explain and improve daylighting. Cockerell in 1850 (a) and Weissman in 1895 (b) established the principle of directing light onto walls. Soane at Dulwich in 1814 (c) had achieved the desired effect intuitively, although the original opaque lantern roof was later glazed. Seager in 1912 (d) developed Cockerell's section, which, reinterpreted by the National Physical Laboratory, was used for the Duveen Wing of the National Portrait Gallery, London (1927) (e).

It was not until the "Conservation Movement", which emerged during the middle of the 20th Century, that it was established how important lighting control was in preventing the deterioration of artworks, and this was to have significant repercussions on the architecture of art galleries.

The two main functions of an art gallery are to display and to protect. In terms of the lighting environment these two aims contradict one another. On the one hand effective display requires a sufficient amount of light to pick out details. On the other hand, the preservation of art demands a minimum amount of exposure, to limit deterioration. Initial detailed research was concentrated on the damaging effects of light, the most important agent of deterioration in art galleries. Research concluded that ultra-violet light was responsible for a great part of the deterioration of organic materials. Furthermore, fixed illumination levels were proposed according to the sensitivity of artwork.

The art galleries that were constructed directly after the conservation movement of the 1950s, are often characterised by total exclusion of natural light. The argument was that by creating an artificial environment the processes of deterioration could be closely controlled. Furthermore, the exhibition designers could have full control over the lighting effects and arrangement of the exhibits, thus offering greater flexibility. This environmentally exclusive approach was particularly prominent in the United States during the 50s and 60s often expressing the protection of art in the "strong box" aesthetic.

The other option was to use daylight in a highly controlled manner to achieve the lighting standards, by the use of adjustable louvres, velaria or prismatic glass. Many galleries have been constructed using this approach but have come under criticism from people in the museum profession and the general public. The main complaint has been that the continuous control of light to a fixed level creates a monotonous environment, with little relief, which will result in museum fatigue. Furthermore the quality of light under the diffusing, translucent ceiling system is unsatisfactory, being likened to an underwater environment.

The segregation in art museums of a mechanistic zone for viewing, from a poetic, exuberant arrival and circulation area, is one manifestation of the dichotomous environmental requirements. Brawne has termed these two parts the "store" and the "temple". An example of this approach is Stirling's Staatsgalerie in Stuttgart. The exhibition spaces are highly controlled environments and the process of entering is more abstract and playful - in short, the difference between the two parts of an art gallery has become more pronounced.

While many new art galleries of the 1960s adopted an environmentally exclusive or mechanised approach, due to concerns about deterioration, to the detriment of the museum-going experience as a whole, other architects were concentrating on issues of display. Scarpa and Albini were designers involved with art gallery design during this period. Scarpa, at Verona's Castlevecchio, respected the existing daylighting in the building, and clearly articulated his modern interventions. The use of "easels" or pivoting display screens allowed the positioning and manipulation of the objects to take advantage of the available daylight.

The poetic settings of these galleries are very appealing, but pressures to preserve art for future generations demand a greater degree of environmental control than they have been able to offer.

A recent example of this integration of display and conservation is Gasson's Burrell Gallery near Glasgow. Here the wooded park provides the background and a setting for the art, reminiscent of the Louisiana Gallery near Copenhagen, Denmark.

Figure 19 - The Staatsgalerie at Stuttgart (Architects: Stirling and Wilford) demonstrates a recent approach to museum design. The public front, entrance and circulation areas are architecturally expressive, while the gallery spaces are restrained and have a closely controlled lighting environment.

Sensitive works of art, such as tapestries and watercolours, are placed at the heart of the building, whereas sculptures for example are positioned near the edge and are bathed in side light. Simple rooflights provide natural illumination along "avenues" and control is provided by blinds. This integrated response, where the nature of the artefact in relation to the need for differing degrees of environmental control and differing display settings is exploited, is at the root of the building's design and success.

There is growing interest in simpler daylighting controls, limiting annual average exposure rather than ensuring fixed illuminance limits. This can lead to energy efficiency and a greater architectural clarity.

The manipulation of light goes to the heart of the architectural enterprise, but in art galleries this is tempered by practical concerns for conservation. A

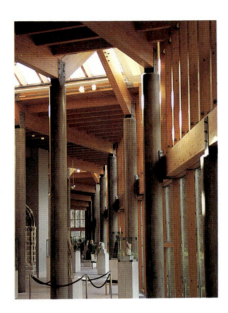

Figure 20 - A view across the north facing glazed wall of the Burrell Gallery near Glasgow (Architect: Gasson) shows the use of sidelighting for sculptures. More sensitive artwork is placed further into the centre of the plan where the environment is more closely controlled.

Figure 21- Lack of planning constraints resulted in buildings packed closely together, as typified by these streets in Paris at the end of the 19th Century. These squalid conditions eventually prompted the adoption of planning laws which explicitly protected rights to daylight and sunlight.

greater understanding of the principles involved in deterioration due to light offers a fresh insight into the extent and form of control that is necessary. New architectural opportunities arise that allow a balanced approach and a greater freedom in daylighting design.

DAYLIGHTING IN UK SCHOOLS

In Britain, prior to the Education Act of 1870, schools provided for the children of the wealthy had been well funded. The modest densities of occupation did not stretch the environmental design skills of the architect unduly, and the model lay somewhere between the country house and the college or seminary. The provision of the Act to make education compulsory for all suddenly created a huge demand for school buildings, and for them to accomodate much larger numbers of children within more constrained budgets.

This in turn created the need to heat, light and ventilate large classrooms, at very low running cost. This requirement was interpreted in almost moralistic terms. The provision of light and air in schools (noticeably absent in the squalid housing prevalent in most of the cities, and in many of the workplaces) became a driving force in the evolution of school design, right up to the present day.

Early progress in daylighting design is illustrated in a book on "School Architecture" written by E. R. Robson, Architect to the London School Board, in 1874. In response to concern for eye health and the avoidance of myopia, he states that "*lighting from the side, especially the left side, is of such great importance as properly to have a material influence over our plans*". He goes on to give rules of thumb - "*a classroom is only well lit when it has 30 square inches of glass to every square foot of floor space*" (about 20% glass to floor area). We must interpret this rule bearing in mind the dismal condition of the urban atmosphere, heavily polluted by coal fires, and the fact that artificial light was still an expensive luxury.

But he was also aware of the issue of visual comfort. In discussing the desirability of sunlight in the classroom he writes - "*It is well known that the rays of the sun have a beneficial influence on the air in the room, tending to promote ventilation, and are to a young child very much what they are to a flower. Acting on this known fact, the builders of some schools have sought to secure as much sun as possible, and produced results of light and glare painful in hot summer weather, either to teachers or pupils or both*". He goes on to give guidance on the appropriate orientation of windows.

The Victorian schools in Britain made a determined move towards deep plans. Light for the central hall was "borrowed" light over the top of the tall classrooms located in the perimeter zone.

This compact planning suited the crowded urban site, and the educational objective of centralised control through head teacher, teachers and assistants.

Figure 22 - Windows for school buildings demonstrate advanced technology of the period, providing daylight and a fine degree of ventilation control. From E. R. Robson.

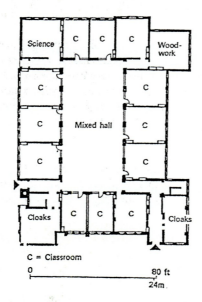

Figure 23 - Deep plan London Board School, late 19th Century. The central hall was lit by "borrowed light" across the perimeter classrooms.

However, in spite of ingenious ventilation devices, using heating coils, roof-mounted cupolas and later, in some cases, fans, air quality was not good. By the early 20th Century, a new plan form was evolving.

The greater space afforded by suburban sites, where most new school building was taking place, no longer necessitated the compact form. The new plan was essentially a row of classrooms connected to the hall and specialist rooms by open corridors or verandahs. This gave opportunity for almost unlimited light and cross-ventilation, and satisfied a renewed obsession with light and air.

In spite of their large area of glazing, these schools did not suffer excessively from overheating since the fabric was massive, and the generous room height permitted useful stratification and ventilation. Moreover the large height-to-depth ratio of the rooms resulted in an acceptable daylight factor variation across the room.

Figure 24 - Earswick School (Architect: Unwin 1911). Large areas of openable glazing were a response to theories concerning the benefit of daylight.

But it was precisely the two problems of thermal discomfort and glare for which the next generation designs, the system-built schools of the postwar boom, were infamous. This type of system building was born out of the need to cater for a rapidly growing school population, under the influence of modernism, both in style and building production. There was also the requirement by the Department of Education and Science that every classroom should have a minimum daylight factor (DF) of 2%.

These factors led to nearly three decades of lightweight, over-glazed and poorly insulated buildings. The 2% DF requirement when applied to a 32-place classroom demands virtually 100% glazing above the work-plane, and up to the suspended ceiling at a height of 2.4m. The variation of DF was excessive and the large area of sky visible from deep in the room led to serious glare problems. Although the 2% criterion was met, artificial lighting was often used in order to reduce the brightness range. Largely as a result of the large glazing area, the thermal environment presented major problems, high room air temperatures in summer being exacerbated by direct sunlight falling on the occupants sitting close to a window. In winter the reverse condition applied, with excessive convective and radiant loss to the single glazing.

Experience with these British system-built schools serves as an object lesson on how the strict

Figure 25 - This secondary school typifies the system designs of the 1970s. The window wall is all glass above the work-plane in an effort to provide a minimum 2% Daylight Factor. This resulted, however, in glare and thermal discomfort. Large glazing areas did not guarantee that electric lighting would not be used, and often the glazing was obstructed due to use as a "pin-up" area.

observance of a single technical criterion can lead down a false avenue of development. Now we are more aware of lighting quality as distinct from quantity - in particular, the significance of uniformity ratio and glare. The experience also illustrates the need for an integrated approach to environmental design.

The epilogue to this story is that these schools are now undergoing major refurbishment. This is often used as an opportunity to improve the daylighting by modifying the façade with reduced glazing areas, shading devices and lightshelves.

By the mid-1970s, some local education authorities were no longer commissioning designs of this type and were beginning to adopt heavyweight systems with small windows and a deep plan, necessitating artificial lighting and mechanical ventilation. New developments in teaching methods, so-called "open plan teaching" required an open plan space, and a strong parallel with the *bürolandschaft* was drawn. In Britain "Integrated Environmental Design" or IED, was the term used by school architects. Undoubtedly these buildings had fewer comfort problems and better energy performance, but they were the first in the history of school building to reject nature's light and air.

Figure 27 - "Integrated Environmental Design" deep plan school at Elmstead Market, Essex, UK. A move away from the over-glazed system schools of the early 1970s, the small sealed windows necessitated artificial lighting and mechanical ventilation.

Not all authorities went down this path. For example, Hampshire County Council rejected the system building approach. The individuality of the

Figure 26 - Daylight model in an artificial sky, testing the reduction of glazing area and the use of a rooflight, during a refurbishment of 1970s system-built school.
(Architect: E. Cullinan, Daylighting: B. Ford)

Figure 28 - Yately Newlands Infants School (Architect: Hampshire County Council) was an important step away from the system designs of the 1970s The absence of a ceiling permits rooflighting to illuminate the white painted spine wall, providing very even illumination to the teaching spaces.

architects' work led to much richer envelope design giving greater opportunity for natural daylight and ventilation. Yately Newlands was the first of a whole series of schools built by Hampshire C.C. with energy conservation on the agenda. As architectural tastes moved firmly away from modernism, the adoption of a more traditional vocabulary, with a more varied envelope design, provided more opportunity for well distributed daylighting. But, despite this trend, in most cases daylighting did not receive special design effort. Daylight was seen as a bonus, part of the ambiance of the building, but not part of the functional brief.

Global issues have recently re-focused attention on both energy use and environmental quality in buildings. This concern has been directed towards a range of building types including schools and we can now find examples all over Europe where conscious design effort has been applied to meet both technical and aesthetic criteria for daylighting. Many of these examples are a product of EC support, and have benefitted from of daylighting design tools and technologies which are described in this reference book.

THE ANALYTICAL APPROACH

The traditional approach to daylighting design has, like most other elements in building design, been by precedent and experience. Probably until well into this century, "daylighting design" would not have been identified by the architect as a particular topic. Window design was closely related to architectural style, the window being such a dominant visual feature of the building. Technical limitations and construction practice also influenced the design.

Although they depended greatly upon precedent and example, that did not prevent some 19th Century architects being very systematic and consciously experimental. Sir John Soane was a good example of this. Practising at the beginning of the 19th Century, he combined his passion for collecting objects of antiquity and paintings with an experimental approach to daylighting design. At his private museum at his home in Lincoln's Inn, London, many fascinating examples can be seen. Looking down on the roof of the main museum, one is struck by the resemblance to an experimental area of a 19th Century building research station ! (See figure 17). His most well known gallery design is the Dulwich Gallery, already described. His imaginative approach did not stop at gallery lighting - the famous Breakfast Room at Lincoln's Inn incorporates mirrors and interreflected light to achieve a complex lighting effect.

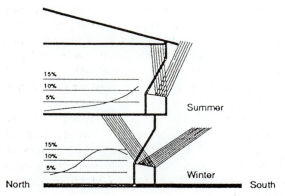

Figure 29 - This school, La Vanoise College, Moderne, France (Architect: P. Barbeyer, Daylighting: M. Fontoynont), has an advanced daylighting system, well suited to the sunny alpine climate of the Savoie region.

Figure 30 - Sir John Soane's Breakfast Room 1812. Innovative use of concealed rooflighting, reflected light and mirrors provided light to this internal room.

It is interesting to speculate just how close Soane came to the use of analytical methods. He built many models and it is tempting to suggest that he would have used them to give a subjective impression of appearance under daylit conditions. However there is no evidence in his many writings that Soane used any mathematical treatment to support his interest in daylighting. It is interesting to note that the current state of knowledge of photometry in the early 19th Century had not reached a stage where even the definition of Daylight Factor was possible, since illuminance had not yet been defined. And it was another 50 years before an instrument for measuring light intensity was invented.

What is certain is that Soane's watercolourist Gandy developed great skill in depicting daylight in his interiors, ensuring that daylight was high on the list of design priorities. We have already seen that the particular functional requirements of galleries and museums had prompted some of the earliest analytical work, as typified by Cockerell and Seager in their work which led to gallery sections for the Tate and the National Galleries in the UK at the beginning of the 20th Century. This period represented the beginning of a period of technical development in daylighting design which brings us up to the postwar, pre-fluorescent period.

Figure 31 - Gandy's water colours of interiors of the Bank of England showed remarkable rendering of daylight and it is interesting to speculate how close it came to being a design tool.

Buildings of this period were becoming larger and deeper in response to the centralisation of business activity. Their growth in size was made possible by increased use of steel and reinforced concrete structures, and the growth in the use of mechanical ventilation. But by the 1930s artificial lighting was still expensive; the luminous efficacy of a typical filament lamp was only about 12 lumens per watt and electricity was at least five times more expensive in real terms than at present, resulting in lighting being about 25 times its present real cost.

Thus daylight was still regarded as the main source of lighting in buildings, with artificial lighting use only at night. Buildings still adopted relatively shallow plans and lightwell forms. This emphasis on daylighting generated a strong incentive to develop daylighting design aids, and this inevitably led to the development of methods of prediction.

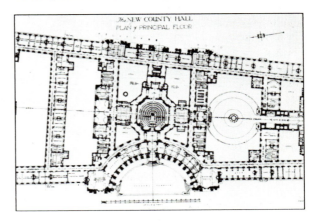

Figure 32 - The plan of County Hall, London, showing the shallow plan of courts and lightwells. (Architect: Knott, 1908)

DESIGN TOOLS

The first problem to be addressed was the prediction of the amount of direct daylight from the sky. As with many matters architectural, Vitruvius got there first. In Book VI of "De Architectura" he says *"On the side from which the light should be obtained, let a line be stretched from the top of the wall that seems to obstruct the light to the point at which it ought to be introduced, and if a considerable space of open sky can be seen when one looks up above the line, there will be no obstruction to the light in that situation"*. This rather enigmatic instruction may have lost something in translation, but must be the origin of the simple "sky-line rule" - that rooms will be well lit if there is a view of the sky from the point of interest.

But more precise methods were largely prompted by the need to adjudicate disputes concerning the obstruction of light by a proposed building. One of the earliest methods in common use was that due to P.J. and J.M. Waldram, dating back to the 1920s.

Known as the Waldram Diagram, it provided a grid upon which the elevation of the window from the point of interest could be drawn. While the grid had a linear horizontal scale, the vertical scale was non-linear to take account of the obliquity (or cosine law) of illumination, and the non-uniform luminance of the sky vault.

1.14 Daylighting in Architecture

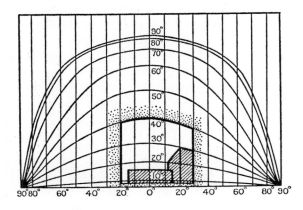

Figure 33 - The Waldram Diagram. The whole diagram represents a projection of half the sky vault. Thus the Sky Component of the Daylight Factor can be calculated from the ratio of the area of the visible sky patch to twice the total area of the diagram. It is drawn for the non-uniform CIE sky and taking account of the variation of the light transmission of glass with angle of incidence.

The Waldram Diagram permitted the vertical edges of windows or obstructing buildings to be plotted as vertical lines. Horizontal edges were plotted as curves, by reference to guide lines known as "droop lines". Other approaches to the same problem included the so-called "pepper-pot" diagrams after Pleijel. Here the sky is represented in stereographic projection and the weighting of the daylighting contribution is indicated by the density of dots.

Perhaps the best know graphical tool is the BRS Daylight Protractor developed by the Building Research Station in the UK.

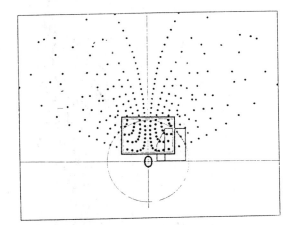

Figure 34 - Pleijel's "pepper-pot" diagrams. Each dot represents 0.1% Sky Component. The advantage of this system is that the geometry is not distorted, thus permitting sunpaths to be superimposed on the same representation of the obstructions and the window aperture.

In all of the above methods, the external reflections from buildings are assumed to make a contribution equal to that of the sky they obstruct, multiplied by the reflectance of the obstruction. This simple approach to the Externally Reflected Component is sufficiently accurate for all but very close obstructions, such as lightwells, where interreflections have a significant effect.

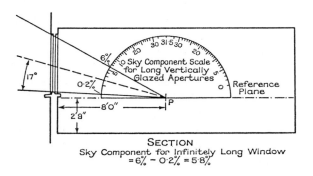

Figure 35 - The BRS Daylight Protractor. The solid angle of the visible sky subtended to the reference point is measured from the plan and section of the room. In this case the scale of the protractor is made non-linear to account for the sky luminance distribution and glass transmittivity.

The Internally Reflected Component requires a different and more complex treatment. Prior to Frühling's work in 1928, the effect of reflected light had been ignored with the design effort being applied to direct skylight. This attitude was probably appropriate when dark natural finishes were common, and a logical way of dealing with the grime resulting from the contemporary artificial light sources. The recognition of the significance of reflected light was a major step towards modern daylighting; almost all limiting cases for sidelighting now involve the manipulation of the Internally Reflected Component (IRC), mainly for the purpose of reducing the variation of Daylight Factor caused by the strong geometric dependence of the Sky Component.

Mathematical treatments had to strike a balance between modelling the complex interreflection between room surfaces, and a formula which gave useful accuracy with a minimum of input data. A successful compromise was the Split Flux Method. This considered light which originated from outside ground reflection onto the ceiling and upper wall, separately from that which originated from above the horizon, but reached the work-plane via reflection from the floor and subsequently the ceiling.

Other interesting work, still in the postwar pre-fluorescent era, included the study of glare problems associated with daylighting which resulted in a glare index prediction nomogram. However, unlike the Daylight Protractors, the input required luminances and special geometric data which would not normally be accessible or familiar to a designer. Thus the work, regrettably had little influence on daylight design.

We have already referred to the influence of the the 2% DF on the design of schools, which was, in retrospect, not a positive influence. Another large scale influence on design has been regulation for the provision of daylight and sunlight at an urban scale. For example in the UK, the Daylight and Sunlight Indicators addressed the old problem of the effect of new buildings on existing buildings and existing sites. The requirement to respect Rights of Light and Rights of Sunlight had considerable influence on the shape of buildings, particularly tall buildings in dense urban situations.

THE POST-FLUORESCENT ERA

The development of fluorescent lighting marked the end of an era where daylighting was at least a design aim, even if in practice it was not attained. Fluorescent lighting gave at least a fourfold increase in lighting efficacy, thus reducing running cost and heat gain sufficiently for designers to abandon daylighting altogether. As soon as this decision was made, glazing areas could be drastically reduced, or clear glazing could be replaced with tinted or reflective glass, so prevalent on many bland, faceless buildings of the 1970s. Large areas of clear glazing employed for reasons which lay somewhere between the pursuit of style and the provision of daylight, had already been identified as the villain, causing massive solar gains and heat losses. Designers happily retreated from the envelope which no longer had to act as an environmentally selective filter. The function of the envelope became to exclude the external environment, and to contain an artificial environment provided by the engineers.

Where daylight was abandoned, the quantity of artificial light as distinct from its quality became an obsession. Studies in the USA had suggested that productivity in offices increases as a function of illuminance, and lighting levels as high as 1200 lux were adopted with no regard for the energy costs. This, together with the move to open plan, led to a monotonous and inhuman working environment of air-conditioned office prairies.

Even in buildings which purported to respond to the cause of energy conservation, daylighting was the most frequently neglected aspect of design. For example, the atrium was generally advocated as an energy-saving feature. But not only did the atrium

Figure 36 - Deep-plan office buildings of the 1970s abandoned the envelope as a source of light and air, providing a totally artificial internal environment.

fail to provide daylight to the surrounding spaces, but in most cases the atrium itself was so badly daylit that artificial lighting had to be provided for the well-being of the plants. This was usually the result of over-cautious solar control, with fixed shading and tinted glass. Reducing unwanted summer solar gains also drastically reduced the daylight.

Where daylighting design in non-domestic buildings is concerned, the last two decades could then reasonably be described as the dark ages. Daylighting design became a neglected if not forgotten art, only a few exceptional cases departing from the norm.

Despite the steady increases in the luminous efficacy of light sources, artificial lighting remains the largest or second largest energy user in most non-

Figure 37 - Tinted glass and permanent shading results in such poor daylighting that the plants require permanent artificial lighting even in the atrium itself.

domestic buildings. However, even the provision of good daylit spaces is not sufficient to guarantee energy savings, since artificial lighting is frequently left on even when not required. Considerable progress has been made in the development of control systems which ensure the use of daylight to displace artificial light, but these systems are not widespread.

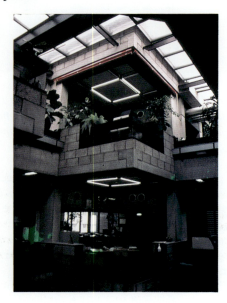

Figure 38 - Hertzberger humanised the open plan at the Centraal Beheer in Utrecht, The Netherlands, mainly by the creation of varied and identifiable spaces, designed to engender social interaction. An important ingredient too was daylighting and the provision of views.

Figure 39 - The Headquarters for the Western Electricity Board at Bedminster Down, Bristol, UK. Designed by Arup Associates. This design adopting a shallow-plan courtyard form and, maximising the use of daylight, was an important departure from the trend in the late 1970s.

Figure 40 - Some atrium buildings did address the problem of daylighting and solar control. This building in Cambridge, UK, has an atrium where clear glazing and light finishes result in a high daylight level, and temperature-operated shading devices and natural ventilation prevents summer overheating. (Architect: The Charter Partnership).

EUROPEAN RESEARCH AND DEVELOPMENT

It is in this last decade that significant progress had been made in Europe on energy conservation in housing design. Much of this progress was a result of National and European Commission (EC) research and development programmes. Attention turned from active solar systems to passive systems, and towards the end of the 1980s, consensus views on effective passive design had been reached, design rules established, and many examples built both to demonstrate the principles and to provide monitored data to refine techniques. However, daylighting received little attention in this work since domestic-sized buildings present little challenge to daylighting design, window design being dominated by solar collection optimisation.

Towards the end of this period, attention began to shift to non-domestic buildings. In 1987 the first EC Concerted Action Programme on Daylighting involving six groups, was started with the main objective of producing this reference book, Daylighting in Architecture.

At about the same time, and following the success of two previous architectural design competitions, a third competition "Working in the City" was organised under the ARCHISOL programme of the CEC. As the name suggests, the subject for design was the workplace, and the constraint was energy conservation by passive means. This naturally focused attention on daylighting since it was by now realised that lighting energy constituted a major component of the overall energy consumption of a building, while heating energy, even in northern European regions, was often quite small.

As part of the design support provided for "Working in the City" a simple energy design tool, the LT Method, was developed. The method evaluated the annual primary energy use for heating, cooling and lighting, taking account of interactions between them, and responded to factors available early in the design development.

Other EC initiatives have also focused on daylighting. The recent Building 2000 programme provided design support for innovative designs in the non-domestic sector, for real low-energy projects. Daylighting design featured high on the list of priorities.

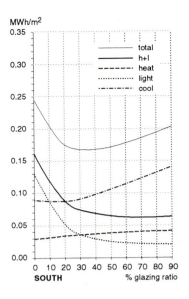

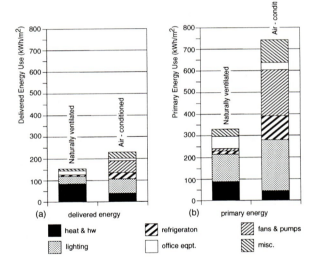

Figure 42 - A set of curves from the LT Method. The graphs illustrate the important contribution that lighting makes to the total energy consumption. The object of the method is not a precision calculation, but rather to indicate the effect and importance of certain design parameters such as plan, section, glazing ratio, etc.

We can see then that the recent intensification of interest in global environmental issues which has focused interest on reducing energy use in buildings has in turn strongly supported a return to the use of daylight in non-domestic buildings. There is also growing support for daylight, stemming from a

Figure 41- Energy end use in 14 UK office buildings. Note that for delivered energy heating energy is the largest component for naturally ventilated buildings, but becomes much less significant in air-conditioned buildings. However for primary energy , lighting is the largest single end use, and constitutes more than 30% of the energy used in environmental conditioning (in individual cases it can be as high as 45%). Primary energy relates much more closely to cost than delivered energy, and for most of Europe gives a good indication of CO_2 production and other environmental impacts.

Many technical developments have occurred in the last decade, which increase the application potential of daylight. These include innovative component design such as light ducts and light-shelves, and also new materials which can be used to control and redirect light.

Figure 43 - This atrium at the Centre for New Businesses, Reze, France (Architect: SETOM, Daylighting: Chauvel) is designed to provide light to the surrounding rooms. It is one of the buildings supported by the EC Building 2000 programme.

Figure 44 - Training Centre Agricultural Bank, Athens, Greece. Light shelves for the classrooms also provide shading. Central clerestory windows (on the right) light a corridor. (Architect: Tombazis).

Figure 45 - St Mary's Hospital, Isle of Wight (Architects: ABK). The use of daylight is part of an overall energy saving strategy, but is also an important consideration for the recovery and well-being of the patients. Here, top-lighting provides good daylighting to positions seven metres away from the perimeter. This removes the need for a large glazing area in the window wall.

concern for the quality of the internal environment from the occupants, point of view. The phenomenon of Sick Building Syndrome(SBS) is particularly associated with deep plan air-conditioned buildings, and lighting quality, particularly spectral composition, flicker and glare, is thought to be one of the cocktail of contributory factors. More specifically, the Seasonal Affective Disorder or SAD syndrome, is directly related to light deprivation. Daylit buildings, due to the non-uniformity of illuminance in both time and space, usually provide sufficient illuminance to trigger the physiological processes necessary to avoid the syndrome. Research is now being carried out to link these physiological requirements with architectural parameters.

These two concerns, the internal and the external environment, are beginning to come together in the designer's mind in concepts such as "green" buildings or "healthy" buildings. In the UK the Department for the Environment is promoting "green labelling" by means of a certification procedure know as BREEAM, the Building Research Establishment Environmental Assessment Model. Originally developed for non-domestic buildings, daylighting scores credits in the assessment both in its role in reducing energy consumption and in its contribution to internal environmental quality.

CONCLUSIONS

Daylighting has been an implicit part of building design for almost as long as buildings have existed. We have seen how the evolution of the window, responding to functional requirement, technological opportunity and cultural influence, has consistently enriched architecture through the centuries, only to lose its sense of direction in the post-fluorescent era of the last forty years.

Now however, concerns for both the global environment and the internal environment have re-focused interest on the art of daylighting. In the meantime scientific analysis techniques, as described in this reference book, have become available, and should allow daylighting design to progress much more rapidly. It is hoped that we are now witnessing the re-emergence of daylighting design from a temporary dark age, to assume once more its vital role in architecture.

Chapter 2
LIGHT AND HUMAN REQUIREMENTS

INTRODUCTION

Luminous Comfort versus Visual Comfort

There are few parallels between luminous and thermal comfort. While optimal thermal conditions are those where the occupant does not feel any need for changes towards warmer or colder conditions, luminous comfort is a much more complex concept.

The luminous environment has more in common with the acoustic environment, in that both are related to receiving messages, rather than just referring to a state of neutral perception of the environment.

Nobody at home or in the work place needs to receive a thermal message or decode a climatic pattern. However, there is a need to interpret acoustic or visual messages, and so people are concerned to improve their reception through the reduction of interferences of transmission. Luminous comfort must be interpreted as the clear reception of visual messages from the visual environment. It thus is in this respect, more appropriate to refer to visual comfort, rather than luminous comfort.

Traditionally, visual comfort in places of work has been associated simply with providing illuminance levels adequate for the intended task, while minimising, as much as possible, all other stimuli from the environment. Some recent studies in environmental psychology and ergonomics have emphasised the need for a more interesting environment in the work place, with the benefit of improving productivity. This has in some cases been achieved by redecorating tedious grey office walls.

The aesthetic aspects of architecture, which can enhance luminous stimuli in non-domestic buildings, are central in achieving visual comfort. There is a need for complexity in architecture as a means of reaching true visual comfort; uniform building interiors are boring.

Visual Perception

Visual perception is an essential part of the cognitive process. Cognitive activities, important for the understanding of complex systems, constitute a central issue of current scientific research (1).

We can recognise within the field of cognitive research, and therefore in the study of perception, two contrasting positions: one considers the cognitive process as a problem-solving activity, related to analysis; the other sees it as an autonomous self-creative action, characterising living systems as sense producers.

The former position predominates. Here, the processes of visual perception are represented as an information flow that starts from the retina and runs towards the visual cortex to reconstruct, through continuous successive information processing, the mappings modelled upon the images received from the external world.

This "representationist" interpretation of visual perception recalls the idea of the mind as a mirror of nature, related to a Cartesian concept of knowledge.

However, the study of visual system anatomy raises questions concerning this interpretation because it has been shown that only a part (1/5) of the information acquired comes from the retina, and thus from the external world. Instead of being represented as a chain of instructions transmitted from the retina to the visual cortex, the perceptual process seems to be a more complex interaction, with feedback and crossing interactions. We thus find a system organised in a reticular way: in place of an input followed by a transformation and then an output, it emerges as a simultaneous convergence of all the involved parts (2). This theory is characterised by the mind creating its own, closed and self-referencing model of the external environment (3) (Figure 1).

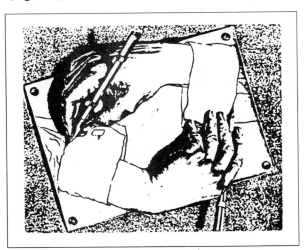

Figure 1 - A cognitive metaphor (Escher).

We can say that this system is determined more by internal computations than by information flows coming from outside. But it presupposes that one must suspend the concept of the objective external world. Instead of having a system that projects the world in which it is living and interacting, we have a system that through its closure is producing a world. The world as we look at it is, in this sense, more the result of a cumulative historic construction based on our experiences than an objective scene projected on a *tabula rasa,* freed from all the preconceptions of previous knowledge.

Comparing the two concepts, Varela defines those systems which are determined from the outside as heteronomous (they correspond to the former position), and those systems whose behaviour is determined from the inside as autonomous (they correspond to the latter position) (4).

This implies that the colours and shapes of objects we see are determined not only by features of the light we receive from these objects. We should understand that these experiences of shapes and colours correspond to a specific state and pattern of activity of the nervous system. This is demonstrated by the fact that we can correlate the names of colours with states of neuronal activity, not just with wavelengths. What states of neuronal activity are triggered by the different light perturbations are determined in each person by his, or her, individual structure and not only by the features of the perturbing agent. We do not see the "space" of an external world; we live our field of vision. We do not see the "colours" of that world; we live our chromatic space. The circularity which connects action and experience, the inseparability between a particular way of being and how the world appears to us, tells us that every act of knowing brings forth a world; every act of knowing a lit built environment brings forth an architectural world.

Everything seen is seen by someone, whose competence in visual language allows their luminous world to be distinguished. What is shared is the language, not the world, and as language brings forth its world we tend to mistake one thing for the other. The circularity taking place in visual perception is not a subjective one, it belongs to a linguistic community.

In the interpretation of an architectural knowledge of the luminous environment, one can consider the distinction between a part of the world whose state should be modified and another whose state should be preserved. The part not within design control becomes a constraint, the context of design, on which one has to base the part under design control. The former part becomes the uncontrollable variable, about which a designer must obtain all the needed information to prepare the design programme. The latter part becomes the controllable variable on which decisions must be taken.

The world is thus a given constraint to a designer which can only be described as a fact, distinguished from the other modifiable world that must be prescribed, and becomes a matter of choice and responsibility, and a question of value.

The world described within the design programme, representing the uncontrollable part of the design problem, follows the language of physics in the description of site and skies, and that of biology in the description of the eye and its behaviour. The world described in the design process, meaning the controllable part of design, follows the language of architecture.

Parameters and Variables of the Design System

Synthetically, the designer's task consists of choosing the controllable causes and adjusting them in such a way that, under the circumstances defined by the uncontrollable causes, desired effects are obtained. These desired controllable effects (the luminous performances of the built environment) constitute the architect's goal, whereas it is in choosing and adjusting the controllable causes (the shapes of apertures in the building envelope) that the architect exercises discretion.

To clarify the fundamental design distinction between the uncontrollable and the controllable parts of the environment, it will be convenient from now on to talk of measures of the system, which is what an architect does when using terms such as luminance or brightness. Let us then define:
- design variables as measures of the controllable causes, the design configurations or the architectural types;
- independent variables as measures of the uncontrollable causes (the site skies) and effects (the eye's responses to light stimuli);
- dependent variables as measures of the controllable effects (the luminous environment).

The handling of a building's parameters must be driven by considering the effects/consequences to improve the visual environment of the building. In order to control these effects/consequences one must know how visual perception operates and what requirements are necessary to improve the visual built environment.

DESIGN CONSTRAINTS

The Human Eye

Vision is a whole perceptual system. One sees the environment not just with the eyes, but with the eyes - in the head - on the shoulders - of a body - that moves about. Vision does not have a seat in the body in the way that the mind has been thought to be seated in the brain. We look at details with the eyes, but we also look around with the mobile head, and we go on looking with the mobile body. The perceptual capacities of the organism do not lie in discrete parts of the body but in systems with nested functions.

From a strictly physical point of view, the human eye is a complex sensory organ which converts the light energy it receives from the spatial and temporal relationships of objects in visual space into electrical signals for processing by the brain. The human eye system can be considered as being structured into two specialised interacting sets of components (Figure 2).
- the optic components (cornea, crystalline lens, pupil and intra-ocular humours)
- the neural components (retina and optic nerve).

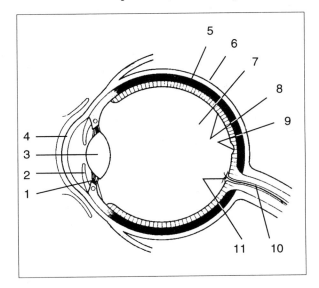

Figure 2 - Horizontal cross-section through human eye.
1 ciliary ligament, 2 iris, 3 lens, 4 cornea, 5 choroid, 6 sclera, 7 vitreous body, 8 retina, 9 macula lutea, 10 optic nerve, 11 blind spot

When light rays from an object pass through the cornea, the lens, and the vitreous body, they are refracted so that an inverted image is formed on the retina, which is a light-sensitive film. The rays are focused on the *macula lutea,* the retinal region where cones are numerous. Their name derives from their flask-like shape. Cones contain pigments which make them sensitive to colour. In dim light, we depend more on rods for vision. These cylindrically-shaped receptors are distributed throughout most of the retina. Both rods and cones contain photosensitive pigments whose chemical structure alters in the presence of light. The changes occurring in rods and cones in turn trigger electrical impulses in nerve cells in the retina, which are then transmitted to the optic nerves of the brain

Central (foveal) vision permits one to see much finer detail than peripheral vision; it represents the detailed visual acuity of the eyes. Foveal vision also provides the most acute colour discrimination because of the concentration of cones. The ability to discriminate among wavelengths of light is believed to be due to a combination of photo-chemical and neurological processes. Signals from three cone types are coded in the retina and the lateral geniculate body (in the brain) into chromatic and achromatic information. The chromatic information is a result of a subtraction of incoming signals, while the achromatic information (luminance) is a result of an additive mechanism.

The outputs of the middle and long wavelength cone systems (receptor levels) are summed to provide luminance information; the short wavelength cones are believed to contribute negligibly to luminance information.

Chromatic information is derived from defining differences in the output of the three cone systems and the combined perception is a mixture of the chromatic and achromatic channels (Figure. 3).

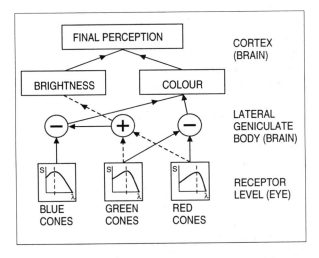

Figure 3 - A model of the biological colour vision system.

The cornea and lens focus light on the multi-layered retina which transmits impulses through the optic nerve to the brain. The size of the pupil is controlled by the iris - the larger the pupil, the greater the amount of light admitted into the eye. Under conditions of high luminance (eg bright skies outdoors), the iris reduces the size of the pupil so less light is admitted. Luminance perception is normally limited to within a brightness range up to a factor of 1000.

This ability of the eye to control the amount of light it admits and to change the sensitivity of the retina is called adaptation. The ability of the eye to focus light on the retina from one distance to another by changing the shape of the lens is called accommodation. For near vision (6 m), the curvature of the lens is increased, the pupil is constricted by the iris, and the lines of sight of both eyes converge on a common point (Figure. 4).

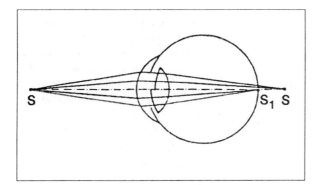

Figure 4 - Change in form and position of the lens during accommodation.

Contrast detection is the basic task from which all other visual behaviour is derived; we can say that we see by contrast. The visual system gives virtually no useful information when the retina is uniformly illuminated, but is highly specialised in informing about luminous discontinuities and gradients in the visual field.

When an obstruction is placed in the light path, as in Figure 5, all neurones of the lower layer remain inactive, except the one at the edge of the obstruction, for it receives two excitatory signals from the sensor to the left (3).

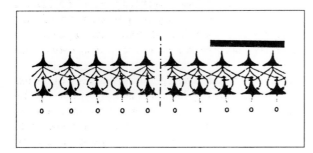

Figure 5 - Contrast detection. Obstruction placed in the light path illuminating the layer of receptors (3).

Human visual perception is based on the existence of contrasts of light and shade and contrast of colour. The term contrast is used in both a physical sense and a perceptual sense. It involves the assessment of the difference in appearance of two or more parts of the visual field seen simultaneously or successively, for example brightness contrast, colour contrast, successive contrast in dynamic lighting situations involving movement.

Optic nerves from the two eyes follow the paths to the visual cortex. At the optic chiasma, fibres from half of one retina cross to join fibres from the corresponding half of the other retina at the lateral geniculate body. The ability of the brain to perceive the images from both eyes as a single image is called binocular vision (vision by one eye alone is called monocular vision). The visual fields of the two eyes overlap to some extent (5).

The visual field is a small area, roughly oval in shape, with a clear centre and soft boundaries, seen when both the head and the eyes are motionless (6). We can define:
- central field - within a visual angle of 2 °
- background - within a visual angle of 40 °
- environment - up to 120 degrees vertical and 180 degrees horizontal visual angle.

The visual world is the larger area covered when the head and eyes are moving and thus has no centre of sharp definition but is clear everywhere.

While contrast detection may be considered the simplest visual function, visual acuity is somewhat more complex. The word "acuity" is often used to describe the visibility of fine details involved in various kinds of displays. Several different kinds of acuity can be recognised, such as resolution acuity and recognition acuity.

Visual acuity means the sharpness of vision, measured by the smallest size of detail which can be seen at a given distance. Visual acuity, like detection, varies with exposure duration and luminance. The operation of the law of diminishing returns can easily be seen in the relation between visual acuity and surface luminance of the task.

The graph in Figure 6 (7) shows that less illumination is needed when the object size is increased (eg using large type, magnifying lenses or enlarging spectacles).

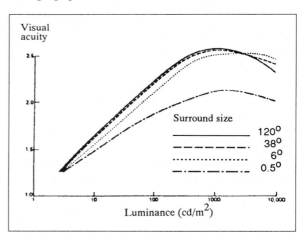

Figure 6 - Visual acuity against task luminance for surround fields of different angular subtends. The luminance of the task and surround areas were equal (7).

Vision and Age

Visual performance decreases from the late twenties onwards. Old eyes have reduced visual acuity because of yellowing of the lens; they require more time for adaptation, a higher illumination level, and an optimum contrast of the visual task to achieve the best visual performance (Figure 7) (8).

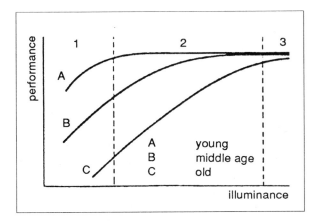

Fig. 7 - Model of general relationship between visual performance and illuminance for three age groups, young, middle age and old. At high levels of illuminance the visual performance of the three groups is very similar. At low illuminances there are very large differences in performance (8)

The increase in the absorption and scattering of light in the eye and a recession of the near point, that is the closest point to the eye where clear images can be achieved, becomes increasingly more evident around the age of 40. This effect can be overcome by the use of spectacles.

Light and Health

Eye health is sometimes put forward as a justification for high levels of illumination, with the implication that low levels of illumination will damage the eye.

Recent research studies developed in the USA have demonstrated that there is no generally accepted evidence that poor illumination results in organic harm to the eyes, any more than indistinct sounds damage the ears or foul smells damage the nose.

The need for wearing glasses arises only from organic causes, not from inadequate illumination levels.

Complaints which occur whenever a task is visually difficult, no matter whether the difficulty arises from poor lighting, from the inherent features of the task, from inadequacies in the individual visual system or from some combination of these elements, are all temporary, so that with rest, the symptoms disappear.

The most common complaints caused by visual work in poor conditions are eyestrain in various forms, muscular aches and pain, and more general reactions such as fatigue, irritability and headaches. Eyestrain may indeed result from the effort of trying to overcome difficult viewing conditions such as those presented by Visual Display Units, but the strain is only a temporary discomfort and does no damage to the eye. Eyestrain can be caused by glare - excessively bright light in the field of view - as well as by inadequate illumination, but eye damage can only be caused by over-exposure to light.

Today, with the more widespread diffusion of television and VDU, visible light itself is viewed as potentially harmful to the eye and, although relatively high retinal illuminances are required to cause damage, the new high efficiency sources potentially provide enough energy to be damaging. For the first time, safe viewing limits are being proposed for visible radiation.

As well as its ability to stimulate vision, light acts on the body in many other ways. In recent years these non-visual effects have become of interest and concern to the illuminating engineer because of the rapid expansion of photo-biological techniques in medicine and because of the controversy surrounding the safety of some lamps.

The most prevalent notion is basically that light is for seeing and has no effects on people, other than ultraviolet light which is harmful. Normal glass eliminates most ultraviolet light. Plastics manufacturers put special UV filters into translucent products to prevent damage to the material. Artificial lights use various segments of the spectrum but rarely UV.

One of the most controversial issues in the lighting community is the statistical relationship between fluorescent lighting and malignant melanoma, a particularly dangerous form of skin cancer. Unless data derived from research studies on the matter become more consistent, controversy will remain.

Light, besides being indispensable for visual perception, also regulates metabolic processes in the human body, and exerts an influence on its immunological state, i.e. the body's resistance to unfavourable agents, such as pathogenic organisms. Lighting conditions indoors also have a considerable influence on the state of mind, and so affect the psychological, psycho-emotional and general health of human beings.

What some researchers are now suggesting is that there seems to be a layer of the retina which has no function in vision, but serves as a receptor for light-waves, which are then carried along non-visual-connected fibres of the optic nerve to the master endocrine glands - the pituitary and pineal glands in the brain that control the entire metabolic system. The components of light taken in through the eye may turn out to be as critical as the kinds of foods eaten.

So light that is significantly different in spectral nature from sunlight may cause the metabolic equivalent of malnutrition. Alterations to the light spectrum taken in could be a factor in metabolic disorders (9).

Daylight is also involved in the setting of the "biological clock" and its associated rhythms. A lack of light for long periods, particularly such as during the winter seasons in the polar regions, can also manifest itself as seasonal affective disorder (SAD) where a general lethargy and depression may set in at the onset of winter. Of particular significance to the architect, this effect could be present in occupants of deep-plan artificially lit buildings where, although adequate for visual tasks, the artificial illumination is insufficient to trigger the necessary physiological response.

The Physics of Light

In describing the optical parameters of the visual system we will distinguish light as physical energy, light as a stimulus for vision, and light as information for perception.

The human environment is exposed to a wide variety of natural and man-made energy sources that emit energy within various bands of the electromagnetic spectrum. Radiant energy propagates through empty space at great velocity. Such energy can be treated either as particles or as waves, but it travels in straight lines or rays.

The human eye, considered as a photo-chemical device, can only perceive light with wavelengths between about 380 nm and 770 nm. Precise limits cannot be set because they depend on the intensity reaching the retina and the visual acuity of the observer.

TABLE I - The optical radiation spectrum

Category	Wavelength Range[nm]	Action
Ultraviolet:		
UV-C	200-280	Germicidal
UV-B	280-315	Actinic (tanning)
UV-A	315-380	Ocular Effects
Visible	380-770	Light, Vision
Infrared:		
IR-A	760-1400	Thermal only
IR-B	1400-3000	
IR-C	>3000	

The light-adapted eye has its greatest sensitivity at 555 nm, which, when expressed in terms of perceived colour, is in the green-yellow region. The sensitivity to red light beyond 700 nm is very low. The perceptual sensitivity of the eye to violet radiation with wavelengths below 400 nm is also very low.

Under conditions of very low illumination, the eye, after a period of time, adapts to give an enhanced response based on rod- as opposed to cone-dominated vision. The peak response then occurs at a wavelength of 507 nm. This visible portion is a function of the sensitivity curve as shown in the graph of Figure 8 (10).

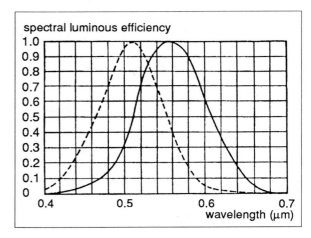

Fig. 8 - The visible spectrum and colour perception: normal vision (unbroken line) and vision adapted to low illuminance levels (dashed line)

Often people are inclined to think about lighting design, and for that matter buildings, in terms of a black and white visual environment. It is important to consider the importance of colour as the characteristic of light by which a human observer may distinguish the difference between two structure-free patches of light of the same size and shape. Furthermore, colour can affect human tension, brain-wave function, heart rate, respiration, and other functions of the nervous system.

Light and pigment create similar visual stimuli. Pigment colour is usually specified in terms of three characteristics: hue, value and chroma. Coloured light is described by three analogous terms: hue, brightness and saturation.

Hue defines the basic colour. Dominant hues include red, yellow, green, blue, and purple. Value or brightness is the subjective sensation of reflectance or brightness, the light or dark appearance of a colour. Chroma or saturation is the intensity of a hue (11). The Munsell system of colour specification enables any colour to be identified in terms of the above three main attributes of colour that are significant for a designer.

The term "colour temperature" relates the colour of a completely radiating (black body) source at a particular temperature, to light sources whose colour match such a body. The qualities of light emitted by heated objects depend on the temperature of the radiating object and this fact is used to describe the colour of light. The perceived colours of "black bodies" at different temperatures depend on the state of adaptation of the observer (Figure 9).

Light and Human Requirements

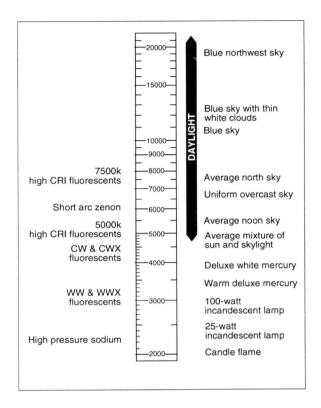

Fig. 9 - Correlated colour temperature (°K), of several electric light and daylight sources

The degree to which the perceived colours of an object, illuminated by a source of light, conform to the colours of the same object illuminated by a standard source of light is defined as colour rendering . The colour rendering index (CRI) is a value used to determine how closely a light source matches the standard source, daylight, at a given colour temperature.

Direct and Indirect Sources of Light

Some material bodies emit light, and others do not. Light comes from sources such as the sun in the sky and from other sources close at hand on the earth such as fires or lamps. They "give" light, as we say, whereas ordinary objects do not. Non-luminous objects only reflect some part of the light that falls on them from a source, and yet we can see the non-luminous bodies along with the luminous ones. In fact most of the things that need to be seen are non-luminous, they are only seen by reflected light.

We will refer to daylight and artificial light as direct sources of light, and to all surfaces reflecting light as indirect sources .Where the average daylight illuminance of an interior is inadequate for the activities taking place within it, artificial lighting is used to supplement the task lighting and the lighting of the room in general. Historically, artificial light has evolved from fire, to candles and oil lamps up to electric light. For a more extensive description of artificial light characteristics, refer to Chapter 6.

Skies and their Light Variables

Over millions of years the only principal source of light has been daylight. From a strictly physical point of view, daylight is defined as the part of the energy spectrum of electromagnetic radiation emitted by the sun within the visible wave-band that is received at the surface of the earth after absorption and scattering in the earth's atmosphere. When considering the direct, diffuse, and ground-reflected components, sunlight represents the direct component of daylight . Since sunlight covers the entire spectrum of solar radiation, designers must carefully consider the shorter wavelengths (UV) that can damage sensitive materials and produce heat, and the longer wavelengths (IR) that also produce heat. The spectral distribution varies with the time of the day, the season, the height above sea level and atmospheric conditions.

The problems of daylighting involve the assessment of lighting generated from natural sources (sun and sky) at a certain reference point. It is therefore a matter of analysing the light flux generated by these sources and the illumination they produce.

Apart from the geometrical correlations that exist between light sources and reference points, the magnitude of the light flux depends upon the radiance of the source.

The illumination produced by the sky depends on its luminance. The intensity of illumination from direct sunlight on a clear day varies with the thickness of the air mass it passes through. It is less intense at sunrise and sunset at any latitude; and at noon it is less intense at high latitudes because the sun is lower.

The masters of architecture through history have demonstrated a very good understanding of sky conditions but the advent of electric lighting has distracted designers from considering the sky under which they have to build.

Sky luminance varies according to a series of meteorological, seasonal, and geometrical parameters that are difficult to codify. In view of this problem some models of standard skies have been worked out, and simplified references can be made to some other limited conditions.

The simplest model is the Uniform Luminance Sky Distribution (Figure 10), which represents a sky of constant luminance (that does not vary with geometrical parameters), and which pertains to a meteorological situation corresponding to a sky covered with thick, milky-white clouds, an atmosphere full of dust and where the sun not visible.

Another model is called the CIE Standard Overcast Sky Distribution (Figure 11), where the luminance is not uniform but varies in accordance with geometrical parameters, and which pertains to a meteorological situation that corresponds to a sky covered with light cloud in a clear atmosphere,

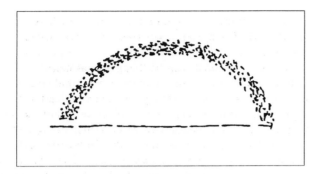

Figure 10 - Uniform Luminance Sky Distribution model.

where the sun is not visible. Briefly, the sky is three times more luminous at the zenith than at the horizon.

A third sky is the Clear Blue Sky Distribution (Figure 12) that represents a sky of variable luminance (that is to say, variable in accordance with geometrical parameters and the position of the sun) and which pertains to a meteorological situation that corresponds to a sky with a clear atmosphere. This model takes into account the effects of the varying position of the sun but direct sunlight is not taken into account.

In as much as the absolute values of luminance vary according to latitude, season and the time of day, all these models express luminance in relative terms with respect to the zenith value.

A good architect should attempt to relate the design to the moods of daylight found in different seasons and at specific places. The development of an appropriate "feel" for regional daylighting climates is important for successful architecture, as there remain so many aspects of daylighting design which it is not possible to incorporate into any formal quantitative design process. The following section will show the dynamic variability of daylight. A qualitative description of cloudless, overcast, and partially cloudy real sky conditions in terms of luminance, brightness, colour and shadows is given.

Real Sky Daylight Quality under Cloudless Conditions

The blue colour of the cloudless sky is not constant, but is strongly influenced by the height of the site above sea level, and also by the amount of atmospheric pollution above the site. The bluest skies are found in mountainous regions with clean atmospheres. As the site level decreases, the total amount of scattering increases and the sky becomes both brighter and less blue. Man-made pollutants like smoke, and natural pollutants like dust, increase absorption and scattering, and alter the amount of energy at different wavelengths. The sky becomes less blue and the sun's light tends to become redder, especially when the sun is low. Some polluting gases absorb light in the blue end of the spectrum. For example, nitrogen dioxide generated in considerable quantities in towns by traffic, makes the colour of the atmosphere brown. This colour is often very evident when looking towards an urban area from the surrounding countryside, or from an aeroplane. High water vapour levels, in the absence of pollution, tend to give the sky a milky-white appearance. Ozone,

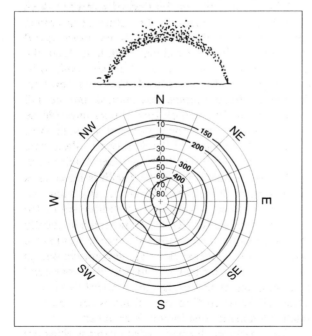

Figure 11 - CIE Standard Overcast Sky model showing an even brightness distribution with respect to azimuth.

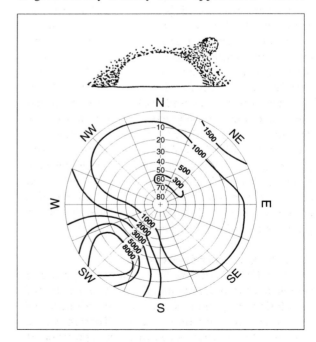

Figure 12 - Clear Blue Sky Distribution model showing anisotropic brightness distribution due to the sun's position, with reference to azimuth and altitude.

because it absorbs the yellow part of the spectrum, as well as the ultraviolet region, influences the colour of both the solar beam and the sky. This imparts a coldness to the quality of light. The type of weather also affects sky colour. Air from polar regions tends to be cold, dry and clear, and gives a high beam transmission and strong blue skies. Air masses from the south that are received in Europe from tropical maritime regions are usually associated with considerable quantities of water vapour, and tend to give rise to milky-blue skies, and much redder sunshine, especially evident when the sun is low. The daylight thus varies in colour from day to day, and invariably has strong regional characteristics. The light of Venice is very different from the light on the west coast of Ireland or drier dustier parts of Spain.

The altitude of the sun exerts a particularly strong influence on the colour of the direct beam, and also on the balance between beam illumination and diffuse illumination. This is especially true when the pollution level is high.

The cloudless day regional characteristics in different parts of Europe can be inferred, at least to some degree, from data given in the European Solar Radiation Atlas, Vol. 2 (12) for over a hundred cities in Europe.

Cloudless skies have a strongly non-uniform brightness distribution. The brightest area of the sky is in the zone immediately around the sun (Figure. 12). The area of lowest brightness is at a point at right angles to the sun in the line of the solar azimuth. Provided that the sky is not too polluted, there is also a "horizon brighting" effect. The colour of this band of scattered light is more white than the rest of the sky. The bluest part of the sky is found in the region where there is the least brightness. The precise distribution of the sky luminance depends on the atmospheric turbidity, and on the solar altitude. The luminance range of different parts of the cloudless sky may exceed 40:1, but the perceptual response of the eye compresses the range of apparent brightness. The CIE has recommended standard luminance distributions of the clear sky for cloudless sky design purposes. These standard distributions are given in Chapter 3. It must be recognised that these standard distributions are simplifications of a complex process of atmospheric scattering in the presence of different pollutants.

Daylight Quality of Overcast Skies

The CIE Overcast Sky model (Figure 11) represents the most commonly used reference for simulation programmes and for the definition of Standards and Recommendations. Therefore care must be taken to avoid an incorrect interpretation of the results obtained by using this sky model, for example, in southern European countries with clear blue skies.

Overcast sky models are representative of northern countries, where most of the glazed Modernist European architecture has been developed.

In this case clouds can be thought of as elements floating in the sky, which transform beam radiation and blue sky radiation arriving from above to certain daylighting characteristics. Clouds also act as reflecting surfaces which can re-reflect light from their undersides towards the earth. This effect is particularly important when there is a complete cloud cover and snow-covered ground. This leads to an appreciable enhancement in light available from the sky on overcast days, which helps offset some of the light disadvantages of high latitude locations in winter.

The transfer process, in the absence of severe pollution, produces white light by colour mixing. If the atmosphere is heavily polluted, changes in overcast sky colour may be perceived, usually in the form of "yellowing" of the light. The amount of light penetrating is very dependent on cloud type and cloud water content. Rain clouds can to some extent be identified by darkness. Low stratus clouds tend to produce the greatest darkening over long periods.

Vertical thunderclouds, because of their great depth and high water content, can make conditions very dark, but usually only for limited periods. Considerable amounts of daylight penetrate thin high clouds and such skies can be very glaring. Information about the relative frequency of different cloud types during different seasons provides important clues to overcast daylighting climates, and can help designers identify regional design priorities.

The luminance distribution of overcast skies depends to some extent on the nature of the cloud cover. Thin high cloud tends to show more visual structure than dense low cloud. With thin cloud, the area of the sky in the direction of the sun tends to be the brightest. As the cloud becomes thicker, directional effects become less marked, but are still often present. Even thick low clouds may not be uniform in brightness. The most stable skies, in terms of relative luminance distribution, occur when there is more than one cloud layer present. Under such conditions, the luminance at the zenith is three times the luminance at the horizon, and the sky shows no directional effects in relation to the direction of the sun above the clouds. The CIE has mathematically standardised the recommended relative luminance distribution of overcast skies for daylighting design purposes, as described in Chapter 3. In practical measuring situations, using real overcast skies, it is often difficult to find occasions when the CIE relative luminance distribution is achieved.

The shadows formed under overcast skies are very weak compared with conditions when the sun is shining. The lack of azimuth directionality implies lack of information about the directional structure of the visual scene. The flatness of the daylighting produced reduces visual stimuli. Vertical surfaces are not only poorly illuminated *per se*, particularly because of a lack of reflected ground illumination, but their surface textural details are also suppressed. Buildings tend to take on a form of dark silhouette against a brighter sky. Because the luminance of the ground is low, due to the lack of reflected sunshine, the eye's attention tends to be drawn to the brighter sky, in spite of its low visual information content and its consequent lack of variety.

A very dominant type of weather in Europe is partially overcast. It is useful to distinguish between frontally developed clouds, orographically developed clouds and convectively developed clouds.

Convectively developed clouds arise from vertical convection of warm moist air to higher colder levels in the atmosphere. Such clouds are usually triggered by solar warming of the ground. The clouds produced are of considerable vertical thickness. There are largish gaps through the clouds, through which the sun shines. In daylighting terms, the colour mixture received is direct sunshine, some sight of blue patches, cloud with some very bright areas and others less bright, and of course the whole scene changes continually. The rate of variation depends to some extent on the horizontal wind speed at cloud level. On the whole, provided excessive glare can be controlled, the variety of the daylighting is stimulating, the visual scene is changing all the time, and there is quite a lot of visual interest in the sky as compared with the somewhat boring overcast sky. As convection is the dominant process, such cloud types reach their greatest frequency in the afternoon. The back scatter to easterly façades from such clouds raises illuminances on such façades. The clouds, in a sense, act like a reflective courtyard wall. In the morning, westerly façades do not benefit in the same way, due to lack of convection in the early part of the day.

Frontally generated clouds are caused by the interaction of moving air masses of different temperature, water content and density. If the cloud masses are vaporous and moving quickly, their shadows move quickly too, producing a very dynamic pattern of illuminance. Such clouds are often described as scudding clouds in English, running before the wind. The existence of such variable illumination conditions can make big demands on electric light control systems designed to respond to external lighting conditions through photo-sensors. One can therefore distinguish "lazy" daylighting climates from "active" daylighting climates. Coastal maritime climates are often very active compared with inland continental areas. Relative wind conditions do affect daylighting climates involving partially clouded skies. A problem arises in trying to use standard hourly meteorological data to describe such factors, because hourly summarised data cannot reveal the true dynamics of daylight.

Orographic clouds are generated when warm air rides up a slope and condensation occurs. Often such clouds are highly site specific and frequently attached to specific landscape features as fixed clouds for long periods. In hilly areas, the choice of daylighting may need to take such features into account.

DESIGN RESPONSE

Architectural Parameters

Based on the multi-scale organisation of the architectural grammar described in Chapter 11, parameters involved in the daylighting design will refer to the different typological levels. First, we consider the interaction between the building and the lighted open space with its lighting sky, and secondly the interaction between the building and the internal lighted room which receives its natural light from the external environment.

The Urban Space Daylighting Parameters

The daylight received at the apertures of buildings comes from a combination of three sources: direct illumination from the sky dome, direct illumination from the sun, and reflected illumination, both from the ground and from other naturally illuminated external surfaces (Figure 13).

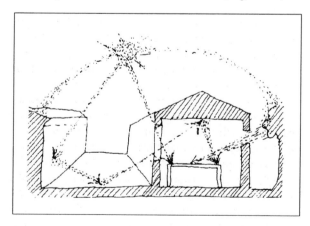

Figure 13 - Daylit open space.

External surfaces may be directly illuminated by the sky and ground-reflected light alone, or receive additional direct beam illumination from sunlight. Such external sunlight patches are an important source of indoor daylight for apertures facing away from the sun. The colour of daylight under a clear blue sky derives from the additive mixing of coloured light from four sources: the blue sky, the rather more yellow-coloured sunlight, the ground, which if covered with growing vegetation is green, and finally the other reflecting external surfaces.

TABLE II - Luminance (cd/m²) of materials under different sky conditions (6)

Material	Reflectance (%)	Overcast Sky 5145	Sunny Sky Solar Altitude 25° 13620	45° 25930	80° 32480
Green grass	6	309	816	1557	1948
Water	7	360	953	1797	2274
Asphalt	7	360	953	1797	2274
Moist earth	7	360	953	1797	2274
Slate (dark-grey)	8	411	1091	2075	2600
Gravel	13	669	1770	3372	4222
Grandolite pavement	17	875	2315	4407	5522
Bluestone, Sandstone	18	926	2452	4665	5848
Macadam	18	926	2452	4665	5848
Vegetation (average)	25	1286	3403	6483	8119
Cement	27	1389	3677	7001	8670
Brick - dark red glazed	30	1543	4082	7779	9744
- dark buff	40	2058	5447	10372	12993
- light buff	48	2470	6537	12447	15593
Concrete	40	2058	5447	10372	12993
Marble (white)	45	2315	6126	11669	14615
Paint (white) - old	55	2830	7488	14262	17836
- new	75	3859	10211	19448	24360
Snow - old	64	3293	8712	16594	20786
- new	74	3807	10077	19187	24037

In order to control the quality of ambient light, the designer should manipulate a set of relevant parameters including the following:

1. The ground surfaces around the building. In Table II, luminance variations under overcast and sunny skies for different reflectances are reported (6). Analysis of the effect of ground cover on daylight quality is important for daylighting design, particularly under cloudless conditions, because a significant proportion of the light falling on façades is light reflected from the ground. This is especially true of surfaces facing away from the sun. Earth-based materials, like brick, terrazzo and mud, reflect more red light than green and blue light, and add "warmth" to the colour of the reflected light. The reflectance of green vegetation in the visible spectrum is relatively low, but becomes higher if the vegetation dries up. The quality of daylight is strongly changed by the presence of snow. Water reflects proportionally more light when the sun is low than when the sun is high. The rippling water surface imparts a dynamic quality to the light which can be used to good architectural effect, but may cause a potentially disturbing flickering effect. This effect can also be caused by the location of trees in front of south-facing windows.

2. The external surfaces of buildings interact with one another in such a way that surface parameters are a design variable for one building and become design constraints for a neighbouring building (Figure 14). The colour of any light reflected from building surfaces is influenced by the colour of those reflecting surfaces.

3. The choice of aperture and orientation is the third factor essential for the control of the quality of natural light. A designer is always, either consciously or otherwise, selecting a colour mix, whose proportions will vary according to orientation, arrangement and properties of the external reflecting surfaces and the detailed fenestration design. For example, rich-blue north light, coming from that part of the sky which is associated by human experience with cold radiative conditions and with absence of sun, and hence with direct beam deprivation, is described subjectively as "cold", even though its colour temperature is higher than that of the direct solar beam (Figure 9).

The Building Space Daylighting Parameters

Although we look upon daylight as our principal source of light, we must remember that, especially in southern countries, it may be difficult to use directly in everyday working life because of its high intensities and constant variation due to sunpaths and meteorological changes. Taking advantage of different surface configurations and of a number of

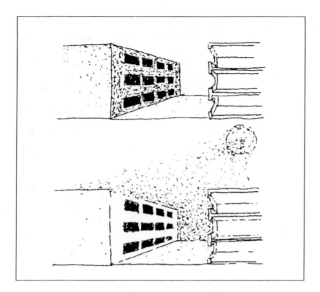

Figure 14 - On an overcast day an adjacent building is always darker than the sky (top) ; on a sunny day an adjacent building can be brighter than the sky (bottom).

physical phenomena, such as reflection, refraction, diffusion, absorption, etc, daylight should be used in a diffused form, ideally as an indirect light.

It is important therefore that designers envisage the visible environment as a highly structured three-dimensional light field. Architecture becomes the shaping of this luminance field. Its structure is made of reflecting and transmitting "objects", acting as lighted and lighting operators.

The most important controllable causes influencing daylight within a room, for a specific building type, are:
- the envelope shape
- the geometry, location and orientation of apertures
- the surface characteristics.

The surface texture, ranging from mat (light reflected diffusely, equally in all directions) to specular (light reflected only in one direction), affects the nature of reflected light.

Surface reflectance, and thus its luminance, is dependent on the angle of incidence and the surface characteristics; reflectance is one of the most important factors in daylighting design.

The luminous reflectance of spectrally non-selective grey objects remains constant for all light sources, but the luminous reflectances of coloured objects will differ in accordance with the spectral power distribution of the light source. For example, under illumination from incandescent sources that are relatively rich in the red and yellow portions of the spectrum and poor in the blue end, yellow objects will appear lighter and blue objects darker than they do under daylight illumination. Under a blue sky the reverse would be true.

VISUAL COMFORT REQUIREMENTS

Often studies on lighting have considered artificial and natural light as equivalent, addressing merely the illuminance levels (see Table III). This attitude leads to a reductive interpretation of measurable, dependent variables, and at the same time under-evaluates the advantages of daylighting.

Several studies have been conducted that illustrate to what extent daylight is preferred over artificial lighting sources, making its quality as an illuminant an important reason for using it in buildings. Daylight, the combination of sunlight and sky light, is the light source that most closely matches human visual responses. Over millions of years, the human eye has evolved using this full light spectrum as the source against which all other light sources are compared. Daylight thus is likely to provide the best visual environment.

To see objects in a space properly, a person must be able to discern foreground from background, while surfaces and objects that are not critical to a given visual environment need to be visually subdued. A poor visual environment is one in which the visual information cannot be readily discerned because there is not enough contrast to enable the viewer to distinguish what it is important to see. Often designers are unable to determine whether a space is sufficiently illuminated or whether it appears dark. It is often assumed that visual capacity is limited only by the strength of the task illumination, a statement which suggests that better vision is simply a matter of increasing the incident illumination. However, it is very important in designing a good visual environment to consider that the way surfaces are illuminated is often more important than how much light falls on them.

Nevertheless, the visibility of a task may be limited by either visual acuity and object size (sharpness of vision) or contrast sensitivity (contrast between objects and/or background), both of which vary with task luminance. Within the design process it is important to relate different visual tasks to the best viewing conditions.

Despite the common misconception that surface illuminance is the only relevant parameter for visual comfort, a more comprehensive attitude to visual perception is needed. This should include notions of:
- spatial distribution of daylight illuminance
- luminance ratios
- shape from shadows
- colour rendering
- glare
- visual noise.

Daylight Illuminance

Illuminance from daylight, especially under clear sky conditions, can vary in direction and intensity throughout the day. Since visual effectiveness is

TABLE III - Lighting recommendations in workplaces

Activity/Space	Building Type	Artificial Lighting:		Daylighting:		
		Illuminance (Lux)	Glare Index	Type of Daylighting*	Average Daylight Factor (%)	Glare Index
Formal teaching and seminar spaces	Schools Colleges Hospitals, etc	300 to 500 (300 on desks, in hospitals)	16 formal 19 seminar	A B	5 2	21 formal 23 seminar
Deep (open) plan teaching spaces	Schools Colleges	300 to 500	19	A B	5 2	23
Lecture theatres and examination halls	Schools Colleges Hospitals	500 (300 on desks, in hospitals)	16	A B	5 2	21
Music rooms and music practice rooms	Educational and recreational buildings	300	19	A B	5 2	23
Art Craft Needlework (studios)	Schools Colleges Factories Offices Recreational buildings	300 to 500	16	A B	5 2	21
Woodwork Metalwork Engineering (teaching)	Schools Colleges Training centres Recreational buildings	500	16	A B	5 2	21
Laboratories	Educational buildings Hospitals Offices Research establishments Factories	500 to 750 (300 to 500 on bench, in hospitals)	16	A B	5 2	21
Staff rooms Common rooms	Educational buildings Hospitals Offices Factories	150 to 300 (100 average in hospitals)	19	A B	5 2	23
Offices (enclosed)	Offices Educational buildings Factories Hospitals Banks Insurance buildings Post offices Libraries	500 (300 on desks, in hospitals)	19	A B	5 2	23
Deep (open) plan offices Landscaped offices	Offices Colleges Banks Insurance buildings, etc	500 to 750	19	A B	5 2	23
Typing Business machines Punch card	Offices Colleges Banks Post offices, etc	500 to 750	19	A B	5 2	23
Computers	Offices Banks Educational buildings Hospitals	500 to 750 Limit illuminance where VDUs are used	19	A B	5 2	23
Drawing offices Design offices	Educational buildings Offices Factories	500 to 750 plus local lighting to 1000 on boards	16	A B	5 1 (in supple- mented area)	21
Workshops Machine shops Processing Production plane	Factories Offices Hospitals, etc	Rough work 300 Medium 500 Fine 750 to 1000 Very fine 1000 to 1500 (300 to 500 on bench, in hospitals)	19	A B	5 1 (in supple- mented area)	23

* A - Full daylighting, B - Supplemented daylighting.
Source: *Basic Data for the Design of Buildings: Daylight.* Draft for Development, DD 73: 1982, British Standards Inst.

influenced by this variable character of daylight, it would appear that any daylight measurement should apply to particular conditions at a particular moment. Fortunately, these changes are relatively slow and do not affect the visual impression of light quantities; in fact, the human eye adapts itself continuously to changing light patterns. However, due to daylight changes as a function of sky conditions, absolute measurements of illuminance are not directly indicative of the actual building performance.

Comfortable Luminance Ratios

The achievement of comfortable luminance ratios in any space requires a careful study of all the factors involved, including not only the light sources but also the reflectances of ceiling, walls, floors and furniture. For best results it is necessary to create a balance between the luminance of the immediate task and that of adjacent surfaces in the field of view, avoiding both excessively dark backgrounds and distracting bright surroundings.

When surfaces of strongly contrasting luminances are present at a work station, the eyes are constantly adjusting to different luminances. If the changes are major and happen often, eyes become tired and the ability to work is reduced.

When the eyes are adapted to 100 cd/m^2, a surface with a measured luminance of 100 cd/m^2 should have a perceived luminance of 100 cd/m^2 to the observer. However, if the observer's eyes are adapted to only 1 cd/m^2 (eg under dark shadow conditions), this surface could have a perceived luminance equivalent to 400 cd/m^2.

Thus, although the measured luminance would be the same in both of these situations, the surface could appear four times brighter due to the luminance adaptation. For example, under identical illumination grey paper viewed on a black surface appears brighter than grey paper viewed on a white surface yet its measured luminance would be the same.

In common usage the term brightness usually refers to the strength of sensation from viewing surfaces or spaces. The sensation is determined in part by the measurable luminance, and in part by conditions of observation such as the state of adaptation of the eye.

The dynamic range of the visual system is large but finite. In daylight, luminances below about 1 cd/m^2 are seen as black, while those above 500 cd/m^2 are said to be glaring (7).

The range of brightness which the eye can appreciate and above which we still see well, is more than 1000 to 1. Apparent luminance is thus relative to the state of adaptation, subjective, and dependent on contrast.

We must remember that what we can see is limited by the contrast between detail and background, and by the size of the detail. As the background luminance of the task increases, visibility also increases up to a point. But the curve reaches a point of diminishing returns above which large increases in background luminance are required to produce even small increases in visibility.

When one is outdoors on a bright day, one can see very little when looking into a room through a window or an open door. The range of brightness from outside to inside will be very high. If, however, one comes into the room and allows the eyes to adapt, then one can see perfectly well, because adaptation has changed the range of useful brightness discrimination.

Visual performance is aided by lighting that gives the correct brightness balances in the indoor environment. It is desirable to make the visual task the brightest object in the normal field of view. Ideally the following brightness ratios within the field of view should be aimed at:
- between task and darker surrounding 3 : 1
- between task and remote darker surfaces 10 : 1
- between light sources and surroundings 20 : 1
- maximum contrast (except if decorative) 40 : 1
- highlighting objects for emphasis 50 : 1.

These limits on contrast, however, do not imply that uniform or unchanging lighting conditions are preferred or are even desirable.

The control of contrast is mainly achieved by an appropriate choice of surface reflectances, and an appropriate design of the window system or light fittings. Studies on visual performance suggest that the ideal situation is a background luminance equal to task luminance. Since this condition is seldom achieved, a luminance contrast of task to adjacent darker surroundings of 3 : 1 is usually acceptable. Ratios no greater than 10 : 1 anywhere in the field of view are desirable, between 20 and 40 : 1 being commonly considered as the maximum that is permissible within the space itself.

When describing the physiology of the eye we noted that visual performance improves with increasing contrast definition (Figure. 5). The ability to perceive luminance differences between adjacent areas of the visual field is called contrast sensitivity.

Contrast sensitivity plays an important role in determining how well we can see; the higher the contrast, the less contrast sensitivity will be required to perceive accurately a given visual task. In the immediate task area, the eye perceives detail through quick eye movements that scan the boundaries separating areas of different luminance and colour. For example, black ink on white paper provides sufficient contrast for the eye to separate detail from the background; there must be contrast between the two.

Shape from Shadows

Shadows are important in determining the shape and position of an object in space, if there are no other references. Directional light accentuates the three-dimensional qualities of an object more clearly than diffuse or shadowless light. However, shadows created by obstructions between the light source and the task are bad for visibility, since contrast is drastically reduced.

Colour Rendering

Daylight is considered the best source of light for good colour rendering; many people have had the experience of matching clothes of different colours under electric light only to discover that they do not match when seen in daylight. This is because a light source may alter the perception of colours to match the spectral composition of the light. Therefore, objects may vary in colour appearance depending upon the light source.

Fading

Daylight has another, but negative, effect upon colour, namely fading. This occurs because natural light often contains UV light, which, by photo-chemical reaction, can rapidly fade the pigment in coloured objects. This negative effect can be avoided by the use of UV filters in the glazing.

Glare

The CIE defines glare as the "condition of vision in which there is discomfort or a reduction in the ability to see details or objects, caused by an unsuitable distribution or range of luminance, or extreme contrasts" (see Glossary). Hopkinson et al. (13) define glare, in general terms, as a condition of eye adaptation that is unfavourable to good sight. It causes discomfort or impairment of vision.

Glare is a complex phenomenon which involves the understanding of many issues, such as the length of time that the glare source is present, the luminance ratio between the glare source and its surroundings, and the requirements of the visual tasks in the room. It is typically caused by lamps, windows and surfaces appearing too bright in comparison to the general background.

It is possible to distinguish between direct and indirect or reflected glare. Direct glare is affected by the characteristics of the room and light sources (natural or artificial) directly visible within the field of view. For instance, when direct sunlight enters the field of vision (which extends 180° horizontally and about 60° up from the horizon) it will be noticed; but when it enters the centre of the field of vision (an area defined by a cone extending about 40° from the centre) glare will be caused.

Since the human eye cannot accept high levels of brightness directly into the foveal vision area, a considerable difference exists between the allowable level of brightness directly in a person's line of vision and in the peripheral area (Figure 15) (14).

Indirect glare involves two forms: reflected glare and veiling reflections. Reflected glare is caused by

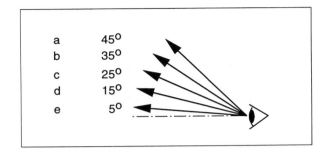

Figure 15 - Acceptable luminance levels as a function of position in the field of vision: (a) 2500 cd/m², (b) 1800 cd/m², (c) 1250 cd/m², (d) 850 cd/m², (e) 580 cd/m².

shiny or glossy surfaces reflecting images of light sources into the eyes. Veiling reflection occurs when small areas of the visual task reflect light from a bright source (windows or light fixtures), reducing contrast between the task and the immediate surroundings. Pencil handwriting is highly susceptible to veiling reflection, as pencil graphite can act as tiny mirrors (Figure 16) (15).

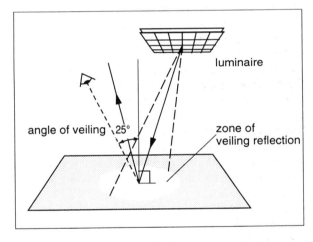

Figure 16 - Veiling reflections on horizontal surfaces.

Veiling reflection can occur when the angle of incidence of light on the horizontal work surface is within the observer's viewing zone. The offending zone of potential veiling reflection light sources is seen if a mirror is substituted for the surface of the reading task. The most frequent viewing angle for reading and writing tasks on flat desk surfaces is 25°.

To reduce veiling reflections:
- locate light sources away from the mirror angle (side-lighting, notably using daylight, can improve visual performance equivalent to providing ceiling lights at three times the illumination)
- use light fixtures with low surface luminance
- provide relatively uniform illumination throughout the room (eg by indirect light sources)

- tilt task or work surfaces away from mirror angles (eg drafting tables with adjustable drawing surface)
- use mat work-surface finishes with a reflectance of 35 to 50% (avoid using white desk surfaces with 85% or more reflectance).

Glare can be experienced as disability glare or discomfort glare, either one of which can occur without the other.

Disability glare is the glare that lessens the ability to see detail. It does not necessarily cause visual discomfort. For example, excessive reflections from shiny white paper can cause disability glare while reading. Such conditions also occur when a person has a direct line of view to a bright object such as a window or a light fixture. The eye must constantly adjust from the task to the higher luminance and resulting in eye fatigue.

Under daylight conditions the influence of the sky through a window can have a disturbing effect. Such conditions can be experienced when looking at a wall surface adjacent to a window or when attempting to see details of an object set against a highly reflective surface in which it is mirrored by the sources of light.

The disabling effects of glare can be reduced by reducing the luminance of the light source, while at the same time increasing the luminance of the object or surface by a better light distribution, and by the use of lighter colours.

Discomfort glare occurs when, even with no significant reduction of the ability to see, the presence of excessively bright sources in the field of view causes a state of discomfort. The sources may be too bright with respect to darker surroundings, or be uncomfortably bright in absolute terms. An unshielded light bulb is a common example.

A quantitative assessment of glare discomfort can be given by the expression of the "glare constant", G:

$$G = K \cdot P \cdot (L_s^{1.6} \cdot \omega^{0.8})/L_b$$

where:
K is a constant depending on the units employed
P is a "position factor" depending on the position of the source with respect to the line of sight
Ls is the luminance of the source
Lb is the field luminance
ω is the solid angle subtended by the source

Table IV (13) gives the values of P as a function of the horizontal and vertical displacements of the glare source from the horizontal line of view.

For artificially-lit rooms, the IES (Illuminating Engineering Society) has defined a glare index GI:

$$GI = 10 \cdot \log_{10} \cdot G$$

The IES has suggested limiting values of GI for different environments. Table V gives some values.

TABLE IV - Values for the "position factor" P (13)

Horizontal angle ($\phi = \tan^{-1} L/R$)

V/R \ °	0°	6°	11°	17°	22°	27°	31°	35°	39°	42°	45°	50°	54°	58°	61°	63°	68°	72°	θ
1.9	—	—	—	—	—	—	—	—	—	0.02	0.02	0.02	0.02	0.02	0.02	0.02	0.02	0.02	62°
1.8	—	—	—	—	0.02	0.02	0.02	0.02	0.02	0.02	0.02	0.02	0.02	0.02	0.02	0.02	0.02	0.02	61°
1.6	0.03	0.03	0.03	0.03	0.03	0.03	0.03	0.03	0.03	0.03	0.03	0.03	0.03	0.03	0.03	0.03	0.03	0.03	58°
1.4	0.04	0.04	0.04	0.04	0.04	0.04	0.04	0.04	0.04	0.04	0.04	0.04	0.04	0.04	0.04	0.04	0.03	0.03	54°
1.2	0.05	0.05	0.06	0.06	0.06	0.06	0.06	0.06	0.06	0.06	0.06	0.05	0.05	0.05	0.05	0.04	0.04	0.04	50°
1.0	0.08	0.09	0.09	0.09	0.10	0.10	0.10	0.10	0.09	0.09	0.09	0.08	0.08	0.07	0.06	0.06	0.05	0.05	45°
0.9	0.11	0.11	0.12	0.13	0.13	0.12	0.12	0.12	0.12	0.11	0.10	0.09	0.08	0.07	0.07	0.06	0.06	0.05	42°
0.8	0.14	0.15	0.16	0.16	0.16	0.16	0.15	0.15	0.14	0.13	0.12	0.11	0.09	0.08	0.08	0.07	0.06	0.06	39°
0.7	0.19	0.20	0.22	0.21	0.21	0.21	0.20	0.18	0.17	0.16	0.14	0.12	0.11	0.10	0.09	0.08	0.07	0.07	35°
0.6	0.25	0.27	0.30	0.29	0.28	0.26	0.24	0.22	0.21	0.19	0.18	0.15	0.13	0.11	0.10	0.10	0.09	0.08	31°
0.5	0.35	0.37	0.39	0.38	0.36	0.34	0.31	0.28	0.25	0.23	0.21	0.18	0.15	0.14	0.12	0.11	0.10	0.09	27°
0.4	0.48	0.53	0.53	0.51	0.49	0.44	0.39	0.35	0.31	0.28	0.25	0.21	0.18	0.16	0.14	0.13	0.11	0.10	22°
0.3	0.67	0.73	0.73	0.69	0.64	0.57	0.49	0.44	0.38	0.34	0.31	0.25	0.21	0.19	0.16	0.15	0.13	0.12	17°
0.2	0.95	1.02	0.98	0.88	0.80	0.72	0.63	0.57	0.49	0.42	0.37	0.30	0.25	0.22	0.19	0.17	0.15	0.14	11°
0.1	1.30	1.36	1.24	1.12	1.01	0.88	0.79	0.68	0.62	0.53	0.46	0.37	0.31	0.26	0.23	0.20	0.17	0.16	6°
0	1.87	1.73	1.56	1.36	1.20	1.06	0.93	0.80	0.72	0.64	0.57	0.46	0.38	0.33	0.28	0.25	0.20	0.19	0°
L/R	0	0.1	0.2	0.3	0.4	0.5	0.6	0.7	0.8	0.9	1.0	1.2	1.4	1.6	1.8	2.0	2.5	3.0	

Vertical displacement (V/R) on left axis; *Vertical angle ($\theta = \tan^{-1} V/R$)* on right axis; *Lateral displacement (L/R)* on bottom.

Position factors: V = vertical distance from horizontal line of view, L = lateral distance, R = horizontal distance eye-source

TABLE V - Limiting values of GI (IES)

Building Type	Limiting Glare Index
Factories	
Rough work	25
Engine assembly	25
Fine assembly	22
Instrument assembly	19
Farm buildings	
(where people work)	25
Jewellery assembly	10
Laboratories	19
Museums	16
Art galleries	10
Offices	
General	19
Drawing	16
Schools	
Classrooms, etc	16
Needlework rooms	10
Hospital wards	13

The most important functions of a window are to admit light into an interior and to provide a view out. However, glare discomfort could arise from a direct view of the sky. To prevent this, the first aim should be to reduce excessive contrasts by controlling the direct light sources and by raising the luminance of surrounding surfaces. For example, a bright sky seen through a window in an under-lit or darkly finished room can cause discomfort glare. It is possible, at the design stage, to solve the problem by ensuring that the view of the open sky through the windows is kept to a minimum by devices such as louvres, and using parts of the building itself as screens. One of the more effective and interesting design strategies is to introduce a secondary window in another wall to increase the general light level.

The luminance of a window can be controlled by the use of adjustable blinds or lightweight curtains, the same as those used to screen windows from solar radiation or to provide privacy.

The surfaces immediately surrounding the sources of light are important in terms of glare. If these areas can be given an intermediate luminance between the source and the general environment, such contrast grading will be helpful. The design of the window itself may eliminate some of the problem by for example using splayed reveals. Glazing bars and any element seen against the light should be light in colour to reduce the contrast.

With artificial sources the ceiling is often the background and some upward distribution of light, especially on to a ceiling of light colour, will assist contrast grading.

A study by Chauvel et al. (16) deals with the issue of daylight as a source of discomfort glare, with reference to sky light but not to direct or reflected sunlight.

The main conclusion of that study is that *"discomfort glare from a single window (except for a rather small one) is practically independent of size and distance from the observer, but is critically dependent on the sky luminance"*. This luminance can be as high as 10000 cd/m^2, even on overcast days, and much higher if bright sunlit clouds are visible. The work is based on laboratory studies using artificial light sources with large areas, which were carried out in England and the USA, and also on studies using daylight seen through real windows, which were carried out in France and England independently.

Discomfort glare has been studied for relatively small-sized sources, that is, for solid angles on the eye of around 0.01 steradian. Increasing the size of the source above this level does not seem to increase glare significantly.

The degree of discomfort glare due to the sky seen through a window can be predicted from a Glare Index based on the "Cornell large source glare formula" (see Appendix F):

$$G = K \cdot (L_s^{1.6} \cdot \Omega^{0.8})/(L_b + 0.07 \cdot \omega^{0.5} \cdot L_s)$$

where:
K is a constant depending on the units employed
L_s is the luminance of the source
L_b is the field luminance
ω is the solid angle subtended by the source
Ω is the solid angle subtended by the source, modified to take into account the position in the field of view.

Ω can be determined by a procedure developed by Petherbridge and Longmore (17); the glare source must be plotted on the diagram, its area being expressed as a fraction of the total area (see Appendix F).

As before:

$$GI = 10 \cdot \log_{10} G$$

A Daylight Glare Index (DGI) can be introduced which is related to the IES Glare Index for artificial lighting (GI) by the equation:

$$DGI = 2/3 \cdot (GI+14) \text{ for values up to 28}$$

This equation expresses the observed fact that there is a greater tolerance of glare from the sky, as seen through windows, than from a comparable artificial lighting situation, provided the Glare Index is not too high (see Table VI).

TABLE VI - Comparison of Glare Indices for artificial light (IES GI) and daylight (DGI)

Glare Citerion	IES GI	DGI
Just imperceptible	10	16
	13	18
Just acceptable	16	20
	19	22
Just uncomfortable	22	24
	25	26
Just intolerable	28	28

Figure 17 shows the results of a study on the influence of window size and sky luminance in a room with average internal reflectance equal to 0.4. These results show that the level of glare discomfort is almost constant (+/- 0.5 Glare Index Units) for all window sizes greater than 2% of the floor area. Therefore, a room has a relatively fixed glare character.

Page (18) has suggested a seven-point scale, called the Daylight Glare Perception Scale, calculated as:

DGPS = (DGI - 16)/2

If the value is negative, it is set equal to zero.

Table VII shows a comparison between the two indices.

TABLE VII - Comparison of two Glare Indices

Glare Citerion	DGI	DGPS
Imperceptible	Below 16	0
Perceptible	16 - 20	0 - 2
Acceptable	20 - 24	2 - 4
Uncomfortable	24 - 28	4 - 6
Intolerable	Above 28	Above 6

Sky luminance variations from 1600 to 8900 cd/m² lead to variations of the Glare Index by four units.

In order to provide guidelines for the design of windows, a standard value for the sky luminance equal to 8900 cd/m² has been chosen (this luminance is exceeded in only 25% of the annual working hours in Southern England). With this luminance, the DGI is equal to 26 for an internal reflectance equal to 0.4, and has a value of 24 for a reflectance of 0.6.

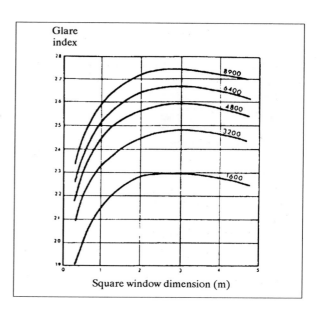

Figure 17 - The relationship between daylight glare index and window size for sky luminances between 1600 and 8900 cd/m². The window is viewed from a perpendicular distance of 6m in a room of dimensions 12m x 12m x 6m. Results for average reflectance of 0.4 (17).

The principal means to reduce discomfort are:
- to reduce sky luminance
- to reduce contrast of sky and surroundings
- to reduce visibility of the sky by obstructions.

The sky luminance as seen through a window can be reduced by fixed or adjustable components. Obviously, fixed devices (louvres, fins, overhangs, etc) restrict daylight penetration in all circumstances. The same applies to low transmission glass. Moveable screens, curtains, awnings, etc, can be very useful, as their position can be adjusted according to the sky conditions.

Light-coloured diffusing walls and ceilings produce a high general luminance, thus reducing contrast and glare. A further way to reduce discomfort glare is to increase the luminance of the walls surrounding the window, thus reducing the contrast. This effect can be achieved by artificial lighting of the walls adjacent to windows. Similar effects result from light-coloured diffusing window frames and sills.

Visual Noise

To understand luminous comfort performance, we must recognise that perceptions are interpretations of information, not sensations. The information sought - what we want to see - should be our focus. Anything that interferes with this is considered visual noise. Therefore, we should illuminate what we want to see, be it a task or a pleasant visual rest area. Conversely, we should not create or accentuate visual noise.

REFERENCES

(1) Neisser, U., *Cognitive Psychology*, Prentice-Hall, Inc. Englewood Cliffs, New Jersey, 1967, and;
-. *Cognition and Reality, Principle and Implications of Cognitive Psychology*, W. H. Freeman and Co, San Francisco, 1976.
(2) Varela, F., "Complessità del cervello e autonomia del vivente" in Bocchi, G., Ceruti, M. (ed), *La sfida della complessità*, Feltrinelli, Milano, 1985.
(3) Von Foerster, H., *Observing Systems*, Intersystems Publications, Seaside, California, 1984.
(4) Varela, F., "The Creative Circle: Sketches on the Natural History of Circularity" in Watzlawick, P. (ed), *The Invented Reality*, W. W. Norton & Co., New York, 1984.
-. *Principles of Biological Autonomy*, North Holland, New York, 1979.
(5) Egan, M. D., *Concepts in Architectural Lighting*, McGraw-Hill, New York, 1983.
(6) Gibson, J. J., *The Ecological Approach to Visual Perception*, Houghton Mifflin Co, Boston, 1979.
(7) Canter, D. (ed), *Environmental Interaction, Psychological Approaches to our Physical Surroundings*, Surrey University Press, UK, 1975.
(8) McGuiness Boyce, P. R., "Age, Illuminance, Visual Performance and Preference", *Lighting Research and Technology*, Vol. 15.
(9) Ott, J. N., *Health and Light*, Pocket Books, New York, 1976.
(10) Barducci, I., *Fotometria e Colorimetria*, ESA, Rome, 1982.
(11) *IES Lighting Handbook*, Reference Volume, Section 3, "Light and Vision," IES of North America, New York, 1981.
(12) Commission of the European Communities *European Solar Radiation Atlas*, Vol. 2, Verlag Tuv Rheinland, 1984.
(13) Hopkinson, R.G., Longmore, J., Petherbridge, P., *Daylighting*, Heinemann, London, 1966.
(14) Robbins, C. L., *Daylighting Design and Analysis*, Van Nostrand Reinhold, New York, 1986.
(15) Lam, M. C. W., *Perception and Lighting as Formgivers for Architecture*, McGraw-Hill, New York, 1977.
(16) Chauvel, P., Collins, J.B., Dogniaux, R., and Longmore, J., "Glare from Windows: current views of the problem", *Lighting Research and Technology*, Vol. 14, no. 1, 1982.
(17) Petherbridge, P. and Longmore, J., "Solid Angles Applied to Visual Comfort Problems", *Light & Lighting*, 55,146 (1962).
(18) Page, J.K., Private communication, 1989.

Chapter 3
DAYLIGHT DATA

INTRODUCTION

Daylight data are essential at various levels during the design process and performance assessment of buildings. Basic sky data, in the form of simple descriptions, make an architect aware of overall conditions with respect to location of the site. The discussion in Chapter 2 about the differences between southern and northern European skies has already been shown to be important in terms of an appropriate architectural response. More detailed information, in the form of sky brightness distribution curves or daylight availability data, gives basic quantitative details that can be used in simple analyses of design options. Accurate data and techniques to determine sky conditions are important to judge the final design's performance in terms of both visual comfort and energy use over time. It is clear that the use of daylight to offset the need for artificial lighting can result in important energy savings and reduced carbon dioxide emissions, particularly in non-domestic buildings such as offices, libraries, schools, etc. Because of the dynamic nature of daylight and the resultant variations of energy and lighting conditions, daylight information is required that is climatically and temporally accurate and easily available in order to assess detailed design solutions.

This chapter presents techniques for determining daylight data from limited information. Various models are discussed and related to measured data. The results of this work are presented in Appendices A and B in the form of standardised probabilities of the occurrence of various sky types, and the availability of daylight on planes of different orientations, for 33 sites in Europe.

There is a lack of luminous data and especially energetic data, the main source being meteorological files. One of the most direct means to obtain luminous values from radiation data is the use of luminous efficacy of daylight, which is a measure of the luminous efficiency of radiant flux. Most meteorological radiation data are recorded simply as the global irradiances on an horizontal plane at certain time intervals. Few weather stations provide measurements of diffuse horizontal irradiance and global irradiances on tilted surfaces (vertical in most cases) for a range of orientations. Obtaining diffuse illuminances from basic meteorological data on various tilted surfaces requires knowledge of the luminance distribution of the sky. Normalised luminance distributions are expressed as the ratio between the luminance of a point of the sky and the zenith luminance. Therefore, it is necessary to know the variation of this zenith luminance according to easily accessible parameters in order to obtain absolute values of the luminance in every point of the sky.

The normalised distributions, expressed as simple equations, are available only for extreme sky conditions, namely overcast sky and clear sky. There are no normalised formulae for intermediate skies, such as partially cloudy skies and more generally real skies, but many research teams who have worked on the matter have proposed expressions of luminance distributions for different sky conditions. Some luminance models, based on monitored data for various types of sky, are discussed.

REVIEW OF SKY MODELS

Overcast Skies

Two formulae can be used to describe the luminance distribution of the following overcast sky conditions:
- the uniform sky
- the Moon and Spencer overcast sky.

In the case of the uniform sky, the relation linking the value of the luminance L_u to the illuminance obtained on a horizontal plane E_H is:

$$L_u = E_H / \pi$$

The Moon and Spencer sky, standardized by the International Commission on Illumination (CIE), is a sky with altitudinal asymmetry, featured by luminance at the zenith three times greater than on the horizon. For a point P at an angular height θ, the luminance is expressed as follows:

$$L_P = L_z \cdot (1 + 2 \sin \theta)/3$$

L_z is the luminance at the zenith and is linked to the illuminance E_H by:

$$L_z = 9 \cdot E_H / 7 \pi$$

Clear Skies

The equation for a clear sky as standardized by the CIE (1), takes into account mean real conditions and fundamental solar light diffusion and refraction effects in a perfectly clear and cloudless atmosphere. The distribution of the luminance is expressed in terms of three angles (Figure 1):

- η: angular distance between the point P of the sky and the sun
- ξ: angular distance between the sun and the zenith - zenithal distance from the sun
- $\pi/2 - \theta$: zenithal distance of the point P

$$L_P(\pi/2 - \theta, \alpha) / L_Z = f(\eta) \cdot \phi(\pi/2 - \theta) / f(\xi) \cdot \phi(0)$$

where:
$f(x) = 0.91 + 10 \exp(-3x) + 0.45 \cos(x)$ (standard diffusion)
$\phi(x) = 1 - \exp(-0.32 / \cos(x))$
$\eta = \arccos[\cos(\xi) \cdot \cos(\pi/2 - \theta) + \sin(\xi) \cdot \sin(\pi/2 - \theta) \cdot \cos(\alpha)]$
α is the azimuth angle

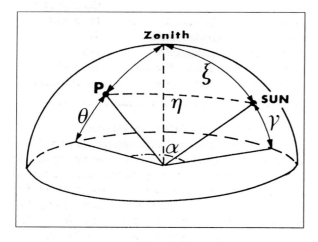

Figure 1 - Definition of sky dome angles.

With polluted atmospheric conditions, as may be found over large towns or industrial zones, the function $f(x)$ becomes:

$$f(x) = 0.856 + 16 \exp(-3x) + 0.3 \cos^2(x)$$

Cloudy Skies - Real Skies

Several intermediate sky luminance distribution formulae have been proposed:

- Tregenza (2): The formula is based on the characteristics of the clouds and on statistical nebulosity data.
- Gillette and Treado (3): Weighting between the two extreme skies (clear and overcast) is based on the use of the Cloud Ratio (CR), a ratio between the diffuse horizontal irradiance (D) and global horizontal irradiance (G):

$$L_P = \eta \, L_{P_C} + (1 - \eta) \, L_{P_O}$$

where:
L_{P_C} is the luminance of the clear sky
L_{P_O} is the luminance of the overcast sky
$\eta = [1 + \cos(CR \cdot \pi)] / 2$

A similar formula is proposed by Winkelman and Selkowitz (4) for calculations of natural lighting in the DOE 2 software.

- Nakamura and Oki (5): A mathematical formula (that can only be used with $\gamma < 80°$) of the intermediate average sky is proposed from measurements made over a long period:

$$L_P / L_Z = L(\gamma, \eta, \pi/2 - \theta) / L(\gamma, \theta/2, \xi)$$

- Pierpoint (6): An expression identical to that of the CIE clear sky, but with different f and ϕ functions:

$f(x) = 0.526 + 5 \exp(-1.5x)$
$\phi(x) = 1 - \exp[-0.80 / \cos(x)]$

- Littlefair (7): Using the results of measurements corresponding to a very broad range of real skies, the distribution of the average sky luminance is provided by:

$$L_P = \alpha e^{-\eta/40} + d[5 - 2\sin(\pi/2 - \theta)] / 3 \quad (kcd/m^2)$$

where:
$\alpha = 0.1 + 0.42 \gamma - 0.7 \sin(7.2 \gamma)$
$d = (0.3 + 0.434 \gamma - 0.0042 \gamma^2) \, 9 / (11 \pi)$
(α and η are in degrees for this equation only)

This average sky comprises three components:
- direct sun illuminance
- circumsolar zone
- remainder of the sky.

LUMINOUS EFFICACY OF DAYLIGHT

The luminous efficacy of daylight is a main parameter required to make predictions about daylighting. Efficacy is defined as the quotient of the luminous flux by the radiant flux, expressed in lumens per watt. It is dependent on solar altitude, cloud cover and water vapour in the atmosphere. Experimentally, values are given by measurements of illuminance and irradiance on a specified plane. In most cases this is for the horizontal plane.

The following details on luminous efficacy are extracted from a review by Littlefair (8).

Luminous Efficacy of Global Radiation

Figure 2 shows various efficacy values (for a solar altitude greater than 10°) for clear sky conditions and Table I for overcast sky conditions.

Luminous Efficacy of Diffuse Radiation

For the diffuse component most authors find a variation of luminous efficacy with solar altitude. The values are predominantly between 84 and 173 lm/W (Table II).

TABLE II - Efficacies of diffuse radiation for clear skies

Author	Place of Measurement	Value Obtained (lm/W)
Drummond	Pretoria, South Africa	132 (average)
Blackwell	Kew, UK	130
Bartenava and Poljakova	Repeteke, USSR	118
Krochmann	Washington, USA	130 - 133
Kuhn	Plateau Station, Antarctica	122 - 156 (increases with γ)
Evenich and Nikol'skaya	Moscow, USSR	95 - 115 (function of γ)
Liebelt	Karlsruhe, Germany	113.3 ± 8.0
Chandra	Roorkee, India	84
Arumi-Noe	Golden, USA	140
Petersen	Vaerlose, Denmark	146 ± 14
CSTB	Nantes, France	132 (average)

Luminous Efficacy of Direct Radiation

There are great differences between values produced by various authors. This can be explained by the assumptions relating to the measurements used to estimate this efficacy: in some cases the direct component is obtained by subtracting the diffuse component from the global value. The results are between 50 and 120 lm/W for a solar altitude greater than 10° (Figure 3).

RESULTS OF MEASUREMENTS

Measurement Instrumentation and Procedures

The luminance measurements carried out in Nantes, France, were monitored over 15 months (9, 10). Sets of sensors measured the following:
- global and diffuse horizontal illuminance
- global and diffuse horizontal irradiance
- four global vertical illuminances (N, E, S, W)
- five luminances (zenith and N, E, S, W at 42° altitude).

A camera fitted with a fish-eye lens was also used.

The luminance meters used consist of lux meters with a tubular fitting to limit the aperture to 15°. The inner surface of the tubes are coated with a mat black paint so as to eliminate spurious reflections as far as possible. Despite such precautions, the

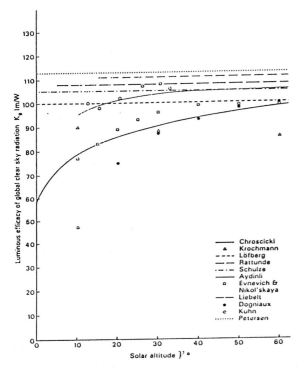

Figure 2 - Graph of global (sun and sky) clear sky luminous efficacies according to various authors, as a function of solar altitude (8).

TABLE I - Efficacies of global radiation for overcast skies

Author	Place of Measurement	Values Obtained (lm/W)
Drummond	Pretoria, South Africa	106
Krochmann	Washington, USA	115
Blackwell	Kew, UK	120 ± 5
Blackwell	Kew, UK	115
Bartenava and Poljakova	Repeteke, USSR	103
Dogniaux and Lemoine	Belgium	110
Evenich and Nikol'skaya	Moscow, USSR	60 - 92 (function of γ)
Lofberg	Stockholm, Sweden	111 ± 18
Rattunde	Berlin, Germany	116 ± 10
Petersen	Vaerlose, Denmark	121 ± 7
Page	UK	112 - 128 (function of γ)
CSTB	Nantes, France	110 - 115 (increases with γ)

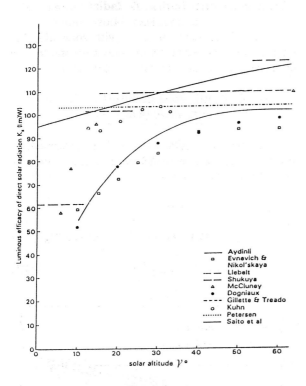

Figure 3 - Graph of direct solar luminous efficacies according to various authors, as a function of solar altitude (8).

presence of the sun near the cone of measurement covered by the luminance meter results in an over-estimation of luminance values due to the very high direct solar illumination of the inner wall of the tube. Although the reflectance is low, the tube contributes to inaccuracies.

From comparisons with measurements made using a "precision" luminance-meter, it has been possible to determine a correction factor to be applied to the data acquired. Only when direct solar radiation is within the measurement cone of the measuring cell will data not be used.

Classification of Real Skies

One of the main results of the measuring campaign carried out at CSTB Nantes is the classification of the skies into five types (Table III). This classification, made by analysis of data on all the parameters recorded, is based on a single index : the nebulosity index I_N, defined as follows:

$$I_N = (1 - CR_M) / (1 - CR_T)$$

where:
CR is the "Cloud Ratio", an index notably used by the National Bureau of Standards (Washington, DC). CR is the ratio between the horizontal diffuse irradiance and the horizontal global irradiance.
CR_M is the measured value of CR.
CR_T is the mean theoretical value of CR for a clear sky condition (11).

TABLE III - Nebulosity Index for five sky types

Type of Sky	Nebulosity Index
Overcast (O)	$0.00 < I_N < 0.05$
Intermediate Overcast (IO)	$0.05 < I_N < 0.20$
Intermediate Mean (IM)	$0.20 < I_N < 0.70$
Intermediate Blue (IB)	$0.70 < I_N < 0.90$
Blue (B)	$0.90 < I_N < 1.00$

Luminance Distribution: Zenith Luminance of Overcast Sky

The value of the luminance at the zenith L_Z contingent on the diffuse horizontal illumination E_H lies, on average, between those of the two standardised overcast skies (Figure 4).

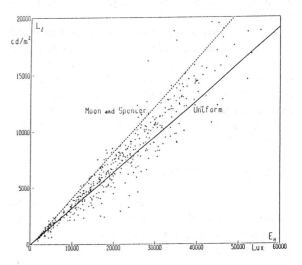

Figure 4 - Zenith luminance (L_Z) under overcast sky condition as a function of horizontal diffuse illuminance (E_H).

The different formulae, defining L_Z as a function of the solar altitude γ, listed by Matsuura (12), are all of the following basic form, with A and B as constants:

$$L_Z = A + B (\sin \gamma)^C$$

The measurements made both of overcast skies (O) and intermediate overcast skies (IO) are also well represented by such a formula (in cd/m^2):

$$L_Z = 90 + 9630 (\sin \gamma)^{1.19} \qquad \text{(O, see Figure 5)}$$

$$L_Z = 100 + 7580 (\sin \gamma)^{1.36} \qquad \text{(IO)}$$

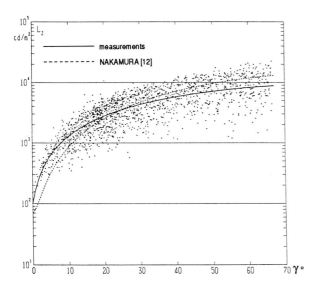

Figure 5 - Zenith illuminance (L_Z) under overcast sky condition as a function of solar altitude (γ).

Luminance Distribution: Zenith Luminance of Blue Sky

In the considerable literature devoted to zenith luminance under clear sky conditions, in most cases a variation of L_Z is determined as a function of the tangent of the solar altitude and with the turbidity. It is not feasible to determine turbidity from measurements; however, an expression proposed by Nakamura, Oki, et al. (13) can be adopted, expressing L_Z in the following form:

$$L_Z = 100 + 600 (\tan \gamma)^{1.1}$$

It has not been possible to approximate the measured results obtained under a blue sky by such an expression, but a fifth degree polynomial achieves a close fit (Figure 6).

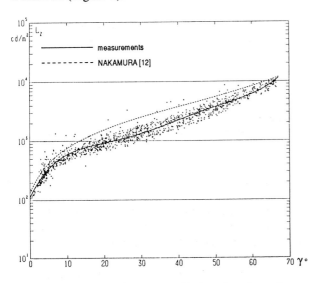

Figure 6 - Zenith luminance (L_Z) under clear sky condition as a function of solar altitude (γ).

Luminance Distribution: Zenith Luminance of Intermediate Sky

The Nakamura et al. (14) formula for intermediate sky zenith luminance is a combination of their formulae for overcast and clear skies:

$$L_Z = 0.07 (100 + 600 (\tan \gamma)^{1.1}) + 0.93 (100 + 220 (\sin \gamma)^{1.8})$$

The measurements made with an intermediate sky, mean (IM) and blue (IB), can be represented by the following expressions:

$$L_Z = A + B (\tan \gamma)^C$$

$$L_Z = 100 + 5290 (\tan \gamma)^{1.19} \quad \text{(IM, see Figure 7)}$$

$$L_Z = 100 + 4150 (\tan \gamma)^{1.18} \quad \text{(IB)}$$

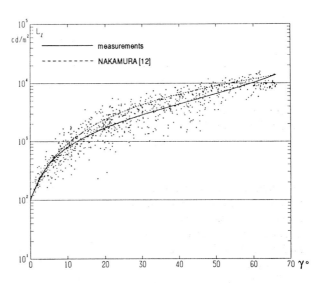

Figure 7 - Zenith luminance (L_Z) under intermediate sky condition as a function of solar altitude (γ).

Luminance Distribution: Overcast Sky

From measurements by the five luminance meters and five lux meters (horizontal + 4 vertical), it has been observed that, for the measurement site of Nantes, France:
- the distribution of the luminances conforms to that of the Moon and Spencer overcast sky expression in only 5 to 10% of the cases
- in 20 to 30% of the cases, the distribution is similar to that of the uniform sky
- for all the other cases, the luminance is greater in the area of the sky near to the sun, concealed by clouds.

Luminance Distribution: Blue Sky

In the great majority of blue sky cases, the theoretical expression of the CIE clear sky is representative of the luminances measured at Nantes. An example is provided in Figure 8 where the four ratios L/L_Z for the four north, east, south and west luminance meters, together with the theoretical values of these four ratios, are shown over a day.

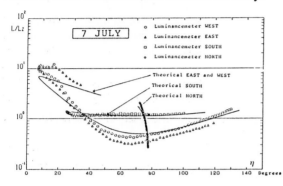

Figure 8 - Ratio L/L_Z as a function of the angular distance (η) for four orientations.

Models

The choice of the formula of the sky models is determined by the comparisons between measurements and standardised models for the extreme sky classes indicated in the previous paragraphs. Sky luminances in both overcast and clear conditions are higher in the area of the sky near the sun, whether obscured or not. It is thus important to express, for the five types of sky defined in Table III, the luminance at each point of the sky with a formula of the same type as that of the CIE clear sky.

As opposed to the CIE clear sky, which provides the values of the ratio L/L_Z, the value of the luminance at a point P is required in the following form:

$$L_P = A\ f(\eta)\ g(\pi/2 - \theta)\ h(\xi)$$

where:
$f(\eta)$ characterises the distance from the sun
$g(\pi/2 - \theta)$ defines the zenithal distance of point P
$h(\xi)$ is linked to the zenithal distance from the sun
A is a scale factor (linked to the turbidity)

The f and g functions above have the following forms:

$$f(\eta) = a_1 + b_1 \exp^{-k\eta} + c_1 \cos(\eta/2)$$

$$g(\pi/2-\theta) = a_2 - b_2 (\cos \pi/2-\theta)^{0.6}$$

where:
a_1, b_1, c_1, k, a_2 and b_2 are coefficients defined in Table IV.

The presence of a specific h function of the distance from the sun is made necessary by the fact that the aim is to obtain an absolute value of the luminance, as opposed to a comparative ratio. This h function is expressed as follows:

$$h(\xi) = a_3 + b_3 \cos \xi + c_3 \sin \xi$$

where:
$a_3, b_3,$ and c_3 are coefficients defined in Table IV.

Two expressions for L_P are considered, corresponding to two "extreme" cases. In the first, the diffuse horizontal radiation D is unknown. It is then impossible to have access to the scale factor A and the model provides only a mean value of L_P. In this case:

$$L_P = f(\eta)\ g(\pi/2-\theta)\ h(\xi) \qquad (cd/m^2)$$

where f, g and h are the expressions indicated above.

It may seem contradictory to consider the case where diffuse horizontal radiation is unknown, when the type of sky is known (depending upon, amongst other things, D). In fact, these models will be used only in the case where the type of the sky is only known statistically.

In the second expression of L_P the diffuse horizontal radiation is known. It is then possible to obtain a more precise value of the luminance defined as follows:

$$L_P = D\ f'(\eta)\ g'(\pi/2 - \theta)\ h'(\xi) \qquad (cd/m^2)$$

where:
$f'(\eta) = a'_1 + b'_1 e^{-k'\eta} + c'_1 \cos(\eta)$
$g'(\pi/2-\theta) = a'_2 - b'_2 (\cos \pi/2-\theta)^{0.6}$
$h'(\xi) = a'_3 + b'_3 \cos \xi + c'_3 \sin \xi$

The coefficients of the f, g, h and f', g', h' functions are indicated in Tables IV and V respectively.

Comparison Between Models and Measurements: Luminances

Comparisons have been made at two levels: on the luminances themselves and on the diffuse horizontal illuminance.

A solar altitude γ and five values of the angular distances (one for each luminance meter) correspond to each of the recordings. Only two values of θ are available (both for the development of the models and for comparison with the measurements):
$\theta = 90°$ for the luminance meter for the zenith
$\theta = 42°$ for the four other luminance meters

From Figures 9 and 10 it is possible to evaluate the accuracy of each of the two models proposed.

TABLE IV - Coefficients of f, g and h functions

Coefficient	Sky Type:				
	O	IO	IM	IB	B
a_1	1.10	0.60	0.70	0.71	0.21
b_1	4.20	11.16	21.24	26.24	32.73
k	3.00	3.00	3.00	3.00	4.00
c_1	5.52	3.83	3.95	2.65	2.60
a_2	1.22	1.70	1.99	2.17	2.29
b_2	0.28	0.89	1.26	1.49	1.64
a_3	0.35	-0.18	-1.04	-0.90	-0.91
b_3	1.58	1.92	2.04	1.74	1.60
c_3	-0.16	0.37	1.25	1.18	1.36

TABLE V - Coefficients of f', g' and h' functions

Coefficient	Sky Type				
	O	IO	IM	IB	B
a'_1	32.33	17.82	14.41	13.05	12.89
b'_1	13.16	23.99	69.70	124.96	243.38
k	3.00	3.00	3.00	3.00	3.00
c'_1	3.24	13.35	10.18	7.49	3.26
a'_2	1.18	1.70	2.03	2.21	2.25
b'_2	0.23	0.89	1.31	1.54	1.59
a'_3	0.76	0.45	0.83	-0.83	1.04
b'_3	0.13	0.10	-0.29	-0.28	-0.41
c'_3	0.20	0.59	0.38	0.42	0.20

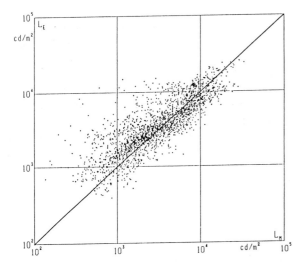

Figure 9 - Comparison between measured luminance (L_M) and estimated luminances (L_E) when horizontal diffuse irradiance is unknown. Results are for the five sky types.

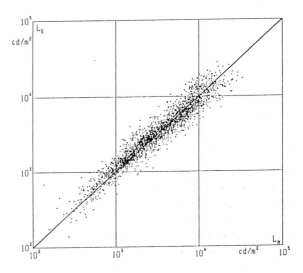

Figure 10 - Comparison between measured luminance (L_M) and estimated luminances (L_E) when horizontal diffuse irradiance is known. Results are for the five sky types.

Comparison Between Models and Measurements: Illuminances

The estimate of the diffuse horizontal illuminance is obtained by integrating luminance models over the whole sky. The comparison between estimates and measurements for the two models shows the advantage of knowing the diffuse radiation (Figure 12) as compared to the opposite case (Figure 11).

The estimates with a intermediate mean sky (IM) when horizontal diffuse irradiance (D) is known are less accurate owing to the problems involved with the actual measuring of luminances, discussed earlier, and the difficulty in determining a correction factor for these sky conditions.

The deviations between calculations and measurements are also due to the small number of luminance measuring points used to produce the models. This is above all considerable for the g and g' (zenithal) functions for which only two references values were available. The expressions of these functions that have been adopted, based on the form of the CIE clear sky function, are not accurately representative of reality.

Relation Between Daily Irradiation and Sky Type

In order to be able to estimate luminous quantities when only a daily value of a parameter such as global irradiation or relative sunshine duration is known, a correlation between sky type and this parameter needs to be established. An example of the relationship between these factors is shown in Figure 13 where daily probabilities P of each sky type have been drawn as a function of the daily relative sunshine duration F. For a value of F the graph provides the five probabilities of occurrence of the sky types.

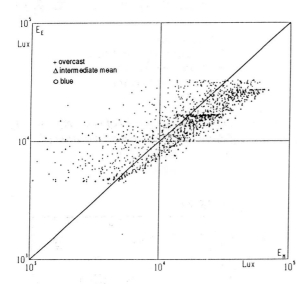

Figure 11 - Comparison between measured illuminance (E_M) and estimated illuminances (E_E) when horizontal diffuse irradiance is unknown. Results are for the three sky types.

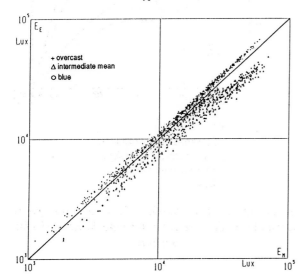

Figure 12 - Comparison between measured illuminance (E_M) and estimated illuminances (E_E) when horizontal diffuse irradiance is known. Results are for the three sky types.

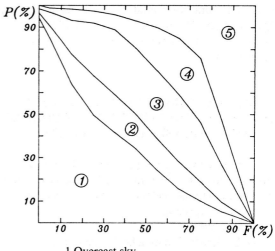

1 Overcast sky,
2 Intermediate overcast sky,
3 Intermediate mean sky,
4 Intermediate blue sky,
5 Blue sky

Figure 13 - Daily probability (P) of each sky type as a function of daily relative sunshine duration (F).

PROBABILITY OF OCCURRENCE OF EACH SKY TYPE AT EUROPEAN METEOROLOGICAL STATIONS

The data used to obtain these probabilities are gathered from the following research:
- development of the Test Reference Years (TRY) (15), developed for the Commission of the European Communities. TRYs have been produced in Belgium, Denmark, France, Eire, Italy, the Netherlands and the UK, and data for 29 sites are available
- data from four German stations (16).

Probabilities have been obtained by using hourly values of irradiances (diffuse, global and direct) when solar altitude is greater than 3°.

The different steps (for each hour) taken to determine the probability of the occurrence of each sky type are:
- establish "measured" Cloud Ratio
- calculate Nebulosity Index I_N
- determine sky type.

Each day is characterised by its relative sunshine duration and the five sky type durations. For German stations, the relative sunshine duration has been estimated from horizontal daily global irradiation. Figures 14 and 15 present typical examples of results for two sites. The graphs for all the 33 sites are in Appendix A.

ALGORITHMS FOR DETERMINING LUMINANCE DISTRIBUTIONS ACCORDING TO AVAILABLE ENERGETIC DATA

Amongst the energy parameters influencing the luminous quantities, the most widely measured in the network of national meteorological stations are:
- sunshine duration
- cloud cover
- global irradiation on a horizontal plane.

The diffuse irradiation on a horizontal plane is only measured at a few stations. Depending on the data available and the accuracy of the calculations to be made, the accuracy of determining the sky luminance distribution varies.

In the best cases, where hourly global and diffuse irradiation data are available at hourly or three-hourly intervals, the whole sky luminance distribution can be determined as a function of sky type and diffuse horizontal irradiance (D). The Nebulosity Index I_N determines the sky type, and the luminance distribution is defined by the established equation: $L_P = D\, f'(\eta)\, g'(\pi/2 - \theta)\, h'(\xi)$.

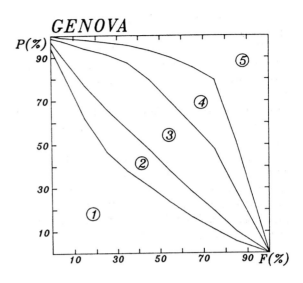

1 Overcast sky,
2 Intermediate overcast sky,
3 Intermediate mean sky,
4 Intermediate blue sky,
5 Blue sky

Figure 14 - Daily probability (P) of each sky type as a function of daily relative sunshine duration (F) for Genoa.

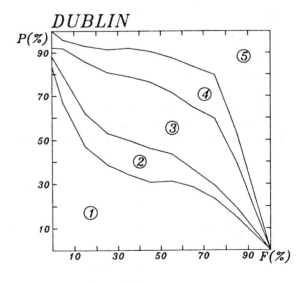

1 Overcast sky,
2 Intermediate overcast sky,
3 Intermediate mean sky,
4 Intermediate blue sky,
5 Blue sky

Figure 15 - Daily probability (P) of each sky type as a function of daily relative sunshine duration (F) for Dublin.

If only sunshine duration data are available, either as global irradiation or relative sunshine duration, then only the most probable sky luminance distribution can be estimated. This estimate of the sky's luminance distribution (or of the diffuse illumination for any orientation plane) at a given moment can only be made statistically. The statistical relationship used is the one existing between the daily probability of each of the five types of sky and the daily relative sunshine duration (Figure 13). Sunshine duration, which has not been specifically measured, is obtained from the ratio between the daily global irradiation and the theoretical maximum global irradiation. The luminance at point P is estimated by using these statistical results and the formula: $L_P = f(\eta) \, g(\pi/2 - \theta) \, h(\xi)$ indicated previously.

The values of L_P thus estimated can be compared to the measured values (Figure 16).

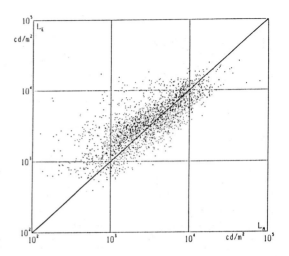

Figure 16 - Comparison between measured luminances (L_M) and estimated luminances (L_E) by using the relation between daily probabilities (P) of each sky type and relative sunshine duration.

Estimation of Illuminances on Various Surfaces

Global and diffuse illuminances have been estimated for the 29 sites of TRY and for the four sites in Germany. Five planes are considered: the horizontal plane and the four vertical planes facing north, east, south and west.

Illuminances are obtained using the method described above. When the sky type is determined, illuminances are calculated using:
- luminous efficacies for the horizontal plane
- luminance distributions for the vertical planes.

Results are given in terms of daylight availability (probability to have an illuminance lower than a chosen level) for four periods:
- Winter: November - February
- Mid-season: March - April and September - October
- Summer: May - August
- Year.

For the horizontal plane and the north and south vertical planes, there is a symmetry between results for the morning (before 12:00 True Solar Time) and results for the afternoon (after 12:00 TST). For these surfaces only one graph is given. Figures 17 and 18 show examples of the curves related to Aberport UK. Equivalent sets of graphs for the 33 stations are presented in Appendix B.

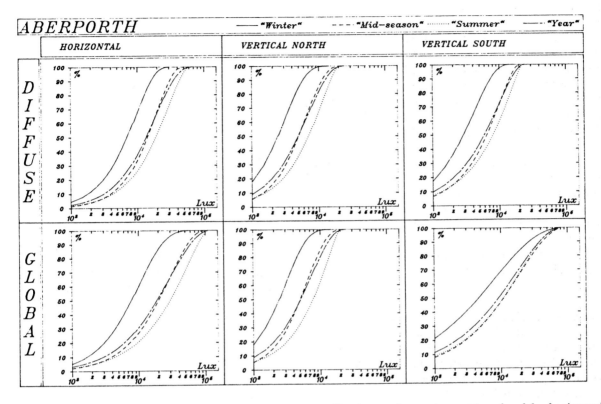

Figure 17 - Availability of daylight: Probability of having an illuminance lower than a given level for horizontal, north vertical and south vertical planes.

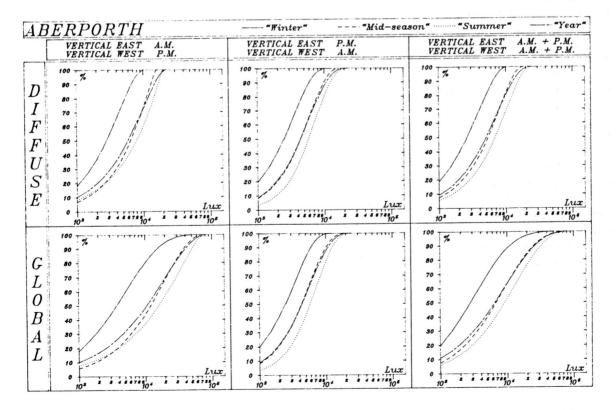

Figure 18 - Availability of daylight: Probability of having an illuminance lower than a given level for east and west vertical planes.

REFERENCES

(1) Commission Internationale de l'Éclairage, "Standardization of Luminance Distribution on Clear Skies", CIE Publication no. 22, Paris, 1973.

(2) Tregenza, P.R., "A Simple Mathematical Model of Illumination from a Cloudy Sky", *Lighting Research & Technology*, vol. 12, no. 3, p. 121-128, 1980.

(3) Gillette, G. and Treado, S., "The Issue of Sky Conditions. Exploring the Issue of Sky Conditions as Applied To Current Daylight Practice", *Lighting Design and Application*, p. 23-27, 1985.

(4) Winkelman, F.C. and Selkowitz, S., "Daylight Simulation in the DOE-2 Building Energy Analysis Program", *Energy and Building*, vol. 8, p. 271-286, 1985.

(5) Nakamura, H. and Oki M., "Composition of Mean Sky and its Application to Daylight Prediction", *Proc. CIE 20th session*, Amsterdam, vol. 1, D303/1-4, 1983.

(6) Pierpoint, W., "A Simple Sky Model for Daylight Calculations", *Proc. International Daylighting Conference*, Phoenix, Arizona, p. 47-51, 1983.

(7) Littlefair, P., "The Luminance Distribution of an Average Sky", *Lighting Research & Technology*, vol. 13, no. 4, p. 192-198, 1981.

(8) Littlefair, P., "The Luminous Efficacy of Daylight: A Review", *Lighting Research & Technology*, vol. 17, no. 4, p.162-182, 1985.

(9) Perraudeau, M. and Chauvel, P., "One Year's Measurements of Luminous Climate in Nantes", *Proc. International Daylighting Conference*, Long Beach, California, p. 83-88, 1986.

(10) Perraudeau, M., "Climat lumineux à Nantes - Résultats de quinze mois de mesures", Rapport CSTB, EN-ECL 86.14. L, 1986.

(11) Perrin de Brichambaut, C., "Météorologie et énergie: l'évaluation du gisement solaire", *La Météorologie*, VI série, no. 5, p. 129-158, 1976.

(12) Matsuura, K., "Luminance Distributions of Various Reference Skies", *CIE Technical Report*, T.C. 3-09 (Draft), 1988.

(13) Nakamura, H., Oki, M., *et al.*, "Preliminary Study on Zenith Luminance", (in Japanese) *Summaries of technical papers of annual meeting A.I.J*, 1985.

(14) Nakamura, H., Oki, M., Hayashi, Y. and Iwata, T., "The Mean Sky Composed Depending on the Absolute Luminance Values of the Sky Elements and its Application to the Daylight Prediction", *Proc. International Daylighting Conference*, Long Beach, California, p. 61-66, 1986.

(15) Commission of the European Communities, *Test Reference Years TRY - Weather Data Sets for Computer Simulations of Solar Energy Systems and Energy Consumption in Buildings*, CEC, DG XII, Brussels, 1985.

(16) Blümel, K., Hollan, E., Kähler, M., Peter, R. and Jahn, A., *Entwicklung von Testreferenzjahren (TRY) für Klimaregionen der Bundesrepublik Deutschland*, BMFT-FB-T 86-051, 1986.

Chapter 4
THE PHOTOMETRY OF MATERIALS

INTRODUCTION

This chapter presents information on the basic principles of the photometry of materials. A classification of translucent and opaque materials is suggested, in order to help designers choose the most appropriate materials to meet the requirements of their selected daylighting techniques. Then, a short review of new products is presented in order to show recent achievements and potential new directions to follow.

A detailed description of the photometry of the materials might appear too complex to manage. Moreover, it would go largely beyond the requirements of a designer. However, the photometry of materials is a key parameter in the performance of daylighting components, discussed in Chapter 5. Values of reflection coefficients, as well as reflection patterns may in some cases eliminate any interest for some applications.

Materials are described in order to:
- distribute the knowledge more widely
- make designers more aware of the impact of their choice
- provide guidelines for the use of materials in efficient daylighting design.

SURFACE PHOTOMETRY CHARACTERISATION

Measurement Procedure

As regards lighting constraints, the transmission through a material can be characterised when the intensity curve and the wavelength distribution are identified for each possible incident beam (angle of incidence, wavelength) (see Figures 1, 2 and 3).

The number of data to record, if a $1°$ angle and a 0.05 micron wavelength region are considered as reasonable steps, would be equal to $(360 \times 90 \times 8)^2$, where:
- 360 is total azimuth angle in the transmission space
- 90 is total angular height in the transmission space
- 8 is number of wavelength steps (0.05) between 0.35 and 0.75 microns
- 2 the square is due to the fact that this has to be known for the incident space as well.

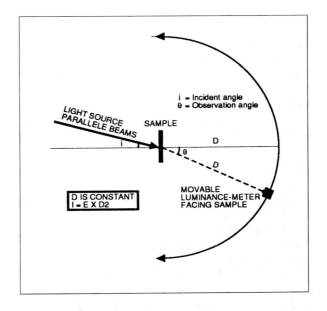

Figure 1 - The characterisation of the optical properties of materials requires the determination of the transmitted, or reflected, intensity for each value of i and θ.

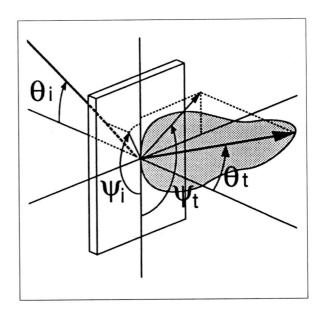

Figure 2 - Light distribution curves can be established when each transmitted or reflected light beam intensity is known for each θ_i and ψ_i angle of incident light.

The total number of data to evaluate in these conditions would be 6.72 x 10^{10}, which is huge. The major difficulty would be to repeat the measurement for each angle of incidence (i) and each angle of transmission (θ) which represents $(360 \times 90)^2 = 1$ billion operations. The equipment able to perform such an analysis would be a spectro-photogoniometer. No such equipment is currently available.

However, photogoniometers are currently used all over Europe by lamp and luminaire manufacturers to determine intensity distribution curves of lamps and luminaires. Their application to material characterisation would require the setting up of a complex system to generate light in a precise manner.

Spectrometers have a widespread use. The range of use covers research (physics, nuclear research, astronomy, biology, etc) as well as industry (characterisation of colours, quality control, etc). Usually, they operate in one single direction, and may work at wavelength ranges and definitions which are highly specific to the use.

It is clear that such a detailed procedure is not adapted to the problems a daylighting system designer may have to face. However, materials must be described since their effect on natural light penetration and distribution is crucial. In particular, it is important to concentrate on:

(A) diffuse reflection (or transmission) coefficients under diffuse light (uniform light)
(B) diffuse reflection (or transmission) coefficients under direct beam light (depending on the incident angle (i))
(C) specular reflection (or transmission) as a function of the incident angle
(D) shape of the intensity distribution curve, which will determine the appearance of the material: mat, glossy, specular or complex
(E) colour change of light after reflection or transmission.

Information on A and C is usually given by manufacturers. Information on A, B, C and D is essential for numerical simulations of daylight and artificial light distribution. Information on E could be determined by the evaluation of the change in colour temperature of light after reflection (or transmission).

Classification

Few manufacturers of wall and floor coverings give information regarding the reflective properties of their materials.

Mat surfaces (such as matt paints) can be described by their diffuse reflection coefficient. With respect to daylighting computations, the major effect is the impact on the calculation of the Internal Reflected Component (IRC) of the Daylight Factor.

The indoor wall, ceiling and floor surfaces are not the only opaque surfaces of interest. In fact,

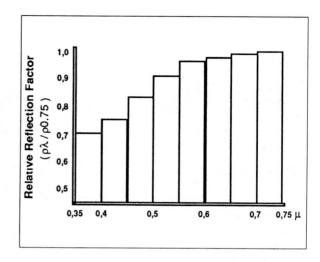

Figure 3 - Relative spectral reflection coefficient of plain concrete (from ENTPE)

designers seem to be looking more and more for a better control of natural light entering a building, particularly with sunlighting techniques. Such solutions tend to use highly reflective surfaces such as mirrors or glossy paint. Mirrors can be characterised through their perpendicular reflection coefficient. Except for cases such as selective coatings, special reflectance values for other incidence angles usually follow a law similar to glazing behaviour (i.e.: $R = R_o (\cos i)^{0.5}$, where i is the angle of incidence).

Glossy surfaces reflect light in a spread around the maximum in the direction of specular reflection. The length of this peak characterises the intensity of this reflection, the width establishes a range between specularity (narrow peak) and diffusion (even spread).

Figure 4 shows a suggested classification of the reflection and the transmission properties of materials. The figure can be directly associated with the phenomena perceived through simple visual observation.

Three parameters are found necessary to describe most common surfaces:
- DIF: Diffuse reflectance
- SPE: Specular reflectance
- SCA: Disperse or scatter reflectance (narrow or wide).

Each of the dispersion levels, described below, is easily observable (see Figure 5):
- Specular reflection, SPE, allows the exact image of the source to be seen
- Low dispersion (Narrow Scatter), SCA N, allows the perception of a bright spot corresponding to a light source such as a lamp; the image cannot be seen.
- In the case of high dispersion (Wide Scatter), SCA W, a bright spot cannot be seen, and the light is reflected in a highly non-uniform manner.

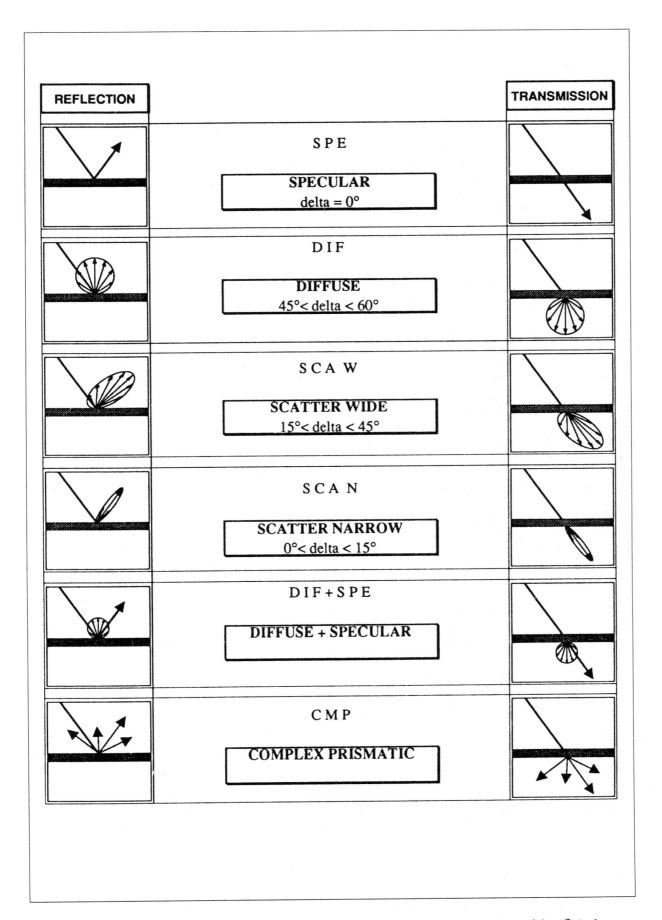

Figure 4 - Suggested classification of both transmission and reflection patterns, where delta (δ) is the dispersion angle for half the intensity.

4.4 Daylighting in Architecture

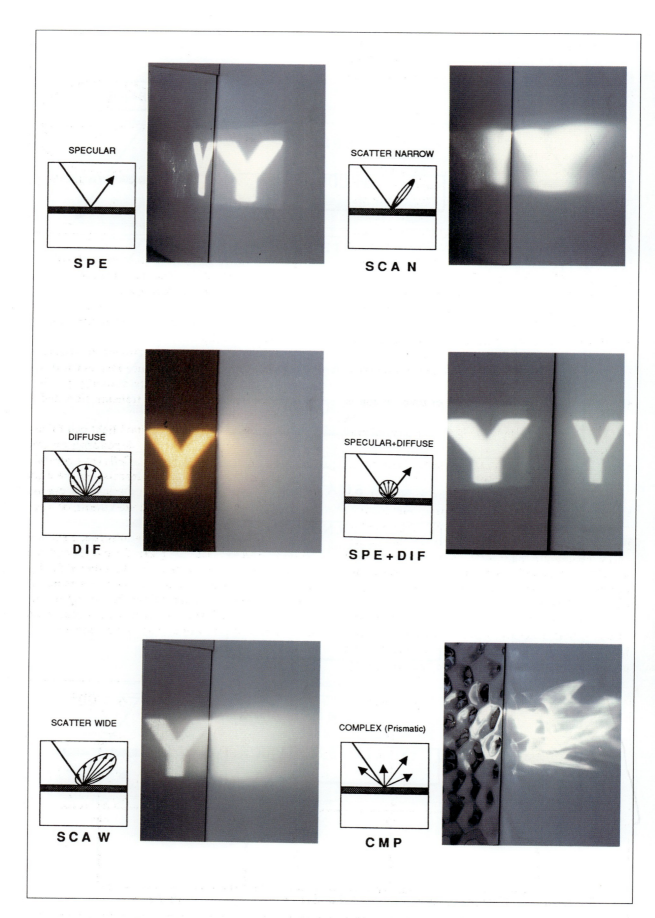

Figure 5 - Examples showing a range of light distribution characteristics.

Low dispersion surfaces reflect light in a softer way than mirrors. High dispersion surfaces allow little control of light reflection but may provide protection from glare.

Dispersion Angle

The following parameter is proposed to characterise dispersion:

The dispersion angle, δ, also called the Half Value Angle, is the angle between the direction of maximum intensity (I_{max}) of transmitted or reflected light, and direction of intensity with a value of $I_{max}/2$, when the intensity distribution curve can be supposed to be symmetrical about the direction of I_{max} (this is typically the case when the incidence angle is zero) (see Figure 6).

The following list represents typical classification values for the dispersion angle (δ) as shown in Figure 7:
- If the reflection or transmission is perfectly diffuse then $\delta = 60°$
- If δ is close to $0°$, the reflection or transmission is considered to be specular
- If $0 < \delta < 15°$, the reflection or transmission is narrow scatter
- If $15° < \delta < 45°$, the reflection or transmission is wide scatter
- If $45° < \delta < 60°$, the reflection or transmission can be considered as diffuse.

The nomenclature can be used in exactly the same way for reflection or transmission characteristics, but the consequent photometric patterns may be very different. Complex photometric characteristics are more common in the transmission mode than in the reflection mode. This is the case for panels of polycarbonate, made with double layers and vertical spacers.

For specular transmittance materials (glass and others), manufacturers provide perpendicular luminous transmittances, and the variation of these transmittances as a function of the incidence angle

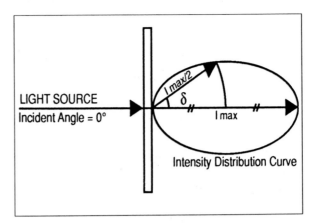

Figure 6 - Definition of the dispersion angle (δ) for transmitted light of normal incidence (Half Value Angle).

can be obtained either from the manufacturer or from other sources (1), (2).

It is rare that diffusing transmission materials are perfectly diffusing (ie the light emission follows Lambert's Law). Manufacturers normally provide only an estimate of the light transmission under perpendicular incidence.

It is known that the transmitted light can follow extremely complex patterns, depending on the direction of the incident light. Cellular materials offer a wide range of such transmission patterns which, depending on their surface treatment, can be specular (images can be seen through them), or offer a low or high dispersion.

When looking through a low dispersion surface (Narrow Spread), one can guess the shapes and see the colours of each element on the other side. By contrast, a high dispersion transmission material (Wide Spread), does not allow the perception of images or forms. Therefore, translucent surfaces are described with the same kind of nomenclature as opaque reflection surfaces.

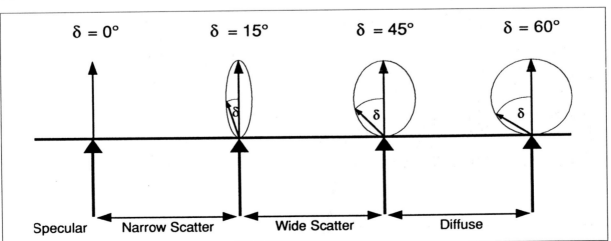

Figure 7 - Suggested classification of light distribution patterns as a function of the dispersion angle (δ).

Corrugated materials have been shown to present complex patterns of transmitted light. Another classification category that can be mentioned is the forward and backward specularity. For these reasons, a final category is proposed dealing with complex photometric properties where reflection or transmission patterns do not follow any standard model.

At the present level of research, it seems that only qualitative considerations associated with light transmission patterns can be determined. The classification presented here corresponds to different applications, depending on the expected trade-off between outdoor environmental perception requirements and the expected distribution of transmitted light.

Experimental Set-up for Classification

To complement the classification of surface photometric properties presented above, a system which allows one to determine easily the way materials distribute light is discussed (see Figures 8, 9 and 10).

The principle is to compare the material to be tested with a reference material, the light distribution characteristics of which are known. The main advantage of this system is that it does not require any measuring instruments. By adjusting the diaphragm until the eye perceives the same transmitted or reflected illuminance, the way a material distributes light is established precisely enough to classify it in one of the categories.

Figure 11 shows two transmitted luminous intensity curves for a translucent laminated glass sample. One of these curves is provided by a photometric test bench, and the other by the Optical Comparative Light-meter.

The precision of this system is satisfactory and errors result mainly from the discrete stepped increase of the diaphragm aperture. The use of a continuously variable diaphragm system will improve the precision of this tool; however, the present system is more than accurate enough to characterise photometric properties of materials.

Colour

The colour of light is associated with the physiological perception of the luminous flux received from different surfaces. This flux can be described by the distribution of the spectral flux over the visual spectrum.

It has been demonstrated that the colour of light as we see it can be characterised by comparing it with a reference coloured light source which uses a mix of three reference colour sources. Various systems to establish colour categorisation have been proposed, and some adopted by the *Commission Internationale de l'Eclairage* (CIE).

With respect to most daylighting design techniques, the colour of a material is not in itself a topic of interest. However, the effect of materials on the colour of the indoor environment should be known since occupants might respond negatively to an inappropriate colour of light.

Colour Temperature

The colour of incoming natural light is the major issue. The first approach described here is to determine the shift of the colour temperature of natural light after transmission or reflection. Using a chroma-meter, and with daylight as the source of illumination, various samples of reflectors and glazing materials were tested under direct and diffuse natural light. The colour temperature of the light source was each time around 6000 K (the sky was overcast). Some results are presented in Tables I and II.

For most materials there is a slight reduction of the colour temperature, which tends to be unnoticeable by observers, especially when they are not offered the opportunity of comparing light sources. Some materials lower the colour temperature significantly down to as low as 4000 K. This effect might be interesting in the case of daylit buildings with low requirements in light levels.

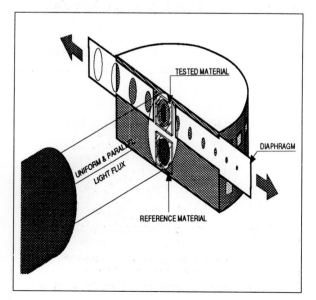

Figure 8 - Experimental set-up to characterise light distribution patterns in transmission mode (Optical Comparative Light-meter, all rights reserved B. Paul and M. Fontoynont, 1987).

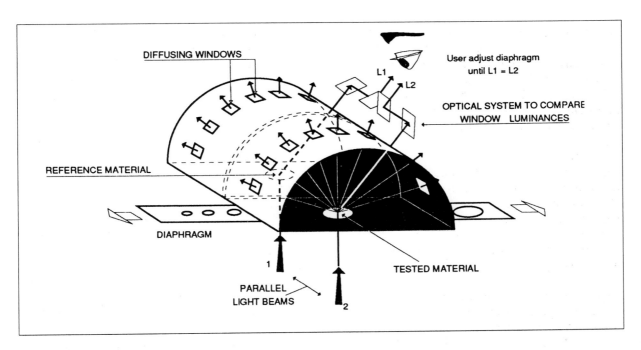

Figure 9 - The Optical Comparative Light-meter in the transmission mode (all rights reserved Paul and Fontoynont, 1987).

SELECTION OF APPROPRIATE MATERIALS

Even if daylighting performance is the major concern in the choice of a material, the photometric properties cannot be dissociated from other considerations such as durability, maintenance, safety, installation constraints or costs. Indoor surfaces are not exposed to the same parameters as outdoor ones regarding deterioration risks, cleaning techniques, or fixture design. Furthermore, fire safety regulations depend on the type of building which is considered, and impose restrictions on many organic materials.

Tables III to IX show important properties of materials which must be considered in conjunction with the photometric parameters in the selection of appropriate materials.

Outdoor and Indoor Reflectors

Table IX proposes a way of comparing reflecting mirror surfaces in a very accessible manner, in order to extract quickly the set of materials appropriate to a problem. For extended and detailed use, this table should refer to a database for more information.

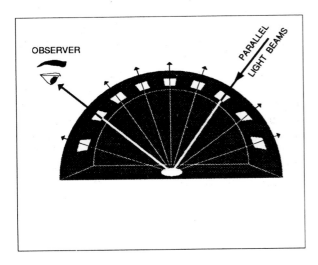

Figure 10 - The Optical Comparative Light-meter in the reflection mode (from ENTPE).

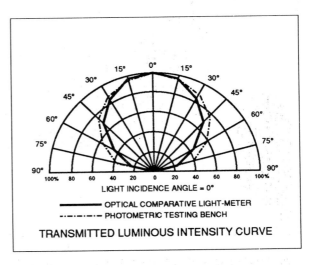

Figure 11 - Comparative transmission pattern from photometric bench and Optical Comparative Light-meter tests on translucent laminated glass.

TABLE I - Shift of natural light colour temperature (K) after transmission

Material	Category	Direct Source			Diffuse Source		
		Source Temperature	Transmitted Temperature	Change	Source Temperature	Transmitted Temperature	Change
Clear Glass	SPE	5130	5065	- 65	6000	5930	- 70
Transparent Acrylic	SPE	5250	5200	- 50	6000	6000	0
'Antelio Clair'	SPE	5250	4615	- 635	6030	5265	- 765
'Antelio Havane'	SPE	5265	4080	- 1185	6050	4530	- 1520
Opalescent White Acrylic	DIF	5250	5115	- 135	6050	5730	- 350
Standard White Paper	DIF	5265	5900	+ 635	6050	7400	+ 1350
Opalescent Laminated Glass	SCA W	5250	5050	- 200	6050	5830	- 220
Embossed Acrylic	SCA N	5250	5250	0	6000	6000	0
Embossed Glass	SCA N	5250	5250	0	6000	6050	+ 50
Polyester Fabric	DIF + SPEC	5250	5000	- 250	6000	5730	- 270
PVC Triple Structured Sheet	COMP	5250	5030	- 220	6000	5850	- 150

TABLE II - Shift of natural light colour temperature (K) after reflection

Material	Category	Direct Source			Diffuse Source		
		Source Temperature	Transmitted Temperature	Change	Source Temperature	Transmitted Temperature	Change
Glass Mirror	SPE	5300	5200	- 100	6000	5675	- 325
Glass plus Chromium Coating	SPE	5250	5100	- 150	6000	5630	- 370
Acrylic Mirror	SPE	5310	5140	- 170	6000	5920	- 80
Polished Steel	SPE	5320	4880	- 440	6000	4950	- 1050
Concrete	DIF	5270	4900	- 370	5670	5230	- 440
Grass	DIF	5350	3470	- 1880	6000	3955	- 2045

TABLE III - Properties of transparent materials (specular)

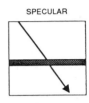

SPECULAR

S P E

Materials	Manufacturer	Reference	Colour	Transmission Factor	Solar Factor	Mechanical Resistance	Resistance to Corrosion	Resistance to Abrasion	Cutting Drilling	Folding	Resistance to Ignition	Stability under Fire	Durability of Optical Properties	Material Cheapness	Installation Cheapness
Clear Glass	St Gobain	Plannilux 4 mm	-	0,90	0,88	○ (1)	●●●	●●●	○ (2)	- (3)	M0	○	●●●	○○○	○
Clear Glass	St Gobain	Plannilux 6 mm	-	0,89	0,82	○ (1)	●●●	●●●	○ (2)	- (3)	M0	○	●●●	○○○	○
Low emissivity Glass	St Gobain	Eko 4mm	-	0,80	0,80	○ (1)	●●●	●●●	○ (2)	- (3)	M0	○	●●●	○○	○
Low emissivity Glass	St Gobain	Eko 6mm	-	0,79	0,79	○ (1)	●●●	●●●	○ (2)	- (3)	M0	○	●●●	○○	○
Colored Glass	St Gobain	Parsol	Varied	0,61 (4) / 0,27	0,78 (4) / 0,48	○ (1)	●●●	●●●	○ (2)	- (3)	M0	○	●●●	○○	○
Fire proof Glass	St Gobain	Contraflam	-	0,85	?	○ (1)	●●●	●●●	○ (2)	- (3)	M0	○○○ (5)	●●●	○	○
Laminated Glass	St Gobain	Stadip 44-2	-	0,89	0,82	○○○	●●●	●●●	○	-	M2	○	●●●	○	○
Laminated Glass	St Gobain	Stadip 44-2	Varied	0,58 (4) / 0,37	0,63 (4) / 0,51	○○○	●●●	●●●	○	-	M2	○	●●●	○	○
Laminated Glass	St Gobain	Stadip 35-2	Silvered	0,45	0,55	○○○	●●●	●●●	○	-	M2	○	●●●	○	○
Reflective Glass	St Gobain	Parelio 6mm	Varied	0,59 (4) / 0,29	0,61 (4) / 0,41	○○	●●●	●●●	-	-	M0	○	●●●	○	○
Acrylic	Altulor	Altuglass 0033	-	0,92	?	○○○	○○	○	●●●	●●●	M4	○	○○○	○	○○
Polycarbonate	Axxis PC	PC 111	-	0,86	0,87	●●●	○○	○ (6)	●●●	○○○	M3	○ (7)	○ (8)	○	○○
Polycarbonate	Axxis PC	PC 121	Bronze	0,50	0,60	●●●	○○	○ (6)	●●●	○○○	M3	○ (7)	○ (8)	○	○○
Double Strength Glass	-	-	-	0,86	0,86	○	●●●	●●●	-	-	M0	○	●●●	○○	○

Key

- ●●● Very good
- ○○○ Good
- ○○ Average
- ○ Low
- - Non-existent
- ? Unknown

- M0 Incombustible
- M1 Very slightly inflammable
- M2 Slightly inflammable
- M3 Fairly inflammable
- M4 Easily inflammable
- M5 Very easily inflammable

(1) Improvement of resistance if hardened or laminated
(2) Cutting only, except if hardened
(3) Thermal forming
(4) Depends on colour and thickness
(5) Keeps its characteristics of mechanical stability, flame resistance, non-emissivity of flammable gases, and limited heating during one hour
(6) Improvement of resistance if special anti-abrasion treatment
(7) Improvement of resistance if fire-proof treatment
(8) Improvement of optical properties' durability if specific anti-UV treatment

4.10 Daylighting in Architecture

TABLE IV - Properties of translucent materials (diffuse)

Materials	Manufacturer	Reference	Colour	Transmission Factor	Solar Factor	Mechanical Resistance	Resistance to Corrosion	Resistance to Abrasion	Cutting Drilling	Folding	Resistance to Ignition	Stability under Fire	Durability of Optical Properties	Material Cheapness	Installation Cheapness
Acrylic	?	?	White	0,50	?	○○○	○○	○	●●●	●●●	M4	○	○○○	○	○○○
Polyester Fabric PVC Coating	Serge Ferrari	Cristal 835	White	?	?	○○○	○○○	○	●●●	●●●	-	○	○○○	○○○	○○○
" "	" "	Precontraint 402	White	?	?	○○	○○○	○	●●●	●●●	M2	○	○○○	○○○	○○○
" "	" "	Cristal 635	White	?	?	○○	○○○	○	●●●	●●●	-	○	○○○	○○○	○○○
" "	" "	630 Automate	White	?	?	○○	○○○	○	●●●	●●●	-	○	○○○	○○○	○○○
" "	" "	Flash 600	White	0,45	?	○○	○○○	○	●●●	●●●	?	○	○○○	○○	○○○
" "	" "	Flash 552	White	0,45	?	○○	○○○	○	●●●	●●●	M2	○	○○○	○○	○○○
" "	" "	" "	Yellow	0,40	?	○○	○○○	○	●●●	●●●	M2	○	○○○	○○	○○○
" "	" "	" "	Red	0,14	?	○○	○○○	○	●●●	●●●	M2	○	○○○	○○	○○○
" "	" "	" "	Blue	0,11	?	○○	○○○	○	●●●	●●●	M0	○	○○○	○○	○○○

TABLE V - Properties of translucent materials (wide scatter)

SCATTER WIDE

S C A W

Materials	Manufacturer	Reference	Colour	Transmission Factor	Solar Factor	Mechanical Resistance	Resistance to Corrosion	Resistance to Abrasion	Cutting Drilling	Folding	Resistance to Ignition	Stability under Fire	Durability of Optical Properties	Material Cheapness	Installation Cheapness
Laminated Glass	St Gobain	Stadip 44 2 opale 1	-	0,57	0,63	○○○	●●●	●●●	○	-	M2	○	●●●	○	○
Glass Paving	St Gobain	Lumax 2210 L	-	0,50 0,60	?	○○○	●●●	●●●	-	-	M0	○	●●●	○○	○○○
Ground Glass	?	?	Varied	?	?	○	●●●	●●●	○	○	M0	○	●●●	○○	○
Triple structured sheet (Polycarbonate)	Everlite	E 609 n°1	Opale	0,68	?	●●●	○○○	○○	○○○	○ (1)	M1	○	●●●	○○	○○○
Triple structured sheet (PVC)	Everlite	E 730 n°1	Opale	0,70	?	●●●	○○○	○○	○○○	○ (1)	M1	○	●●●	○○○	○○○

Key

(1) Can be bent

TABLE VI - Properties of translucent materials (narrow scatter)

SCATTER NARROW

SCAN

Materials	Manufacturer	Reference	Colour	Transmission Factor	Solar Factor	Mechanical Resistance	Resistance to Corrosion	Resistance to Abrasion	Cutting Drilling	Folding	Resistance to Ignition	Stability under Fire	Durability of Optical Properties	Material Cheapness	Installation Cheapness
Embossed Glass	St Gobain	Listral 200	-	0,80 0,90	?	O	●●●	●●●	O	O	M0	O	●●●	OOO	O
" "	" "	Imprime 077	-	" "	?	O	●●●	●●●	O	O	M0	O	●●●	OOO	O
" "	" "	Boreal 108	-	" "	?	O	●●●	●●●	O	O	M0	O	●●●	OOO	O
" "	" "	Goutte d'eau 054	-	" "	?	O	●●●	●●●	O	O	M0	O	●●●	OOO	O
Wire Glass	" "	Uni Arme 698	-	" "	?	OO	●●●	●●●	O	-	M0	-	●●●	OO	O
Embossed Acrylic	Altulor	?	-	0,85	?	OOO	OO	O	●●●	●●●	M4	O	OOO	O	OO

TABLE VII - Properties of translucent materials (specular and diffuse)

SPECULAR+DIFFUSE

SPE+DIF

Materials	Manufacturer	Reference	Colour	Transmission Factor	Solar Factor	Mechanical Resistance	Resistance to Corrosion	Resistance to Abrasion	Cutting Drilling	Folding	Resistance to Ignition	Stability under Fire	Durability of Optical Properties	Material Cheapness	Installation Cheapness
Shading Device Fabric	S. Ferrari	Soltis 92	Alu Sand	0,08	0,05 0,08	OOO	OOO	-	●●●	●●●	M2	O	OOO	OOO	OOO
" "	" "	Soltis 86	Bronze Grey Alu	0,14	0,13	OOO	OOO	-	●●●	●●●	M2	O	OOO	OOO	OOO
" "	" "	Soltis 78	Alu Black	0,22	0,35 0,16	OOO	OOO	-	●●●	●●●	M2	O	OOO	OOO	OOO
" "	" "	Soltis 87	Varied	0,22	?	OOO	OOO	-	●●●	●●●	M2	O	OOO	OOO	OOO
" "	Bat Taraflex	Batyline HM	Varied	0,45	?	OOO	OOO	-	●●●	●●●	M1	O	OOO	OOO	OOO
" "	" "	Batyline HM Calendree	Varied	0,20	?	OOO	OOO	-	●●●	●●●	M1	O	OOO	OOO	OOO

TABLE VIII - Properties of translucent materials (complex)

COMPLEX (Prismatic)

C M P

Materials	Manufacturer	Reference	Colour	Transmission Factor	Solar Factor	Mechanical Resistance	Resistance to Corrosion	Resistance to Abrasion	Cutting Drilling	Folding	Resistance to Ignition	Stability under Fire	Durability of Optical Properties	Material Cheapness	Installation Cheapness
Polycarbonate Structured sheet	Poly-U	450	-	0,83	?	○○○	●●●	○	●●●	○ (1)	M3	○	○○○	○○○	○○○
" "	" "	410	-	0,70	?	○○○	●●●	○	●●●	○ (1)	M3	○	○○○	○○○	○○○
PVC triple Structured Sheet	Everlite	E 109 n°2	-	0,63	?	○○○	●●●	○	●●●	○ (1)	M1	○	○○	○○○	○○○
" "	" "	E730 n°2	-	0,70	?	○○○	●●●	○	●●●	○ (1)	M1	○	○○	○○○	○○○
Polycarbonate Structured Sheet	" "	E 609 n°2	-	0,74	?	○○○	●●●	○	●●●	○ (1)	M2	○	○○○	○○	○○○
Prismatic Panels Acrylic	Siemens	?	-	?	?(2)	○○○	●●●	○	●●●	○ (1)	?	?	○	○	○○
Complex Blinds Aluminium	Koster	Okalux	-	?	?(2)	- (3)	- (3)	- (3)	- (3)	- (3)	- (3)	- (3)	○○○	○	○○

Key

(1) Can be bent
(2) Depends on solar ray incidence
(3) Similar characteristics to double strength glass

TABLE IX - Properties of reflective materials (specular)

Materials	Manufacturer	Reference	Colour	Transmission Factor	Solar Factor	Mechanical Resistance	Resistance to Corrosion	Resistance to Abrasion	Cutting Drilling	Folding	Resistance to Ignition	Stability under Fire	Durability of Optical Properties	Material Cheapness	Installation Cheapness
Glass Mirror	-	-	Silver	0,90	-	○ (1)	○ (2)	●●●	○ (3)	○ (4)	M0	○	○○○	○○○	○
" "	-	-	Bronze	0,45	-	○ (1)	○ (2)	●●●	○ (3)	○ (4)	M0	○	○○○	○○○	○
Glass (Chromium Coating)	Hirtz	-	Silver	0,70	?	○ (1)	●●●	●●●	○ (3)	-	M0	○	●●●	○	○○
Reflective Glass Pyrolitic	St Gobain	Antelio clair	Silver	0,34	0,56	○ (1)	●●●	●●●	○ (3)	-	M0	○	●●●	○○	○○
" "	" "	Antelio havane	Bronze	0,35	0,39	○ (1)	●●●	●●●	○ (3)	-	M0	○	●●●	○○	○○
Reflective Hardened Glass	" "	Parelio clair	Silver	0,37	0,62	○○	●●●	●●●	-	-	M0	○	●●●	○○	○○
" "	" "	Parelio bronze	Bronze	0,36	0,44	○○	●●●	●●●	-	-	M0	○	●●●	○○	○○
" "	" "	Parelio creole	Creole	0,36	0,46	○○	●●●	●●●	-	-	M0	○	●●●	○○	○○
" "	" "	Parelio gris	Grey	0,36	0,45	○○	●●●	●●●	-	-	M0	○	●●●	○○	○○
" "	" "	Parelio vert	Green	0,37	0,41	○○	●●●	●●●	-	-	M0	○	●●●	○○	○○
" "	" "	Parelio corail	Coral	0,37	0,58	○○	●●●	●●●	-	-	M0	○	●●●	○○	○○
Acrylic (Al Coating)	Forum	Oroglass Argent	Silver	0,85	-	○○○	○ (2)	○	●●●	●●●	?	○	○○○	○	○○
Polyester Reflective Film	Sun X	88151 Winter	Silver	0,43	0,11	○○ (5)	○○○	○	●●●	-	?	○○	○○	○○	○
" "	" "	9020 22 brz	Bronze	0,31	0,22	○○ (5)	○○○	○	●●●	-	?	○○	○○	○○	○
" "	Reflectiv	Sol 101	Silver	0,50	0,21	○○ (5)	○○ (6)	○	●●●	-	?	○○	○○	○○	○
" "	" "	Sol 103	Grey	0,37	0,35	○○ (5)	○○ (6)	○	●●●	-	?	○○	○○	○○	○
Anodised Aluminium	Satma	Bandoxal 1499S	Silver	0,92	-	○	○○ (6)	○	●●●	●●●	M0	○	○○○	●●●	●●●
" "	" "	Bandoxal 1080	Silver	0,72	-	○	○○ (6)	○	●●●	●●●	M0	○	○○○	●●●	●●●
Polished Steel	-	-	Silver	0,55	-	○	○○ (7)	○	●●●	●●●	M0	○	○○○	○○○	●●●

Key

(1) Improvement of resistance if hardened or laminated
(2) Specific protection of reflective surfaces and edges required
(3) Cutting only, except if hardened
(4) Thermal forming, except of hardened or laminated
(5) Improves mechanical resistance of glass
(6) Indoor applications only
(7) Risk of tarnishing

Indoor Surface Coatings

Indoor surface photometric properties are often assumed to be diffuse, especially in most lighting simulation equations and programs. In fact, few finishes come close to being totally diffusing. Most mat paints present some dispersion (wide spread).

Diffuse

DIF

There are only a small number of perfectly diffusing materials, among them the following can be mentioned:
- Moquette
- Velvet seen from particular angles
- Mineral fibre panels
- Porous or granular concrete
- Grass.

Wide Scatter

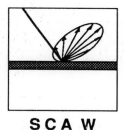

SCAW

Most surface materials which are employed in buildings come under the "wide scatter" heading. Such materials diffuse the light they receive, but the reflected light retains a general direction which depends on the incident light direction. The following materials have this characteristic:
- Mat paints
- Polyester fabrics
- Concrete
- All mat surfaces.

Narrow Scatter

SCAN

Under the heading of "narrow scatter" are classified all kinds of materials which present a glossy appearance, without producing a distinct reflection of the light source:
- Satin paints
- Some types of veneered surfaces
- Some plastic coatings.

Specular and Diffuse

SPE+DIF

All materials and surfaces which can perfectly reflect a distinct image of the light source, without being considered as reflective materials, are considered to be "specular and diffuse" types. The specular reflection component is principally due to surface coating characteristics, whereas the diffuse reflection component mainly depends on the nature of the underlying layer, as for example:
- Lacquered surfaces
- Varnished surfaces
- Highly polished surfaces
- Some sorts of stratified surfaces.

Complex

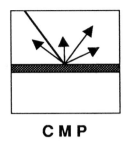

Under the heading of "complex" are all reflective materials which present non-uniform and non-flat surfaces, so the light is randomly reflected and distributed, such as:
- Corrugated surfaces
- Irregular reflective surfaces
- Creased or crumpled metal
- Prismatic surfaces.

NEW MATERIALS

Prismatic Devices

One way to control sunlight penetration into a building is to increase the sensitivity of the transmission factor as a function of the incidence angle, and especially as a function of the solar altitude angle (Figures 12, 13 and 14). This may be achieved by employing Fresnel-type lenses made of glass or acrylic material. Devices can consist of a single prismatic layer (3) or two successive layers (4).

Prismatic devices can be made inexpensively but their poor transparency restricts their use to industrial buildings or some very specific locations, such as the upper parts of façade windows (5).

Such products have been included in the catalogues of large glass manufacturers but have been mostly abandoned owing to a lack of demand. However, recent initiatives by companies such as 3M should be followed closely (6).

Holographic Films

The principle of such films is to intercept sunlight and to diffract part of the radiation in another direction. When applied to a window incoming sunlight can be redirected deeply into a room (Figure 15), and, furthermore, the luminous efficacy of this radiation can be increased (7). Chapter 5 demonstrates the application of holographic components in more detail.

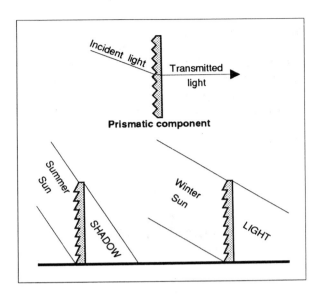

Figure 12 - Solar control with a single-layer prismatic component (3).

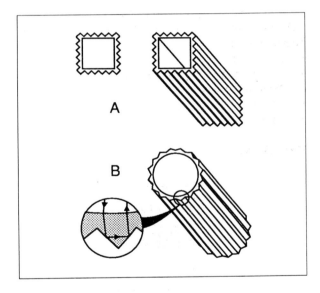

Figure 13 - Details of light guides: rigid wall rectangular guide (A) and circular guide illustrating ray geometry (B) (5).

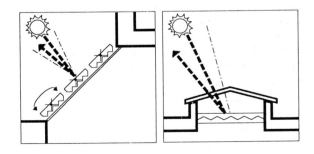

Figure 14 - Prismatic devices used in combination with window material (6).

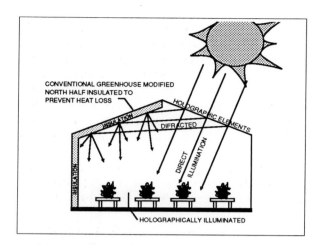

Figure 15 - An example of the use of holographic elements to improve greenhouse efficiency (8).

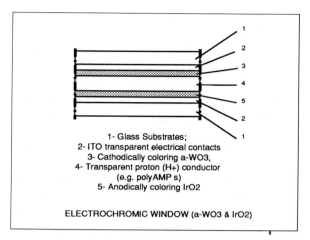

Figure 16 - Detail of electrochromic window (layer thicknesses are not drawn to scale) (13).

Holographic films are generated by a photographic process on a dichromate gelatin, which at present imposes a limitation on the maximum size of the films. Furthermore, such films are wavelength-sensitive and create rainbow reflections.

The major field of applications seems to be currently far removed from the building domain. Holographic devices are used to produce low-cost and light-weight optical systems for "head-up" displays in military jets, or light concentrators (9). The possibility of use in windows offers the best large-scale potential. Although tests and simulations have been performed (8 and 10), it seems that there is no production of such films in large dimensions.

Aerogels

Aerogel is a transparent, low density, thermally insulating solid, 1 to 10% of which is actually in solid state, the remaining fraction being open cell voids (11). The thermal conductivity of the component can be as low as 0.008 W/mK (12). Because the material is extremely lightweight it is fragile and needs to be encased between two plates of glass or plastic. It is transparent since the size of pores is smaller than the wavelengths of visible light.

Electrochromic Devices

These are active systems in that their operation requires an external source of energy (electricity). The principle consists of changing the optical absorption properties of certain laminated materials by an externally applied electric field (see Figures 16 and 17).

The tinting of the material is caused by a small external field, and the opacity remains for a while after the field is interrupted and disappears again when the field is reversed (13).

Typical transmission modulation varies between 70% and 15% in the visible spectrum, and such windows can undergo 4000 to 6000 cycles without degradation (14). In the case of an average of 5 cycles a day, this would lead to a typical usage time of two to four years maximum, which is not long enough for general use in the building industry. The number of successive layers makes the costs of such devices very high. However, the dynamic range is promising and makes this product competitive with movable shades without the complexity of dealing with movable parts. Furthermore, the simple DC operating mode makes it easy to integrate the control of the process in response to heating or cooling requirements as well as outdoor climatic parameters.

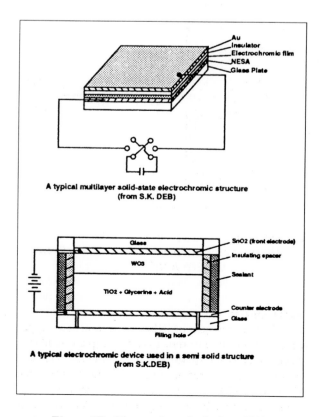

Figure 17 - Electrochromic devices (13).

Thermochromic Devices

The objective is to produce a material that will passively switch between a heat-transmitting and a heat-reflecting state at specific design temperatures within the human comfort range. The composition is usually either tungsten trioxide or vanadium dioxide. On windows, it could have the appearance of a thin film on the glass (15).

This product seems attractive for the reduction of cooling loads, and offers ways to provide sunshading in response to solar gain. The inconvenience may be the blocking of potentially useful solar beams during the heating season. The transmission factor can vary by a range of 1 to 3 (15), which makes this product attractive. Durability seems to be questionable at the moment.

Fluorescent Concentrators

When light comes from a material with an index of refraction higher than the one it is directed to, there can be a specific phenomenon called "Total Internal Reflection" on the boundary plane. When the angle of incidence of the incoming light exceeds a threshold value, the reflection coefficient becomes 1. This effect can be used to concentrate natural diffuse light inside certain materials (see Figure 18). The efficiency is low, since less than 10% of the light can be usefully converted. Furthermore, the dyes have a very specific colour (yellow, green, red) which limits the field of use.

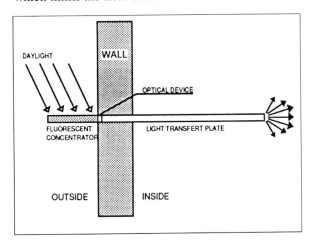

Figure 18 - Working principle of natural lighting system using fluorescence concentrators (from BASF).

Fibre Optics

The technique of fibre optics allows light propagation through thin cylindrical glass or plastic fibres. Losses are minimised due to the phenomenon of total internal reflection. This occurs when the angle of incidence (i), exceeds a certain value i_o. This value is such that $\sin i_o = n_2 / n_1$. The factor n_1 is the index of refraction of the first medium (for example air), and n_2 is the index of refraction of the second medium (for example glass): n_1 should be larger than n_2. For angles of incidence greater than the critical angle i_o, the reflection coefficient of the surface separating the two media is equal to 1. For instance, for the couple Glass/Air, the reflection indices of which are respectively 1.5 and 1, total reflection is achieved for angles of incidence between 42° and 90°.

The technology of fibre optics involves two glass or plastic materials. The material used in the centre has a higher index of refraction than the coating material.

Fibre optics are currently used for specific applications using electric light sources: communication and lighting devices. As regards daylighting, its potential is associated with the desire to bring light into deep plan or underground buildings.

Although light losses are much lower than most light pipe systems, the total flux which can be transmitted is limited. Furthermore, there are difficulties in the collection of natural light.

A technique has been developed in Japan by Key Mori (16), which includes a solar tracking device and a Fresnel-type concentrating optical system. Such a solution is aimed at sunlight alone. For daylight, a different type of light concentrator should be employed. A fluorescent concentrator is a possible solution, but efficiency appears to be very limited.

CONCLUSIONS

The brief overview of materials which has been presented in this chapter indicates the large number of possibilities opened up by the modification of light during its interaction with materials: light is scattered, diffused, sent in specific directions, following simple or complex patterns; it can be carried through guides and its spectral distribution can be greatly modified.

For the daylighting designer, the understanding of the behaviour of materials is essential to meet the luminous specifications of an architectural project. A poor selection of materials might annul the performances of components or daylighting techniques. On the other hand, the selection of an appropriate material can generate a very attractive space. Using inventive combinations of materials is, in itself, a field of exploration for the daylighting designer.

REFERENCES

(1) Groupement des Producteurs de Verre Plat, *White Book,* 1983.

(2) ASHRAE, *Handbook of Fundamentals*, American Society of Heating Refrigerating and Air Conditioning Engineering, New York, 1981.

(3) Nardini and Associates, *Patent Document*, Carimate, Italy.

(4) Ruck, N., "Beaming Daylight into Deep Rooms", *Bâtiment International*, Conseil International du Bâtiment, May-June, 1985.

(5) Saxe, S.G., "Progress in the Development of Prism Light Guides", *Proc. SPIE Conference*, vol. 692, San Diego, California, 1986.

(6) "Tageslichtsystem", *Produkt-Programm*, Siemens A.G., Germany.

(7) Hunt, A., "Holographic Window Coatings for Solar Control and Daylighting", *Assessment Report*, Lawrence Berkeley Laboratory, Berkeley, California, Report LBL-15305, November 1982.

(8) Bradbury, R. *et al.*, "Holographic Lighting for Energy Efficient Greenhouses", *Proc. SPIE Conference*, vol. 692, San Diego, California, 1986.

(9) Hull, J.L. *et al.*, "Holographic Solar Concentrator", *Proc. SPIE Conference*, vol. 692, San Diego, California, 1986.

(10) Ian, R., "Holographic Diffractive Structures for Daylighting", *Advanced Environmental Research Group Report*, Cambridge, Massachusetts, 1985.

(11) Hunt, A. and Loftus, K., "Silica Aerogels: A Transparent High Performance Insulator: Advances in solar technology", *Proc. ISES Conference*, Hamburg, Germany, Sept, 1987.

(12) Caps, R. and Fricke, *Journal of Solar Energy*, vol. 36, no. 361, 1986.

(13) Cogan, S.F. *et al.*, "Optical Switching in 'Complementary' Electrochromic Windows", *Proc. SPIE Conference*, vol. 692, San Diego, California, 1986.

(14) Deb, S.K., "Some Perspectives on Electrochromic Device Research", *Proc. SPIE Conference*, vol. 692, San Diego, California, 1986.

(15) Lee, J.C. *et al.*, "Thermochromic Materials Research for Optical Switching", *Proc. SPIE Conference*, vol. 692, San Diego, California, 1986.

(16) Mori, K., "The Hinawari", *Proc. UK-ISES Conference*, Imperial College, London, April, 1989.

Chapter 5
DAYLIGHTING COMPONENTS

INTRODUCTION

In order to take full advantage of all the benefits offered by daylighting it is necessary to acquire a deeper understanding of the behaviour of light. With this purpose in mind, it is important to analyse all the possibilities available to architects. The first step is to achieve a good approximation of the effects that the use of daylighting components has on architectural design. Then the different types of components available and their different combinations, together with their luminous performance and fields of application must be known.

To analyse all the possible daylighting components, the following areas have been developed in this chapter:
- classification system for daylighting components
- the basic component: the window
- description and performance of daylighting components
- fields of application
- experimental analysis of selected components.

To analyse and classify daylight components it is necessary to differentiate between two main groups, called "**conduction components**" and "**pass-through components**".

Conduction components, which at the simplest level are a space, guide and distribute light towards the interior of the building, connecting pass-through components together.

Pass-through components, of which the window is the most common, are devices designed to allow light to pass from one light environment to another.

Using this approach, a variety of component combinations can be established. Conduction components may be linked by pass-through components located at different points in the building. Each pass-through component can incorporate a set of **control elements** which are devices designed to admit and/or control the entry of light into a building.

GENERAL CLASSIFICATION SYSTEM FOR DAYLIGHTING COMPONENTS

In order to classify the general categories of conduction components, pass-through components and control elements, a table has been established which indicates possible relationships. The table also classifies them in order to allow for analysis of the individual behaviour of each item. The collective behaviour of a combination of components which form a system can be interpreted from this individual analysis.

This classification does not attempt to list all possible cases which are found in buildings. The aim is to present a general plan where specific cases of the more general types can be classified with respect to luminous behaviour.

A graphic presentation has been used to facilitate comprehension of the terminology. Moreover, the presentation offers a visual approach suitable for architects and designers. The range combinations of the various components provides a wide collection of possible design solutions (Figure 1).

Conduction Components

Conduction components can be defined as spaces designed to guide and/or distribute daylight towards the interior of a building, from one pass-through component to another. Two groups can be identified:

Group I: Intermediate light spaces. Conduction components which are part of the perimeter zone of a building, guiding and distributing daylight into attached interior spaces.

Group II: Interior light spaces. Conduction components which are part of the interior zone of a building, guiding and distributing daylight into specific zones of a building separated from the outside.

To identify the characteristics of conduction components they will be categorised according to their shape factor (relationship between the surface area of the space of a conduction component and that of a sphere of the same volume), slenderness (dimension in the light penetration direction, in relation to its width) and optical properties (type of reflection and reflection coefficient) (Figure 2).

5.2 Daylighting in Architecture

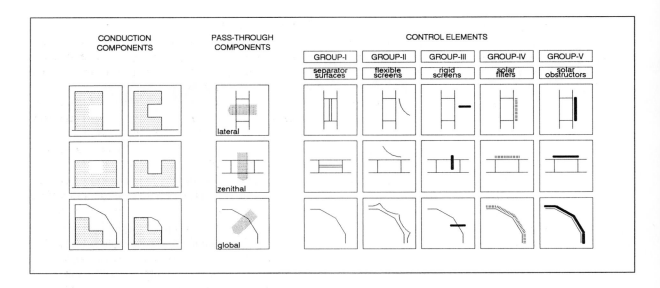

Figure 1 - General classification.

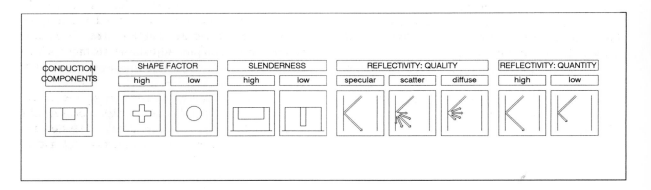

Figure 2 - Characteristics of conduction components.

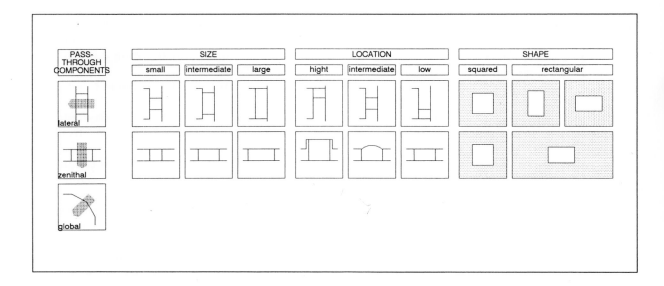

Figure 3 - Characteristics of pass-through components.

Pass-through Components

Devices under this heading are those which link two luminous environments, permitting light to pass from one to another. Three basic types can be identified:

Group I: Lateral pass-through components. Lateral pass-through components are those which are situated in the vertical envelope of a building. They separate two light environments, permitting lateral penetration of light.

Group II: Zenithal pass-through components. Zenithal pass-through components are situated in the roof of a building. They separate two light environments, allowing top-lighting or zenithal entry of daylight to the space below.

Group III: Global pass-through components. Global pass-through components are part of the enclosure of a constructed volume. They surround a space partially or totally, permitting a global (lateral and zenithal) entry of daylight.

These three groups of components may also incorporate control elements such as "separator surfaces", "flexible screens", "rigid screens", "solar filters" or "solar obstructors" in order to control the entry of daylight into the building.

These components will be analysed by their geometric characteristics such as: size, location and shape (Figure 3). On the basis of these geometric characteristics, general light penetration laws can be determined.

Control Elements

Devices specially designed to admit and/or control the entry of light through a pass-through component are referred to as control elements. Five categories are identified:

Group I: Separator surfaces. These are elements of a transparent or translucent material which separate two light environments, permitting light to pass through while not admitting air and sometimes obstructing the view.

Group II: Flexible screens. These elements partially or totally obstruct direct sunlight and diffuse daylight, allowing natural ventilation. They can be opened or closed to control views.

Group III: Rigid screens. Rigid and opaque elements redirect and/or obstruct direct solar radiation falling upon a pass-through component, and are normally fixed structures which cannot be regulated.

Group IV: Solar filters. Such elements cover the entire surface of an opening, protecting the interior zones against direct solar radiation whilst allowing ventilation. They can be fixed or adjustable.

Group V: Solar obstructors. Elements composed of opaque, adjustable surfaces which cover the whole of an opening are called solar obstructors.

All the elements included in these five groups can be analysed according to their location, mobility and optical properties (transparency, diffusion and re-direction), as summarised in Figures 4, 5, 6 and 7.

5.4 Daylighting in Architecture

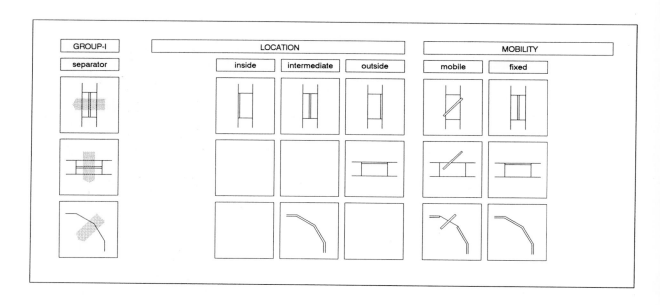
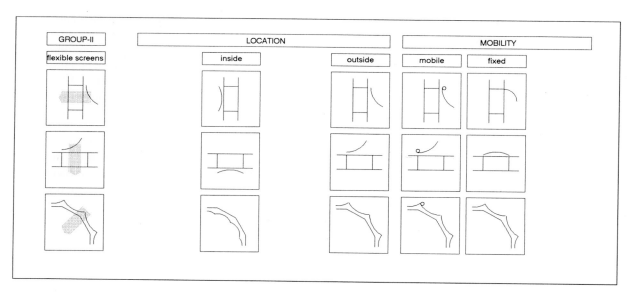
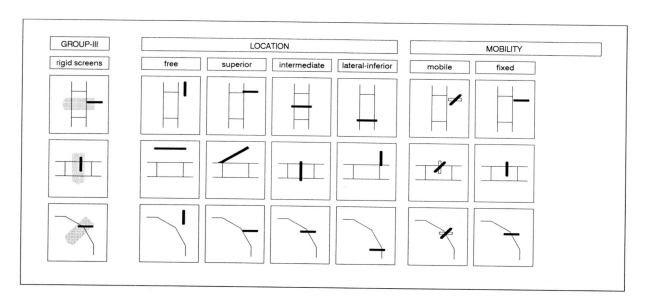

Figure 4 - Location and mobility characteristics of separator (I), flexible screen (II) and rigid screen (III) control elements.

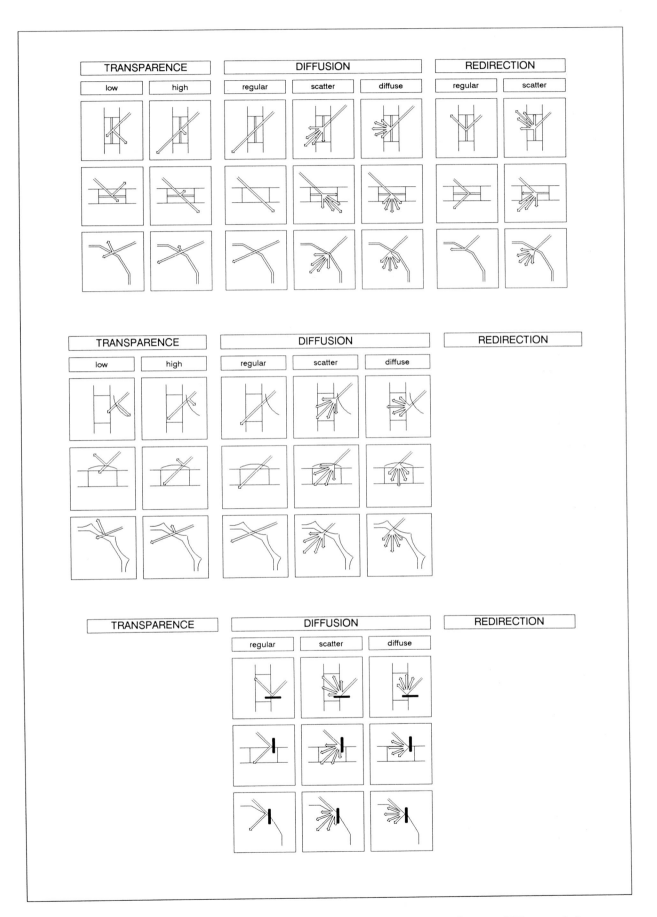

Figure 5 - Optical characteristics of separator (I), flexible screen (II) and rigid screen (III) control elements.

5.6 Daylighting in Architecture

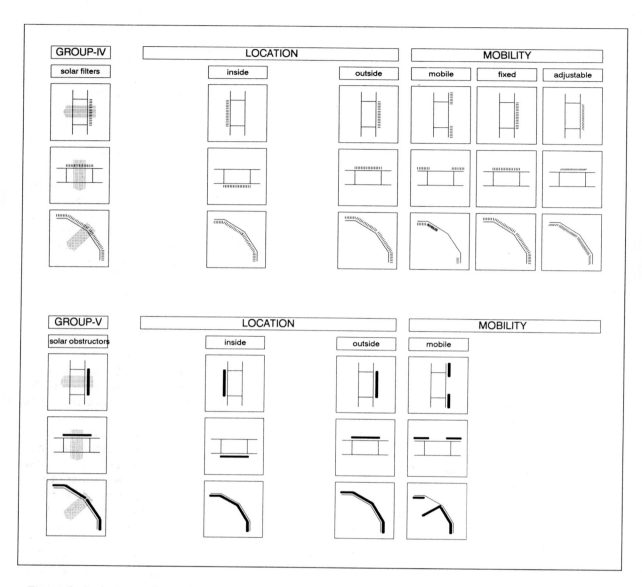

Figure 6 - Location and mobility characteristics of solar filter (IV) and solar obstruction (V) control elements.

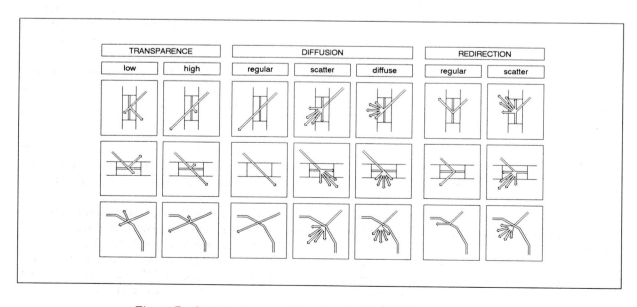

Figure 7 - Optical characteristics of solar filter (IV) control elements.

THE BASIC COMPONENT: THE WINDOW

A window is an opening in the vertical enclosure of a building which allows an interrelationship between the exterior and the interior. A window permits luminous, thermal, and acoustic interchange as well as natural ventilation and view. The correct use of this component can greatly improve the visual aspects of the interior of a building.

A window is characterised by its type, size, shape, position and orientation. In addition, controls may be added to regulate specific interchanges.

Type

To classify windows with regard to type, the following criteria for characterisation will be used: daylighting, exterior view, and natural ventilation. Five main types of windows may be distinguished:
- window for daylighting
- window for natural ventilation
- window for daylighting and exterior view
- window for daylighting and natural ventilation
- window for daylighting, exterior view and natural ventilation.

In defining these types of windows, reference must be made to characteristics such as: size, shape, position, orientation or controls which will be analysed more extensively later.

If the most important function of the window is illumination, it is usually best to locate it in a high position and size it to optimise the entry of natural light. If the ventilation aspect is to be favoured, its position in the wall is more important than its size. For a better exterior view, the size of the window and the height of the sill from the floor are extremely important. The lower the window, the more favourable it will be for views. In practice these three functions are combined in the most common types of window.

The description of a window can be based on its functional characteristics, but there are innumerable window types, depending on the materials used, situation in the facade, style, etc. Since the nomenclature varies according to each country, climatic zone, etc, only the analysis of windows according to their functions will be taken into account in the general classification presented here.

Size

A distinction will be made between the absolute surface area of a window, and fenestration (total window surface in relation to the area of the room which is illuminated by the window, expressed as a percentage).

The "absolute surface" of windows influences only the possibilities of ventilation and vision. "Fenestration" affects the amount and distribution of light.

Absolute surface (m^2). Windows will be classified according to size, taking into account the human scale:
- small: surface less than $0.5\ m^2$
- medium: surface between 0.5 and $2\ m^2$
- large: surface greater than $2\ m^2$.

In general, small-size windows give a limited and specific view to the exterior which intensifies the sensation of isolation from the outside. This kind of window can also create glare.

Fenestration (%). If there is more than one window in the same room, the sum of surfaces of all the windows must be considered from a luminous point of view in relation to the area of the room.

Depending on the relationship between the window surface and the interior space, the following classification can be made:
- very low fenestration: less than 1%
- low fenestration: 1 - 4%
- medium fenestration: 4 - 10%
- high fenestration: 10 - 25%.
- very high fenestration: greater than 25%

As a general rule high or very high fenestration can cause problems of thermal control and glare. To prevent such problems control elements can be introduced.

Low or very low fenestration can produce excessively low illumination levels, especially where predominantly overcast skies, atmospheric pollution or adjacent buildings reduce the availability of daylight. This could be a major problem when the spaces to be illuminated are used for activities which require a high level of illumination.

Whether there is one large window or several small windows with the same total surface area in a room, the amount of light admitted is much the same, but light distribution, view and natural ventilation are affected. Thus, the relationship between fenestration and the mean daylight illumination in a room is approximately linear. However, division of the total window surface into several windows can produce a more uniform distribution of light, but will impede to some extent the view of the exterior. When windows are situated in different walls of the same room they enhance natural ventilation.

Shape

Window shapes differ greatly. A first approximation is to define the relationship between height and width. Windows may thus be classified as:
- horizontal window: shape coefficient of 1/2
- vertical window: shape coefficient of 2
- intermediate window: coefficient from 1/2 to 2.

Window shape principally influences light distribution in the illuminated space, quality of view and the potential for natural ventilation.

With horizontal windows the illumination of the interior is in a band parallel to the window-wall, producing little difference in light distribution throughout the day, with little glare. The relatively large horizontal dimension allows a panoramic view.

With vertical windows the illumination of the interior is in a band perpendicular to the window-wall, thus producing a greatly variable luminous distribution throughout the day. This window shape offers better illumination in the zones farthest from the window; however, there is greater glare. Exterior views are limited horizontally but may contain a greater depth of field, combining foreground, middle distance and long distance views. The greater height dimension of the window offers better ventilation potential.

Position

The position of a window may be described by its horizontal and vertical location in the wall in which it is placed.

By reference to its position with respect to wall height, the window may be classified as:
- high window
- intermediate window
- low window.

The higher a window is, the greater the depth of natural light penetration, producing a better distribution in the illuminated room. Height also encourages the extraction of warm air through natural ventilation. The height of the window sill determines the exterior view, hence a high window generally hinders vision.

By reference to its position with respect to wall width, the window may be classified as:
- central window
- lateral window
- corner window.

A window in a central position produces a greater distribution of light into the interior while a corner window causes less glare.

Orientation

With regard to the orientation of a window, reference is made to the geographical orientation since the sun path can have a great influence on natural illumination. From this point of view, east- and west-facing windows have been considered as being equivalent since the general effects produced are the same, although occurring at different times of the day. Windows are thus classified as:
- south-facing windows: high luminous levels and somewhat variable illumination; high energy gain in winter, medium in summer.
- east- and west-facing windows: both provide medium luminous levels, but the illumination throughout the day differs greatly since the east orientation provides a high level in the morning while that of the west is high in the afternoon; high energy gain in summer and low in winter.
- north-facing windows: low luminous level, however illumination is constant throughout the day; poor energy gain.

Solar radiation admitted through east- and west-facing windows presents control problems, and therefore moveable control elements are required. In the case of north- and south-facing windows it is much easier to use fixed control elements.

Controls

Controls are mechanisms or devices capable of altering the effects of a window. These devices may be:
- fixed: not operable by the user and generally not requiring significant maintenance.
- moveable: adaptable to different conditions and may be directly controlled by the user or operated automatically.

As a general rule, the lighting, ventilation and view characteristics of a window can be controlled, each characteristic requiring an appropriate control:

To control direct light, "separator surfaces", "rigid screens", "flexible screens" and "solar filters" may be used. To completely obstruct radiation, "solar obstructors" may be used.

There are controls which affect ventilation, such as "separator surfaces" or "solar obstructors", in which control is effected by opening or closing. There are other controls, not necessarily manipulated, which permit natural ventilation such as "flexible screens" and "solar filters".

In order to preserve visual privacy "separator surfaces", "flexible screens", "solar filters" and "solar obstructors" may be used to control views.

The control devices referred to above are explained in the general classification given at the beginning of this chapter. The number of devices needed for a window will normally depend on the different functions that it serves in the space.

DESCRIPTION AND PERFORMANCE OF DAYLIGHTING COMPONENTS

Introduction

To use daylighting components and elements effectively in architecture, it is necessary to know their behaviour. It may be difficult to distinguish which characteristics of an element or component are most significant. In many cases, a specific component may produce different effects in terms of the lighting or thermal conditions of a building, therefore, it is necessary to select which of these effects is most important.

A range of examples of components is described in the following pages; the characteristics are outlined, and diagrammatic representations and practical examples are given.

Gallery (GROUP I : Intermediate Light Spaces)

A gallery can be described as a covered light space attached to a building. It may be open to the exterior (open gallery) or closed off by glass (closed gallery).

It is an intermediate living space which admits daylight to the inside parts of a building connected to the gallery by pass-through components. It provides a decreased and less contrasting light level to the inside zones adjoining the gallery.

Usually it is one storey high, although in many cases a single gallery may serve two or more stories. Its depth typically varies from 0.8 to 4 m.

When the gallery is closed to the exterior, the pass-through component consists of a frame holding a transparent or translucent surface that may be single or double glazed.

Porch (GROUP I : Intermediate Light Spaces)

A porch is a covered light space attached to a building at ground level, open to the exterior environment.

It is an intermediate living space which permits the entry of daylight to the parts of the building directly connected to the porch by pass-through components. It provides a decreased and less contrasting light level to the inside zones adjacent to the porch, and protects them against direct solar radiation and rain.

Normally, a porch is one storey high, but sometimes rises to two storeys. Its depth normally ranges from 1 to 5 m.

It may be made of heavyweight construction materials such as brick, concrete, etc., or it may consist of a metal or wooden structural frame, either uncovered or covered with vines, wattle screens, etc.

5.10 Daylighting in Architecture

Greenhouse (GROUP I : Intermediate Light Spaces)

A greenhouse space is attached to a building by one or more of its faces, the others being separated from the exterior by a frame supporting transparent or translucent surfaces.

It permits the entry of light and direct solar radiation towards the interior space through its surfaces. At the same time it protects the interior from cold, wind and rain.

It provides an inside luminous level similar to that of the outside.

Its height may vary from one to several storeys and the enclosure is usually made of clear or acrylic glass supported by a metal or wooden frame.

Courtyard (GROUP II : Interior Light Spaces)

A courtyard space is enclosed by the walls of one or several buildings and is open to the exterior at the top and sometimes in one direction.

Courtyards have luminous properties similar to exterior space but daylighting and natural ventilation are reduced.

The finishes of the enclosing walls influence the illumination performance of the courtyard, with light colours or mirrored surfaces increasing light levels.

Atrium (GROUP II: Interior Light Spaces)

An atrium is a space enclosed laterally by the walls of a building and covered with transparent or translucent material.

It is an inside living space of a building which permits the entry of light to other interior spaces linked to it by pass-through components. It provides a decreased and less contrasting light level to the spaces connected to the atrium.

Its dimensions may vary widely depending on building size. Normally it occupies the total height of the building.

The covering may consist of a metal structure supporting the glazing. The interior finishes should have high reflectances to ensure good daylight penetration into adjacent spaces.

Adjustable control elements may be added to the pass-through component to avoid overheating.

Light-duct (GROUP II : Interior Light Spaces)

A light-duct can conduct natural light to interior zones of a building which are not otherwise linked to the outside but are not far from the exterior. Its surfaces are finished with light-reflective materials in order to direct and diffuse natural light downwards.

Usually the section of the duct is small, between 0.5 x 0.5 m and 2 x 2 m. The photograph below shows a series of such light-ducts separated by structural elements. The length depends on building size, although the luminous performance imposes a limit of about 10 m.

The top of the duct can be openable to permit natural ventilation or closed by transparent materials.

Sun-duct (GROUP II : Interior Light Spaces)

A sun-duct is a light space designed to reflect solar beams to dark interior spaces; it may also provide ventilation.

Surfaces are covered with highly reflective finishes, such as mirror, aluminium, highly polished surface or glossy paint, in order to reflect solar radiation.

Typical dimensions of the section of the duct are between 0.5 x 0.5 m and 1.2 x 1.2 m, although rectangular ducts may be used. The length may be greater than 15 m.

Sun collection requires fixed or movable devices at the top of the duct specially designed to redirect solar rays.

Window (GROUP I : Lateral Pass-through Component)

Windows permit the lateral penetration of light or direct solar radiation, interchange of view, and natural ventilation.

A window increases the luminous level in the interior zone close to the window, but the light level drops off rapidly with distance from the window.

Width and height dimensions may vary from small windows of 0.1 m to large ones over 6 m. The typical range is between 1.2 and 1.8 m in height and between 0.8 and 2.5 m in width.

Openings can be made in most forms of wall constructions. Window frames are generally made of wood, metal or PVC. A range of transparent or translucent glazing materials (polycarbonate sheet, etched glass, prismatic glass, transparent insulation material, etc) may be used individually or forming multilayers (double or triple glazing). Each will have different light transmitting characteristics.

A window can be opened to permit natural ventilation, and it can incorporate elements to control daylighting.

Balcony Window (GROUP I : Lateral Pass-through Component)

A balcony window permits the lateral penetration of light or direct solar radiation, interchange of view, passage of persons and natural ventilation. It increases the inside light level particularly in the zone close to the balcony.

Dimensions vary between 1 and 3 m in width and between 2 and 3 m in height.

The balcony is made of construction materials such as brick, concrete, or it consists of metal or wooden structures. The opening in the wall is closed by transparent or translucent surfaces as those used in normal windows.

Translucent Wall (GROUP I : Lateral Pass-through Component)

Such walls are constructed with translucent materials and make up part of a vertical enclosure in a building.

The surface separates two luminous environments, permitting the lateral penetration of natural light and diffusing it through the translucent material.

A translucent wall modifies the natural light which penetrates into a space, providing a homogeneous diffuse light level in the interior zones close to the wall.

It may occupy the entire lateral area from floor to ceiling and its thickness varies between 5 and 30 cm depending on the construction material.

It is made of glass blocks, acrylic materials, etc, and limits views through it.

Curtain Wall (GROUP I : Lateral Pass-through Component)

A curtain wall typically implies a continuous translucent or transparent vertical surface, with no structural function, that separates the interior from the exterior of a building.

It permits lateral penetration of natural light and direct solar gains, and interchange of view but often does not allow ventilation.

It increases the light level of inside zones close to the curtain wall.

It generally consists of a metal frame, which holds a transparent or translucent surface whose thickness is normally less than 5cm.

Clerestory (GROUP II : Zenithal Pass-through Component)

A clerestory roof window is a vertical or tilted opening constructed on the roof.

It permits zenithal penetration of daylight into the space below, sometimes protecting against direct radiation and/or redirecting it towards lower spaces. It can provide natural ventilation without an exterior view.

It increases the light level in the interior, usually with diffused light.

Its height over the roof may vary between 0.8 and 3 m.

The clerestory is typically constructed of the same materials as are used for the roof. As with windows, the opening can be covered with transparent or translucent glass or plastic, either fixed or openable.

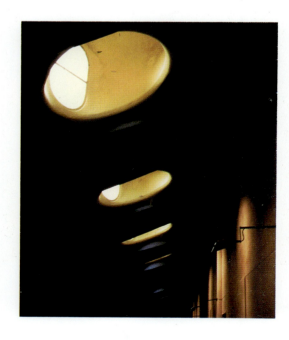

Monitor Roof (GROUP II : Zenithal Pass-through Component)

A monitor roof light is a raised section of a roof, including the ridge, with vertical openings.

North-light Roof (GROUP II : Zenithal Pass-through Component)

A north-light roof consists of a series of successive parallel south-oriented slopes with vertical or tilted linear openings facing north.

From within the space the daylight received is predominantly zenithal, increasing light levels and also allows ventilation through the apertures if openable.

The elevation of a monitor rooflight over the roof varies from 1 to 2.5 m and its length is typically the same as that of the room beneath. It is made of roof construction materials and the openings are closed by a range of glazing materials.

It permits zenithal entry of daylight providing diffuse light and high light levels without contrast to the space below.

The height of the opening varies from 1 to 2.5 m and its length is usually the same as that of the roof.

It is constructed of roof materials with openings which are closed by a range of translucent materials.

Translucent Ceiling (GROUP II : Zenithal Pass-through Component)

A translucent ceiling is defined as a horizontal aperture partially constructed with translucent materials, which separates the inside from the outside space or two inside superimposed spaces.

It permits the zenithal entry of daylight diffused through the translucent material to the lower space providing a homogeneous light level.

Its dimensions may be similar to or smaller than the lower area illuminated, but normally larger than 4 m.

The traditional form of construction employs glass blocks supported by a metal or concrete structure.

Skylight (GROUP II : Zenithal Pass-through Component)

A skylight is an opening situated in a horizontal or tilted roof.

It permits the zenithal entry of daylight into the space under the skylight. It can be opened to permit ventilation.

Standard width and length dimensions are up to 1 m per unit, although in specific cases larger surfaces are used in a variety of shapes and sizes.

The skylight is made of transparent or translucent materials covering the opening in the roof.

Daylighting Components

Dome (GROUP II : Zenithal Pass-through Component)

A hemispherical roof dome may have perforations or can be constructed in its totality with translucent materials.

It permits the zenithal illumination of the space under it.

When translucent it can be made of glass, acrylic, polycarbonate or fibreglass fabric. When perforated, it is made of opaque construction materials and the perforations may be covered by the above-named translucent materials.

Lantern (GROUP II : Zenithal Pass-through Component)

A lantern is an elevated part of the roof, often at the highest point, with vertical openings through which light enters.

It permits the zenithal entry of daylight to the zone directly below.

Its dimensions are usually between 0.5 and 1.5 m in height and between 0.5 and 2 m in diameter. In some specific cases the dimensions may be larger.

The lantern is typically made of roof construction materials.

When openable, it is in an effective location to exploit natural ventilation.

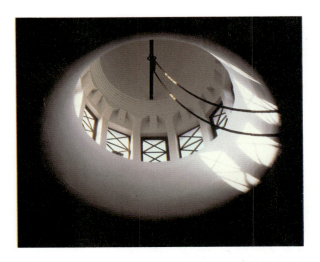

5.18 Daylighting in Architecture

Membrane (GROUP III : Global Pass-through Component)

A membrane envelope consists of a translucent or transparent surface totally or partially enclosing a space.

It permits the global entry of light to the space and provides a low contrast interior light level.

It can be made of polycarbonate, glass, acrylic material, or fibreglass fabric surfaces supported by a metal or wooden frame.

Conventional Division (GROUP I : Separator Surfaces)

A conventional or simple division is defined as a control element, placed in a pass-through component, which shares two environments by allowing view and light to pass through.

Its dimensions vary depending on aperture size, usually covering the whole opening.

It is made of a wooden or metal frame, supporting one or several transparent surfaces. For example: clear glass, laminated glass, polycarbonate, acrylic or polyester surfaces, etc.

It may be opened or closed, allowing natural ventilation.

Daylighting Components 5.19

Optical Division (GROUP I : Separator Surfaces)

The more complex optical division is a control element, placed in a pass-through component, which shares two environments and modifies the characteristics of the radiation passing through it.

Natural light enters the pass-through component, and the division diffuses, redirects, or controls its intensity, depending on the specific treatment of the division.

Its dimensions vary according to size of the aperture it covers.

It is made of a wooden or metallic frame, supporting one or several treated surfaces. For example: permanently coloured glass, mirrored glass, translucent or embossed glass, and glass with thermochromic or holographic films. Some of these treated surfaces may not be variable, in which case they have the disadvantage that light intensity is controlled in the same way throughout the year.

It may be opened and closed, allowing natural ventilation. In some cases views to the exterior are limited or distorted.

Prismatic Division (GROUP I : Separator Surfaces)

A prismatic control element, placed in a pass-through component, redirects light because of its optical and geometrical characteristics.

Natural light is redirected by the prismatic panel, changing the beam direction depending on the angle of incidence.

Its dimensions vary according to the aperture size.

The rigidity of the materials employed permits the installation with or without a supporting frame. It may be made of glass, polycarbonate, acrylic or polyester in various shapes.

It may be openable to allow natural ventilation.

Its poor visual transparency restricts its use.

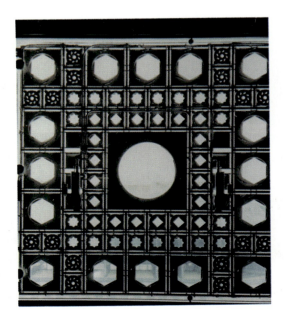

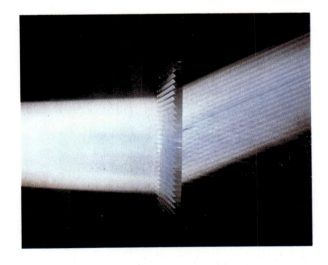

Active Division (GROUP I : Separator Surfaces)

An active control element, placed in a pass-through component, can change its optical absorption properties by applying an external electric field.

Natural light enters the pass-through component and the division controls its intensity according to the interior condition required.

Its dimensions vary depending on the aperture size covered.

It is a complex multi-layer structure made of high technology materials.

For operation at will, an external energy source (electricity) is required.

At present, the high cost discourages its use.

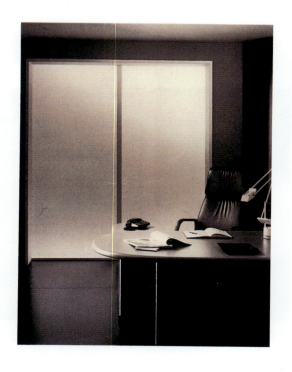

Awning (GROUP II : Flexible Screens)

An awning is a control element made of opaque or diffusing flexible material placed on the exterior of a pass-through component to obstruct or diffuse

direct solar radiation.

It provides a decreased and less contrasting light level in the zone close to the awning, and full or partial shade for the window as required.

It may be drawn down flush with the window-wall surface or protruding to the exterior. It may also roll up or be drawn sideways.

Its dimensions are determined according to the window surface and the shading required.

Awnings are usually made of canvas but may also be made of flexible plastic material.

The possibility of projecting the awning to the exterior permits a selective protection against direct solar radiation, according to the position of the sun, while allowing a view to the exterior.

Curtain (GROUP II : Flexible Screens)

A curtain could be described as a control element, made of opaque or diffusing flexible material, placed inside a pass-through component to protect against view and to protect the interior zones

close to the opening by totally or partially obstructing or diffusing solar radiation.

It may be rolled up or drawn laterally away, leaving the windows open to radiation and view if desired.

It is made of cloth or flexible plastic materials providing full or partial shade for the window as required. It may be made of opaque materials to darken a space completely.

Overhang (GROUP III : Rigid Screens)

An overhang is part of the building itself, protruding horizontally from the façade above a vertical pass-through component.

It protects the zones close to the openings of the building, obstructing high angle direct solar radiation.

It results in a lower interior light level, provides solar shading and partially protects the opening against rain.

Its dimensions are determined according to local and seasonal sun angles, typically protruding 0.4 to 1 m from the façade.

It can be made of a variety of construction materials such as concrete, metal, wood, etc.

An overhang used in a east or west orientation does not provide shading in early morning or late afternoon respectively because of the path of the sun.

5.22 Daylighting in Architecture

Lightshelf (GROUP III : Rigid Screens)

A lightshelf is usually placed horizontally above eye level in a vertical pass-through component, dividing it into an upper and a lower section.

Sill (GROUP III : Rigid Screens)

A sill, described as an element placed horizontally at the bottom of a window opening, can reflect and redirect the natural light which falls upon the sill in order to increase the light level in the interior space.

It protects the interior zones close to the openings against direct solar radiation, and redirects light falling on the upper surface to the interior ceiling. It thus provides shade in summer and makes the interior light distribution more uniform.

It is made of various construction materials, the upper surface having a reflective finish such as mirror, aluminium or a highly polished material.

At different latitudes and orientations, sizes will vary depending on sun angle.

Dimensions are determined by the opening size and thickness of wall.

A variety of construction materials may be used for the sill, the upper surface having a reflective finish such as mirror, aluminium, highly polished surface or glossy paint.

The tilt of the sill may be chosen according to sun angle. Disturbing glare may result when it is located below eye level.

Fin (GROUP III: Rigid Screens)

A fin is a control element placed on the exterior façade of a building and fixed vertically on the sides of the opening.

It reflects and redirects natural light which falls laterally on the fin to the inside. Depending on its location, direct solar radiation may be partially avoided and possible interior discomfort reduced.

It provides a lower, and possibly more homogeneous, internal light level by reducing light at the opening.

Its dimensions are determined according to shading needs and the required reflection to the interior. At different latitudes and orientations its size will vary depending on sun angle. Typically it is as tall as the window. Its projection varies between 0.3 and 1.2 m.

It is particularly appropriate for shading oblique low angle sun received on west and east façades.

Baffle (GROUP III : Rigid Screens)

A baffle is a fixed single opaque or translucent element which protects a pass-through component against direct solar radiation at certain angles, and may reflect daylight to the interior.

It may provide a more homogeneous light level, avoiding the entry of direct solar radiation in the zone close to the opening or increasing lighting in interior zones.

Dimensions are normally determined by the opening area.

Jalousie (GROUP IV : Solar Filters)

A jalousie is defined here as an exterior or interior element composed of slatted screens placed over the whole of a window.

It permits the control of direct solar radiation and regulates the entry of light.

The slats may be fixed or moveable. When moveable they may be adjusted according to sun angle and shading requirements. This device can be moved along the opening, drawn to the side, or rolled up to the top.

Slats are usually made of wood, plastic, aluminium, stainless steel, etc.

A jalousie protects against view from the exterior while allowing natural ventilation.

Louvre (GROUP IV : Solar Filters)

Louvres are a series of exterior slats which may be fixed or adjustable. They usually cover the whole of the opening outside the window but may cover a larger surface including the walls surrounding the opening.

Depending on the orientation of the slats, direct solar radiation which falls upon the louvres may be obstructed and/or reflected and/or redirected to the interior zone.

The parallel slats may be made of painted or galvanised steel, anodised aluminium, PVC, wood, etc.

Horizontal louvres are usually located on southern façades, vertical louvres on eastern and western façades.

If the slats are closed to form a panel, they act as a solar obstructor and may cast complete shade over the openings.

Brise-soleil (GROUP IV : Solar Filters)

A brise-soleil is defined here as an exterior, fixed structure covering the whole pass-through component or a larger area.

The open structure permits the passage of light and air. Depending on the geometric design, the structure will obstruct direct solar radiation at certain sun angles.

A brise-soleil can be made of wood, metal, concrete or other construction materials.

Shutter (GROUP V : Solar Obstructors)

A shutter totally obstructs radiation. It may be exterior or interior, and is openable.

It is a continuous opaque surface which totally blocks daylight and views. It can be folded or drawn towards the side of the opening.

It normally covers the whole opening and is usually made of wood, aluminium or PVC.

When closed, the interior is visually and in part thermally isolated from the exterior.

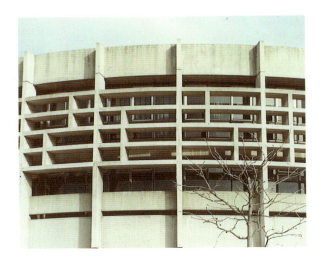

FIELDS OF APPLICATION

Introduction

For buildings inpredominantly daytime use, it seems wasteful to consume a large amount of energy to provide artificial illumination. For this reason daylighting components which provide the necessary quantity and quality of light under normal climatic conditions are being studied. In particular, buildings such as schools and offices, with predominantly daytime occupancy, have their own characteristics of size and shape which up to a certain point are typifiable. The following section attempts to identify the most usual characteristics of these types of buildings and recommends appropriate daylighting components.

General Characteristics of Types of Schools and Offices

The types of school and office buildings are characterised according to the general layout of their plans. The basic characteristics of these types of buildings are divided into three groups: general dimensions, area distribution, and form coefficients. All parameters have upper and lower limits between which the average case may be found, which is considered as the most typical:

- in the general dimensions group, global characteristics of areas and volumes are defined.
- in the area distribution group, area percentages are given according their uses.
- finally, in the form coefficients group, compactness and slenderness of building types are characterised.

From the initiation of an architectural project, the distribution of volumes between perimeter and interior zones conditions the potential for daylighting. This zone distribution depends in great part on the relationship between the enclosure area and the constructed volume. In order to express this relation, a "coefficient of compactness" (enclosure area in a sphere of given volume/enclosure area in the building of equal volume) is used. A low coefficient of compactness would mean that a greater proportion of the total volume may be in contact with the exterior.

The different possibilities of connection with the exterior, by façades or roofing, influences the types of lighting conditions in the interior zones. The "coefficient of slenderness" of the building expresses the relationship between horizontal area and height. Thus, a tall narrow building will be characterised by a high slenderness coefficient, while a low wide building will have a low slenderness. It is evident that the possibilities of zenithal illumination are reduced in the most slender buildings.

Typology of Schools

In Figure 8, large, intermediate and small schools (A, B and C) are analysed with respect to the following types: linear, double linear, comb-shaped, nucleate, module pattern, informal and dispersed.

The linear, double linear and comb-shaped types are derivations of the same model, in which considerations concerning orientation, terrain adaptation, finance, etc modify the final result.

The nucleate and modular pattern types correspond to schemes which group different uses: in the first case, around a space or central nucleus, and in the second case, supported by a framework.

Lastly, the informal and dispersed types have characteristics which are not readily classified.

In all the above cases, the proportions of the classrooms, the most important space in school buildings, must be analysed. In general, classrooms will have different proportions according to the type to which they belong, with a tendency to reduce the possibility of openings in the façade in the linear, comb-shaped and nucleate school buildings.

In terms of "dimensional characteristics", "distribution of areas" and "form coefficients" of the different types of schools, double linear and nucleated types stand out for their advantageous use of area; they present a reduced proportion of circulation area in relation to the total. As a general rule, school buildings are not slender, since they are usually low whereas their degree of compactness has a greater range of variation.

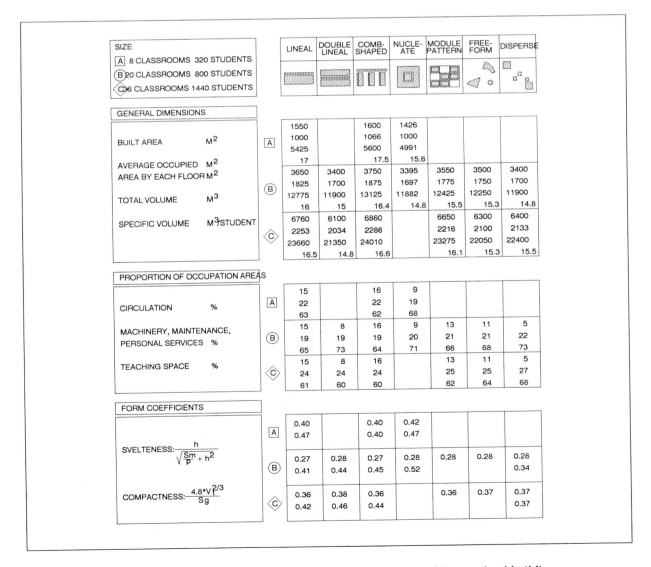

Figure 8 - *The typological characteristics of small, medium and large school buildings.*

Typology of Offices

In Figure 9, large, intermediate and small offices are analysed under the following four types: linear, nucleate, perimeter and radial. In each of them, variants of centralised or decentralised vertical circulation are given. In general, the linear type corresponds to buildings of limited height, whereas tower buildings are very frequently of nucleated types.

The perimeter type, with working areas surrounding an empty central space, is usually of limited height with complex circulation. The radial type, of star or cross shape, is usually a building of great height, sometimes understood as an integration of linear and nucleate building.

Finally, informal and dispersed types have not been analysed since they are difficult to classify and may often be described as subdivisions or deviations of the four types analysed.

All office types are normally characterised by high slenderness, with the exception of small office buildings. The compactness of these buildings is variable, although intermediate values are generally presented.

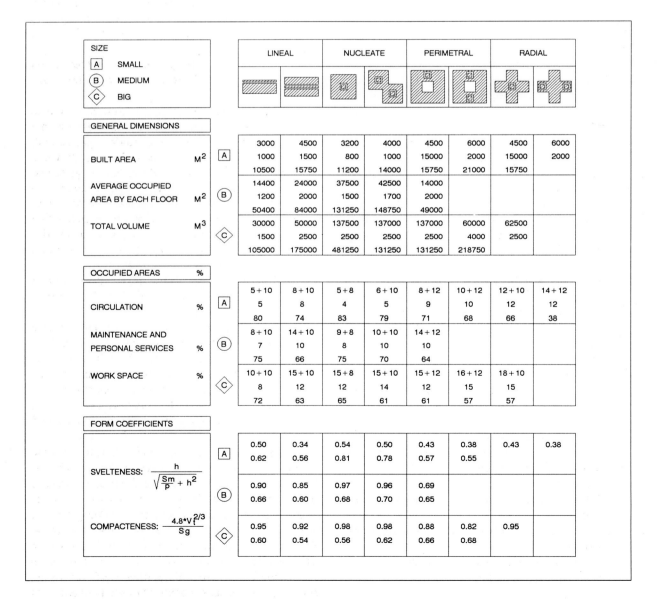

Figure 9 - The typological characteristics of small, medium and large office buildings.

Illumination Conditions in Schools and Offices

The forms of the buildings described above need to be analysed with respect to daylighting. To do this, the two fundamental types of space found in all buildings must first be distinguished. Those of the perimeter zone are situated next to the vertical or horizontal enclosures, and the interior zone is distanced from the enclosures and thereby not in direct contact with the exterior.

In the perimeter zone there are great daylighting possibilities, simply through the openings in the enclosures. The design of these openings is complicated by considerations of light incidence, which is non-uniform in space and time, and by problems of glare. Non-illumination problems are those of privacy, thermal insulation, security, etc. When openings are placed in the vertical envelope (windows), the illumination distribution is irregular and the illumination level falls rapidly away with distance from the window-wall. When light penetrates through roof openings (skylights or clerestories) the light distribution is more uniform; however construction problems occur, as well as those of controlling excessive summer solar radiation.

In the interior zone, problems arise from the absence of light and the difficulty of providing natural ventilation. In these zones, therefore, artificial illumination and ventilation are normally required. In the classical building plans, inner areas are reserved for secondary use, such as circulation, maintenance and storage.

In the following section, the illumination needs of different types of schools and offices are discussed and suitable daylighting components are recommended.

Recommended Daylighting Components

The question of how and when to apply daylighting components as well as control elements is discussed for the following building types:
- schools
- offices
- small buildings.

The third group has been included in order to point out the characteristics peculiar to small schools and offices.

In applying daylighting components and control elements, the following factors must be borne in mind when choosing a specific component:
- building size
- relative proportions of interior and perimeter zones
- light needs in regard to quantity and light uniformity
- maintenance and security requirements
- predominant sky conditions.

A summary of recommended daylight components is presented in Figures 10 and 11. A more detailed discussion of recommendations for the three building types is presented below.

Daylighting Components in Schools

The relatively low compactness of school buildings theoretically allows daylighting illumination of practically all areas. They have almost no interior zone, and present potential problems only in circulation areas and maintenance and personal services areas in the double linear and nucleate types. In buildings of low slenderness, one or two floors in height, these areas may be zenithally illuminated.

It is in teaching zones (classrooms) that lighting requirements are most strict. In all the building types, problems arise when lateral illumination results in non-uniform distribution of light. These problems are accentuated in linear and nucleate schools in which circulation paths are limited, thus resulting in decreasing width and increasing depth of the classrooms.

To improve daylighting behaviour in these types of buildings, the following components are recommended:

The most appropriate conduction components are courtyards, atria and galleries. The first two are used for illuminating adjacent interior spaces, and, depending on climatic and practical needs, they can be open or covered. Rooms described as galleries, which conduct light towards more interior zones, can be used for illuminating hallways and common areas. Light-ducts and sun-ducts may be useful conduction components to reinforce the illumination of interior zones of classrooms and corridors without direct contact with the exterior light conditions.

The most commonly used lateral pass-through components are windows and, less frequently, translucent walls.

Zenithal pass-through components may be used in schools since these are normally low buildings with large roof areas through which light may be admitted. The most common components of this type are clerestories and skylights, which may supplement the illumination provided by windows in classroom zones, and admit light to common zones and hallways.

In order to control the lighting effects of these components, control elements are added. In schools, these elements must usually be fixed and robust to resist wear-and-tear by users. To screen and filter direct solar radiation, as well as to avoid glare and contrast, solar filters are added. These should be fixed or easy to manipulate, such as louvres or a brise-soleil. To improve light distribution in classrooms and to avoid glare, fixed rigid screens are set up. The types most commonly used in classrooms are light-shelves, overhangs and baffles.

Daylighting Components in Offices

In contrast to schools, the usual compactness and large size of modern office buildings means that the volume of the interior space clearly predominates over the perimeter zone. This tendency has often led to the use of daylighting being totally ignored in this type of building. This may even be found in the perimeter spaces, which could benefit from daylight.

Analysis of the different types of offices shows that the linear and radial forms are those which present the least compactness and the greatest possibilities for daylighting. In all the types, exterior contact is given up in the areas of circulation, maintenance and services. The perimeter type may allow complementary illumination through central zones. Lastly, the nucleate type is that which most clearly gives up the advantages of daylighting.

The requirements for quality and quantity of light in working areas are perhaps less demanding than those for schools but nevertheless good illumination levels and low glare are required.

In order to improve the luminous environment in office buildings the following daylighting components are recommended:

The most appropriate conduction components are atria and, in special cases, light-ducts and sun-ducts. Atria illuminate interior pass-through areas and the ducts are used for illuminating specific interior areas.

The most commonly used lateral pass-through components are windows and curtain walls, although translucent walls are often found. Curtain walls illuminate large areas of offices since the whole vertical enclosure is utilised to introduce light. However, the quality of the luminous environment is low, having a poor uniformity with plan depth. Translucent walls offer greater diffusion, while windows illuminate more discrete areas. Large areas of glazed pass-through components frequently present problems of climatic control, especially in temperate climates. Hence their use must be carefully considered and integrated with appropriate control elements.

Zenithal pass-through components are not normally used in office buildings since these are generally several storeys high. These components can however be used to limit the great amount of light in atria, when compared with a completely glazed roof. In some cases translucent ceilings are used.

Global pass-through components are used for covering and illuminating large entrance areas, halls, etc.

In order to control the lighting effects of these components, control elements are added.

Because of the large size of many office buildings, it is feasible to install specialised systems to control artificial lighting. Thus, devices will tend to be much more sophisticated than in schools, and regulation may be automated or motorised.

In office buildings, a wide range of separator surfaces may be used to control radiation or view. Flexible screens such as awnings may be used externally and curtains are commonly used internally.

Rigid screens are also used, the most common being overhangs, fins and baffles. Similarly, solar filters are applied in these buildings and in a great majority of cases are moveable. Blinds are generally regulated by the users, while louvres may be adjusted throughout the day by a centrally motorised control. Brise-soleils can also be adopted, often used as fixed architectural elements.

Daylighting Components in Small Buildings

The characteristics and general needs of small schools and offices are similar to those of residential buildings. Simple low-cost components are normally used which may be controlled and maintained directly by the user.

The need for conduction components in these buildings is limited, because of the reduced building volume. Most often, courtyards and galleries, but occasionally light-ducts and sun-ducts, are used for solving problems arising from the lack of lighting in interior zones.

Lateral pass-through components commonly used are windows and balconies; translucent walls may be useful in a few projects.

To control solar radiation and interior illumination, all types of control elements may be used. These include separator surfaces, flexible screens, solar filters and shutters adjustable by the user. Rigid screens and fixed solar filters are less commonly used.

Summary of Recommendations

Figures 10 and 11 summarise the recommendations about the components and control elements useful in schools, offices and small buildings. When utilising these charts, one must take into account many factors apart from the lighting needs, such as maintenance, safety, etc.

It should be noted that two very different conditions are studied: overcast sky (Figure 10) and clear sky (Figure 11). In these charts, if a component is particularly appropriate it receives three stars, a recommended one has one star, and one which is not recommended or may have counter-effects has no sign at all.

Daylighting Components

	SCHOOLS	OFFICES	SMALL BUILDINGS
CONDUCTION COMPONENTS			
INTERMEDIATE LIGHT SPACE			
GALLERY	*		*
PORCH	*		
GREENHOUSE	*	*	*
INTERIOR LIGHT SPACE			
COURTYARD	*	*	*
ATRIUM	* * *	* * *	
LIGHT-DUCT	* * *	*	*
SUN-DUCT	*	*	*
PASS-THROUGH COMPONENTS			
LATERAL			
WINDOW	* * *	* * *	* * *
BALCONY		*	*
TRANSLUCENT WALL	*	*	*
CURTAIN WALL		*	
ZENITHAL			
CLERESTORY	* * *	*	* * *
MONITOR ROOF	*	*	
NORTH-LIGHT ROOF	*		*
TRANSLUCENT CEILING	*		*
SKYLIGHT	*	*	*
DOME	*	*	
LANTERN	*		
GLOBAL			
MEMBRANE	*	*	
CONTROL ELEMENTS			
SEPARATOR SURFACES			
CONVENTIONAL DIVISION	*	* * *	* * *
OPTICAL DIVISION	* * *	*	*
PRISMATIC DIVISION	* * *	*	*
ACTIVE DIVISION		* * *	
FLEXIBLE SCREENS			
AWNING			
CURTAIN	*	*	* * *
RIGID SCREEN			
OVERHANG			*
LIGHTSELF	* * *	* * *	*
SILL	*		*
FIN			
BAFFLE	*		*
SOLAR FILTERS			
BLIND	*		*
LOUVER			
JALOUSIE			
SOLAR OBSTRUCTORS			
SHUTTER	*		*

Figure 10 - Summary of recommendations for predominantly overcast sky conditions.

5.32 Daylighting in Architecture

	SCHOOLS	OFFICES	SMALL BUILDINGS
CONDUCTION COMPONENTS			
INTERMEDIATE LIGHT SPACE			
GALLERY	*		*
PORCH	*		*
GREENHOUSE	*		*
INTERIOR LIGHT SPACE			
COURTYARD	* * *	*	* * *
ATRIUM			
LIGHT-DUCT	*		*
SUN-DUCT	* * *	* * *	*
PASS-THROUGH COMPONENTS			
LATERAL			
WINDOW	* * *	* * *	* * *
BALCONY		*	*
TRANSLUCENT WALL	*	*	*
CURTAIN WALL		*	
ZENITHAL			
CLERESTORY	* * *		*
MONITOR ROOF	*		
NORTH-LIGHT ROOF	*	*	*
TRANSLUCENT CEILING			
SKYLIGHT	*	*	*
DOME			
LANTERN	*		*
GLOBAL			
MEMBRANE			
CONTROL ELEMENTS			
SEPARATOR SURFACES			
CONVENTIONAL DIVISION	*	* * *	* * *
OPTICAL DIVISION			
PRISMATIC DIVISION	* * *	*	*
ACTIVE DIVISION	*	* * *	*
FLEXIBLE SCREENS			
AWNING	*		* * *
CURTAIN		*	* * *
RIGID SCREEN			
OVERHANG	* * *	*	* * *
LIGHTSELF	* * *	* * *	* * *
SILL	*	*	*
FIN	*	*	*
BAFFLE	*	*	*
SOLAR FILTERS			
BLIND	* * *	*	* * *
LOUVER	*	* * *	* * *
JALOUSIE	*	*	*
SOLAR OBSTRUCTORS			
SHUTTER	*		* * *

Figure 11 - Summary of recommendations for predominantly clear sky conditions.

EXPERIMENTAL ANALYSIS OF SELECTED COMPONENTS

This section deals with the analysis of selected components which improve the quantitative and qualitative characteristics of daylighting conditions in a space.

The components are selected for their high potential to improve the daylit environment. An analysis of two main components representing innovation, changing requirements, the use of VDUs, and international architectural trends is proposed. These two components are:
- the atrium as a source of daylight
- light-directing elements in the façade aperture.

These components are respectively classified as:
- conduction component
- control element.

The analysis of light-directing elements is subdivided into three sections based on different physical phenomena of light manipulation:
- reflection
- refraction
- diffraction

Methodology of Analysis

The scale model technique was used for the analysis (Figure 12). Scale models at 1:50 and 1:20 were built and tested using photometric equipment (Figure 13).

Studies for Overcast Sky Conditions

Initially, scale model testing took place under real overcast sky conditions. Owing to the continually changing luminance distribution of the sky, even for overcast sky, the results of these studies were sometimes not reliable.

These experiences led to the decision to build an artificial sky (mirror box sky). The sky chamber (1.6 x 1.7m) is equipped with fluorescent lamps and provides a luminance distribution close to the standard CIE overcast sky (Figure 14). The horizontal illumination in the artificial sky is 17000 lux. .

Results from real sky measurements and the artificial sky measurements differ slightly. Sidelighting studies showed more differences than toplighting studies.

Figure 13 - Photometric instruments used for scale model testing: (a)- remote photosensor, (b) - daylight meter for exterior illuminations, (c) - sky contrast illumination meter.

Studies for Clear Sky Conditions

Sunlighting studies have been carried out under real sky conditions on clear days. The testing site was virtually free of obstruction.

To determine the Sunlight Factor (SB), the model was placed horizontally and rotated on its vertical axis for different sun angles.

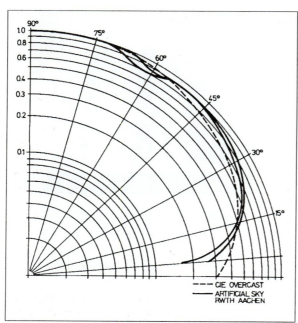

Figure 14 - Luminance distribution of the artificial sky at RWTH Aachen, Germany.

Evaluation

The use of the photometric equipment enabled the illuminations to be measured at specified locations in the model. For overcast sky conditions the results were converted into Daylight Factors (D) and for clear sky conditions into Sunlight Factors (SB).

To allow comparison of results, all studies were based on the same type of office space (9 x 9 x 3 m) with the same reflectances of the interior surfaces (walls 0.6, floor 0.25, ceiling 0.7). The dimensions of the selected space reflect the current architectural tendency of small group offices instead of single office spaces or large office areas. The dimensions of the window were changeable. Glazing material was not used except for the atria studies where the reflected light from the glazing contributes to the illumination of the atrium space.

Ground reflectance is taken to be zero (except for the study on atria) by using black material in front of the scale models.

CONDUCTION COMPONENT: ATRIUM

Glazed atria have become a very popular architectural feature internationally in recent years. Glazing the space between the buildings mainly for climatic reason is an old idea. After a long period of disuse, the idea found a promoter in the architect John Portman who designed Hyatt Regency Hotels, USA, in the early 1970s, with atria to give the designs a distinctive image.

Today, atria are designed for many purposes and for a variety of building types. Glazing the space between the buildings allows an increase in the size of the windows oriented towards that space without significant heat losses, an important consideration in cold climates. The atrium may be used for less thermally sensitive activities, such as circulation or temporary exhibitions, as well as for protection against noisy and polluted surroundings. Furthermore, atria can, with suitable design and landscaping, provide an attractive view from the adjacent buildings.

Atria in commercial buildings provide a potential source of daylight for the illumination of the spaces facing them if they are properly designed. Thus, the combination of being able to increase glazing ratios without significantly affecting heat losses, while improving daylighting into the centre of the plan, displacing the need for artificial lighting, makes the use of atria particularly attractive in terms of energy efficiency.

Outline of the Scale Model Testing

This study focuses on atrium characteristics that influence the admittance and distribution of daylight. Conventional daylighting calculation techniques are inadequate in all but the simplest atrium design.

For this analysis three types of scale models were built with the following shapes:
- square
- triangular
- rectangular.

The scale of the models is 1:50 (Figure 15). The geometry of the section of each atrium space can be manipulated by changing the level of the floor of the atrium. The reflectances of the atrium surfaces (walls and floors) can be altered by replacing the finishes. All scale models are designed with the same width and height of the atrium but with different lengths. The model of the square-shaped atrium represents an atrium of 20 x 20 m. Each storey height is 3 m. Maximum height of all atria is 10 storeys.

Figure 15 - Models for the square and triangular atria.

The following façade conditions have been tested:
- completely white façade (reflectance 0.7)
- façade with a 50% glazing and 50% white wall (referred to as 50% glazing/white walls)
- totally glazed (average reflectance 0.1)
- completely black façade (reflectance 0.05).

A single pane of glass is used in the models, separating the atrium and the adjacent spaces, to take into account the reflection of light from the glazing in an atrium.

The use of remote photosensors allows the measurement of illuminations in the scale models simultaneously at the following locations:
- in the atrium space: horizontal illuminations at the floor of the atrium (centre and four additional locations); vertical illuminations in the centre-line of atrium façades at different levels of the atrium height (quarter, half and threequarters the height of the atrium façades); vertical illumination in front of the tested office space
- in the adjacent space: horizontal illuminations at desk height; vertical illuminations at mid-height of the rear wall; horizontal illuminations at the ceiling. Within a given atrium configuration, the illumination in the room is tested at three different levels: room located at the level of the atrium floor, at mid-height level of the atrium, and at the highest level next to the roofing of the atrium.

All measurements were carried out in the artificial sky, thus providing results for diffuse daylight. The results lead to design recommendations for atrium buildings designed for countries with mainly overcast sky conditions.

Illumination in the Atrium

The illumination in the atrium can be quantified through two components of the Daylight Factor (D) (Figure 16):
- the direct light from the sky reaching the floor and the walls of the atrium, D_s (Sky Component horizontal or vertical)
- the reflected light from the surrounding surfaces atrium walls and floor, D_i (Internally Reflected Component).

The direct light from the sky may also include reflected light from outside the atrium which in this case is considered to be sky light.

Results of Scale Model Studies

The daylight performance of an atrium is very complex although only two components of the Daylight Factor are involved.

The geometrical aspect of the atrium plays an important role in the quantity and quality of daylight in the atrium. In order to describe the geometrical aspects of the atrium, the Room Index (1) and the Well Index (2) can be used:

Room Index (RI) = l w / ((l + w) h)

where:
l = atrium length
w = atrium width
h = atrium height

Well Index (WI) = (h (w + l)) / (2 l w)

Both Room Index and Well Index define the relationship between the light-admitting area (l.w) and the surfaces of the atrium. However, various shaped atria with the same Index can have different light-admitting areas. In order to compare the performances of different shaped atria with the same sized light-admitting area, Room Index and Well Index, the Aspect Ratio is adopted:

Aspect Ratio (AR) = l.w / h^2

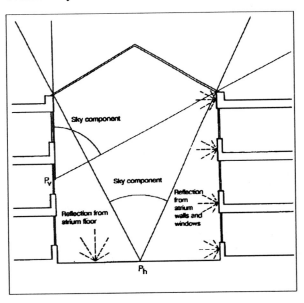

Figure 16 - Daylight in the atrium

Horizontal illuminance at floor level and atrium geometry. Daylight in the centre of an atrium floor depends on the geometrical shape of the atrium. Quantitative illumination at atrium floor level of the square atrium is up to 10% higher for all tested height/width atrium configurations compared to the results for a triangular/rectangular atrium floor plan layout (Figure 17). The main contribution of light is the direct light coming from the sky, with little variation due to the shape of the light-admitting area (Figure 18). The impact of the geometry of the atrium becomes evident in the performance of the internally reflected component (Figure 19): due to the fact that the surrounding surfaces of the square floor plan are minimised compared to the rectangular/triangular floor plan, the number of inter-reflections within the atrium is smaller.

Horizontal illumination at floor level of a square atrium. The total illumination varies from 60 to 80% for atria wider than a cubed volume; for tall atria the total illumination decreases rapidly because both daylight components are decreased with height. For an atrium with 1.5 times the width, only a Daylight Factor of 30% can be reached (Figure 20). All these values are based on a completely white atrium and indicate the maximum contribution of daylight.

The design of the façades will influence the illumination, due to the reduced overall reflectances (Figure 21).

A façade with 50% window openings will reduce the contribution of the IRC by half. A totally glazed wall (curtain wall) as separation between the offices and the atrium reduces the IRC to one third of that of the white walls (Figure 22).

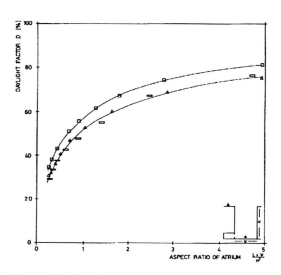

Figure 17 - Daylight Factor at the centre of the atrium floor for three atrium shapes. For a given light-admitting area the square atrium receives more daylight.

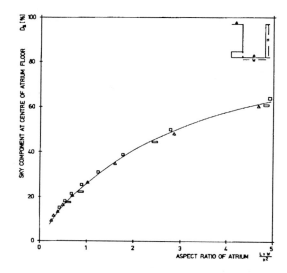

Figure 18 - Sky Component at the centre of atria.

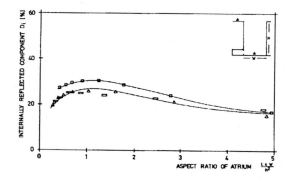

Figure 19 - Internally Reflected Component at the centre of the atrium floor.

Daylighting Components

Vertical illumination at the walls of a square atrium.

The vertical illumination at the atrium walls determines the amount of light that enters the adjacent space.

As with the horizontal illumination, the vertical illumination can also be quantified by two components, D_s and D_i, of the Daylight Factor (D) (Figure 16).

Fish-eye photographs from the window of an office space towards an atrium demonstrate the parameters influencing the vertical illumination at the atrium façade (Figures 23 and 24).

Detailed results of the amount of vertical illumination on atrium walls, as a function of atrium geometry and façade reflectances, are presented in Figures 25 to 30.

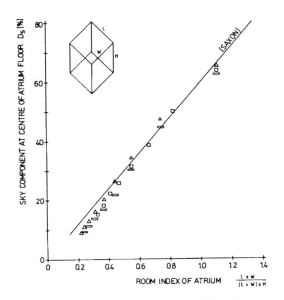

Figure 20 - Comparison of measured Sky Component of different atrium geometries with Saxon (1).

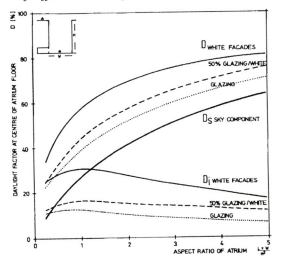

Figure 21 - Daylight Factor at the centre of a square atrium as a function of the Aspect Ratio: (a) white walls, (b) 50% glazing/white walls, (c) 100% glazing.

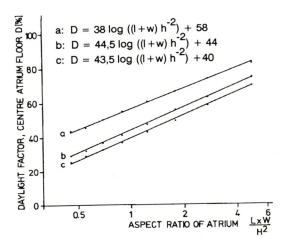

Figure 22 - Impact of different atrium façades on the Daylight Factor at the centre of a square atrium.

Figure 23 - Braas Headquarters, Oberursel, Germany (Architect: Gallwitz). Low contribution of direct sky light.

Figure 24 - Züblin Headquarters, Stuttgart, Germany (Architect: Gottfried Böhm). Little obstruction due to atrium roof structure.

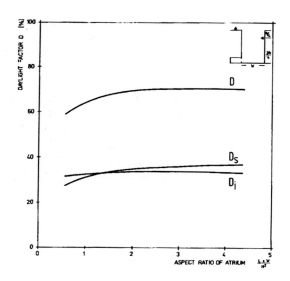

Figure 25 - Components of vertical illumination at three-quarter height of atrium (white walls and floor).

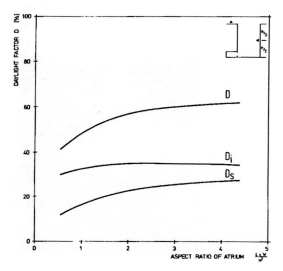

Figure 26 - Components of vertical illumination at mid-height of atrium (white walls and floor).

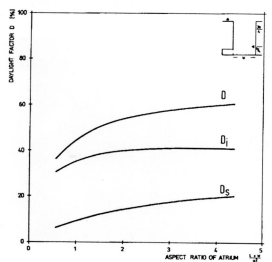

Figure 27 - Components of vertical illumination at quarter-height of atrium (white walls and floor).

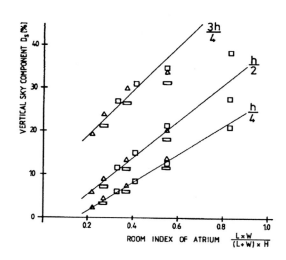

Figure 28 - Sky Components of vertical illumination at defined heights of various atria (white walls and floor).

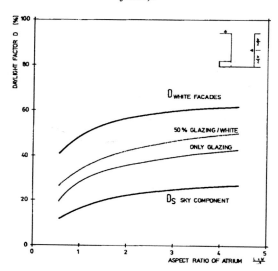

Figure 29 - Vertical illumination at mid-height of atrium due to different façade reflectances.

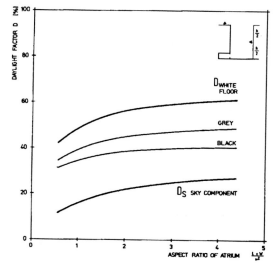

Figure 30 - Vertical illumination at mid-height of atrium due to different floor reflectances.

Illumination of the space facing the atrium. Both the Sky Component and the Internally Reflected Component of the atrium can be considered as light sources for the illumination of an office space; this light has to pass through the thermal and visual separation between the office and the atrium. The amount of light passing into an office space will be reduced by the framing of the window and the transmittance of the glazing. In order to provide as much daylight in the office as possible, optimisation of the framing and the transmittance properties of glazing materials is necessary.

The contribution of daylight to the office illumination varies according to the location of the room within the height of the building (Figure 31). A room located at the atrium floor level is mainly illuminated by light reflected from the floor. A room located near the atrium roof receives most light directly from the sky.

Since many rooms depend mainly on reflected light as the main daylight source, the reflectances of the atrium surfaces should be as high as possible to reflect as much light downwards as possible.

The measured results are presented in Figures 32 to 35. The quantitative improvement of natural illumination is based on the contribution of the reflected light from the surrounding atrium walls and from the atrium floor.

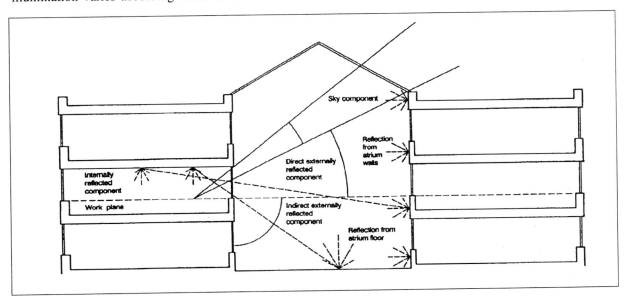

Figure 31 - The atrium as a source of daylight for adjacent spaces.

Although the quantity of light is decreased in atria higher than 0.75 of the width, the quality of illumination in adjacent spaces may be improved. These spaces are illuminated mainly by reflected light, which can both improve the uniformity and reduce glare problems.

The presence of plants on the atrium floor affects the contribution of reflected light, IRC_{floor}, to the illumination of adjacent spaces. Landscaping in the atrium using dark-coloured materials and plants will reduce the reflectance of the atrium floor to around 0.2. The natural illumination of most rooms facing the atrium, except the top floors, depends particularly on reflected light from the atrium floor.

The quantity of light can be improved by using light-directing elements (such as lightshelves or prismatic glazing) in the façade apertures which redirect zenithal light onto the ceiling in the atrium-adjacent spaces. Additional studies (3) show that this measure can increase the quantity of daylight in the space only if the reflectances of atrium walls and floor are very low. When light-directing elements are used in combination with white atrium façades, only the quality of illumination (distribution and glare) in these spaces can be improved. The additional studies also show that the quantity of light in the spaces due to white atrium façades (high contribution of reflected light) is higher compared with the quantity of light having a low contribution due to reflected light and using light-directing elements (Figure 36).

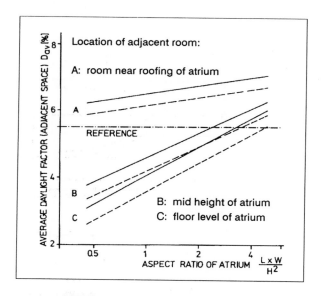

Figure 32 - Average Daylight Factor in atrium-adjacent room as a function of Aspect Ratio for a square atrium with 50% glazing/white walls.

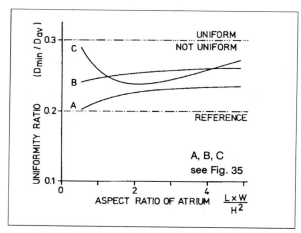

Figure 33 - Uniformity of illumination in atrium-adjacent space for different geometrical configurations of square atria (white façades).

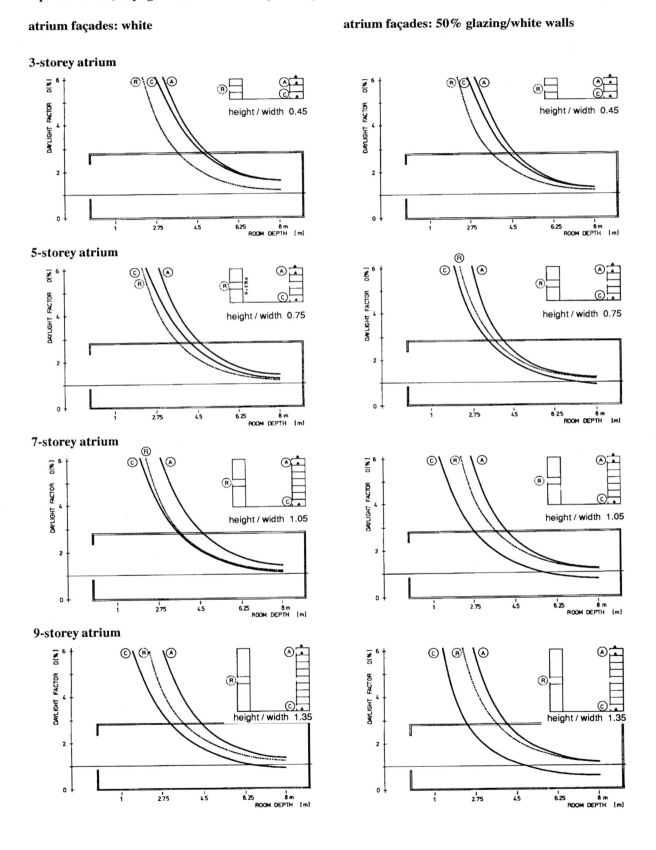

Figure 34 - Daylight distribution in atrium-adjacent rooms A and C for different atrium reflectances - white atrium façades (left column), 50% glazing/white walls (right column). Also for reference room (R - dotted lines) with non atrium-facing glazing.

Square atrium, illumination in adjacent space to atrium

Room located next to roofing of atrium   **Room located at floor level of atrium**

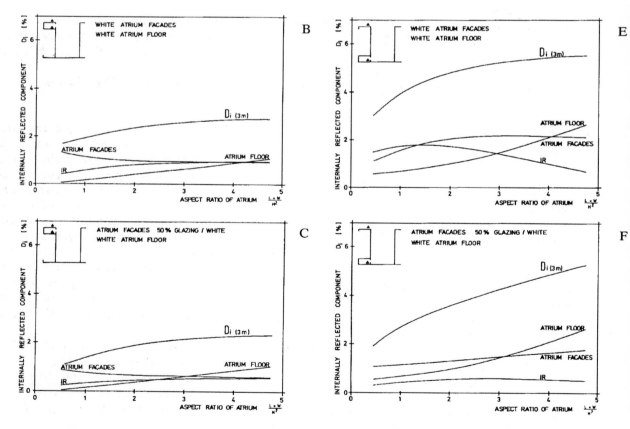

Components (D_s, D_i) of total Illumination (D);
D_i is split into $D_{i\ atrium\ façades}$, and $D_{i\ atrium\ floor}$,
(atrium: white façades, white floor)

Components (D_s, D_i) of total Illumination (D);
D_i is split into $D_{i\ atrium\ façades}$, and $D_{i\ atrium\ floor}$,
(atrium: white façades, white floor)

Components $D_{i\ atrium\ façades}$, $D_{i\ atrium\ floor}$ and IR of Internally Reflected Component D_i
(IR = interreflection between façades and floor)
B: atrium: white façades, white floor
C: atrium: 50%glaczing/white façades, white floor)

Components $D_{i\ atrium\ façades}$, $D_{i\ atrium\ floor}$ and IR of Internally Reflected Component D_i
(IR = interreflection between façades and floor)
E: atrium: white façades, white floor
F; atrium: 50%glaczing/white façades, white floor)

Figure 35 - Illumination at desk level of atrium-adjacent room, at three m from window, related to different square atrium geometries, for top floor rooms (left) and ground floor (right).

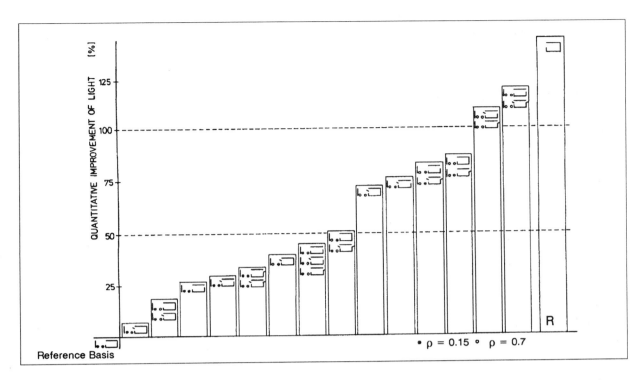

Figure 36 - Comparative study of efficiency of the use of control elements compared with conventional measures to enhance illumination levels in a space facing a three-sided atrium (quantitative improvement of the average daylight on four floors related to a base case: low reflectance atrium walls (low reflectance ρ=0.15, high reflectance ρ=0.70), conventional side window) (3).
The icons in the columns indicate configurations of atrium, room and light-controlling element.

Daylight Requirements in the Atrium

Two major requirements have to be satisfied to achieve a well-lit atrium:
- light level requirements of users in the atrium and adjacent spaces
- light requirements for plant growth.

Good quality lighting and control of glare are expected in atria.

Design Variables

The light system atrium can be subdivided into two major elements:
- light collecting system (atrium roof)
- light guiding system (atrium space).

According to the path of light travelling down through a building and the priority of design decisions by the design team, the following design parameters have to be taken into consideration:
- structural system for the glazing and fenestration
- glazing system and its transmittance
- geometry of the atrium
- surface reflectances of the atrium walls and floor
- glazing and illumination of spaces adjacent to the atrium
- dimensions of adjacent spaces
- interior reflectances of the adjacent spaces
- types and locations of plants within atria.

Structural system for the glazing and fenestration. The fenestration system controls the intensity and spatial distribution of daylight entering the atrium. The net transmittance of this system depends on the glazing system's geometry and structure, the glazing orientation, the type of shading, and the daylight availability (diffuse sky, direct sun). The fenestration system will have a major impact on interior illumination levels and distribution. The illumination level is influenced by the degree of obstruction within the light-admitting area whereas the light distribution is affected by the directional properties of each fenestration system (Figures 37 and 38).

The amount of daylight entering the atrium depends on the structural system for the glazing, the glazing transmittance (solar and visible), the shading devices and the sun controls. The size of the structural system determines the net glazed area (Figure 39).

When designing for climates with mainly overcast sky conditions, the dimensions and type of roof construction must be carefully evaluated. The aim is to minimise obstruction within the light-collecting area, thus maximising the contribution of daylight. However, even the most minimal roofing construction would reduce the light-admitting area by at least 8-10%. Using single-pane glazing material would cause a further reduction of 10%. Covering an open court with any kind of glazing system therefore would reduce the daylight level in the atrium by at least 20%.

As a consequence of these facts, moveable shading devices are preferred to fixed systems.

Glazing system and its transmittance. The optical properties of glazing materials influence daylighting quality and the potential for energy savings due to reduced artificial lighting. The thermal properties affect heating and cooling loads. The two major types of glazing are: transparent and translucent, either colourless, tinted or reflective. Transparent colourless glazing transmits the most daylight and provides the most natural view of the sky. However, these materials admit strong direct beams of sunlight to the building. Beam sunlight may be blocked, redirected, scattered or diffused by interior objects (shades, structural members, etc) if desired, to prevent possible discomfort for occupants.

Translucent glazing materials, which diffuse and distribute sunlight, do not allow a direct view of the sky. Only major differences in weather conditions outside can be detected. To provide a uniform light quality under direct sunshine the glazing material has to be highly diffusing. However, highly diffusing materials tend to have lower light transmittance which drastically reduces the light levels under overcast sky conditions. Diffusing materials with higher transmittance properties are not as good at diffusing the light uniformly (5).

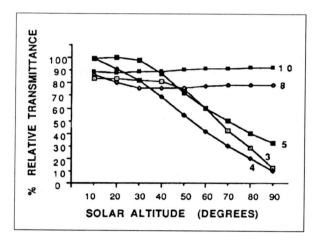

Figure 38 - Transmittance relative to an open atrium, for different roof types (see Figure 37) as a function of solar altitude (4).

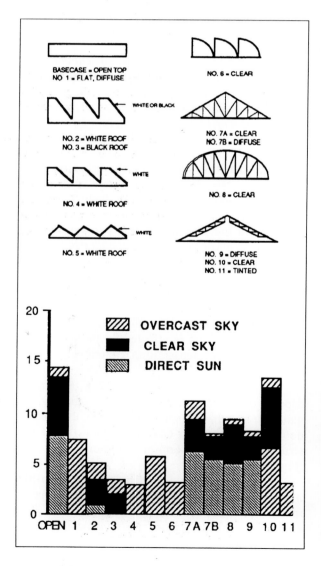

Figure 37 - Daylight Factor at the centre of a nine-storey atrium, with different roof types, for three sky conditions (4).

Figure 39 - Atrium roof constructions showing high and low levels of obstruction in the light-collecting area.

Geometry of the atrium. The proportions of an atrium determine the amount of direct daylight reaching the floor of the atrium. The shallower and wider the atrium space, the better the contribution of direct daylight. Atria that are higher than their width generally result in poor daylight levels for all lower floor levels, and thus require atrium walls with high reflectances. Atrium walls with low reflectances are only suitable for daylighting when the space is wider than height.

Surface reflectances of the atrium walls and floor. The design of atrium walls determines the distribution of light in the space. The quantity of reflected light is the product of the average reflectance of the walls and the type of reflection. Diffuse reflecting materials reduce the quantity of daylight reaching the lower parts of atrium walls and the floor; specular reflecting materials perform better but tend to increase glare for occupants, especially when sunlight strikes the atrium façades (Figures 40, 41 and 42).

Light enters adjacent spaces in two ways, according to the level in the atrium: the upper atrium receives predominantly direct light from the sky and the lower atrium receives predominantly reflected light from the opposite walls and the floor of the atrium. As a consequence, as much light as possible should be reflected to the lower storeys. As glazing does not reflect as much light as white walls, the amount of glazing in the upper storeys should be reduced to the minimum extent that provides well daylit space for the occupants. Theoretically, therefore, each floor should have different window sizes: small windows in the upper floors, large windows in the middle and lower floors, and

Figure 41 - Board of Trade, Chicago: the use of large areas of reflective glazing material in atrium façades creates visual noise.

complete glazing near the bottom of the atrium. Another design option is to alter the ceiling heights of each floor, with the tallest spaces at the bottom of the atrium.

The reflectance of the floor of the atrium influences the light levels of the lower storeys. The greatest benefit would be provided by a glossy floor material (such as marble). However, plants and greenery reduce the reflection of the floor. Trees can also shade the lower floors if they are planted too close to the walls. To maximise the contribution of

Figure 40 - Wiesner Building, MIT, Boston, USA: white enamel panels with glossy surfaces are used for the atrium façades.

Figure 42 - Züblin Building, Stuttgart:, Germany dark precast concrete façade elements with a rough surface reflect little light.

light reflected from the floor of the atrium, plants should be positioned in the centre of the space, with a band of highly reflective floor near the atrium walls.

Glazing and illumination of adjacent spaces. Thermal requirements demand a separating wall between the occupied space and the atrium. In order to maximise the quantity of light entering through this separating wall, glazing should have high transmittance properties. If building codes and fire regulations permit, single glazing is often most appropriate.

The illumination contribution to adjacent spaces can be improved by designing atrium walls in such a way that the skylight and sunlight that enter the atrium can also be used for lighting the occupied space. Interior reflectances in the office space should be as high as possible. Because of the decreasing light levels in an atrium, the floor plans of adjacent spaces further down in the atrium should be less deep than in upper storeys. An advantage of the atrium concept is that double sidelit spaces are feasible: only a small dark zone is then left in the middle of the room which can be used as a circulation zone within a large office space.

Different design strategies can be applied:
- set-back of the atrium walls so that each room has a view of a portion of the sky
- use of a guidance system that directs the light from the atrium to the ceiling of the adjacent space: lightshelves, reflectors, prismatic systems, etc.

In climates with mainly overcast sky conditions, sunlighting strategies may be feasible, but they reduce the light-admitting area.

Plants in the atrium. Four basic lighting factors have to be considered for the lighting of plants:
- intensity
- duration
- spectrum
- direction.

Plants require different amounts of daylight and for a period of time longer than normal office hours, usually more than 10 hours per day, depending on the plant species (Table I). The minimum illumination needed for survival depends on the size and the species of the plants and varies from 250 lux for small plants to 2000 lux for trees and tropical plants.

In daylighting terms, a daylight factor of 5 to 40% will be needed under the 5000 lux standard overcast sky.

Attractive effects can be created with plants; however they do affect the illumination in the atrium and the adjacent spaces by reducing the reflectances of the floor and the walls to as low as 0.2.

TABLE I - Recommended illuminances for acclimatised plants (6)

Species		Examples	Illuminances (lux)
Trees	1.5 - 3.0 meters	Araucaria excelsa (Norfolk Island Pine)	above 2000
		Eriobotyra japonica (Chines Loquator, Japan Plum)	above 2000
		Ficus benjamina Exotica (Weeping Java Fig)	750 - 2000
		Ficus lyrata (Fiddleleaf Fig)	750 - 2000
		Ficus retusa nitida (Indian Laurel)	750 - 2000
		Ligustrum lucidum (Waxleaf)	750 - 2000
Floor plants	0.6 - 1.8 meters	Brassaia actinophylla (Schefflera)	750 - 2000
		Chamaedorea erumpens (Bamboo Palm)	250 - 750
		Chamaerops humilis (European Fanpalm)	above 2000
		Dieffenbachia amoena (Giant Dumb Cane)	750 - 2000
		Ficus elastica Decora (Rubber Plant)	750 - 2000
		Phoenix roebelinii (Pigmy Date Palm)	750 - 2000
		Pittosporum tobira (Mock Orange)	above 2000
		Yucca elephantibes (Palm-Lily)	above 2000
Table or desk plants		Aechmea fsciata (Bromeliad)	750 - 2000
		Aglaonema Pseudobacteatum (Golden Aglaonema)	250 - 750
		Asparagus spengeri (Asparagus Fern)	750 - 2000
		Cissus rhombifolia (Grape Ivy)	750 - 2000
		Citrus mitis (Calamondin)	above 2000
		Dieffenbachia Exotica (Dumb Cane)	750 - 2000
		Dracaena fragans massangeana (Corn Plant)	250 - 750
		Hoya carnosa (Wax Plant)	750 - 2000
		Philodendron oxycardium (Common Philodendron)	250 - 750

Sunlight in Atria

Sunlight entering buildings gives a sense of orientation, time and weather conditions, and creates interior environments that are potentially more comfortable and attractive, resulting in improved occupant productivity.

Atria should be designed to allow the maximum amount of direct and reflected light in to reach the façades. In addition, light courts should shade low-angle sunlight which is difficult to control at the façade and can cause glare problems, and redirect that light into buildings.

Sunlight in atria may cause glare by direct sunlight or sunlight reflected from walls opposite.

Shading becomes necessary: shading devices at the fenestration system can be fixed or movable. Movable shading devices provide great flexibility to cope with daily and seasonal variations in daylight availability: fixed shadings are simpler but less flexible (Figures 43 and 44).

Two major techniques can be employed to shade an atrium:
- diffusing the direct sunlight to create a bright diffuse source
- guiding the sunlight by using light-directing elements such as reflectors, mirrors (Figure 45), prismatic or holographic elements

Figure 43 - Colonia Building, Cologne, Germany: white flexible fabric shading devices used for solar control.

Figure 44 - Board of Trade, Chicago, USA: glazing materials with different transmittance properties are used for fixed solar control.

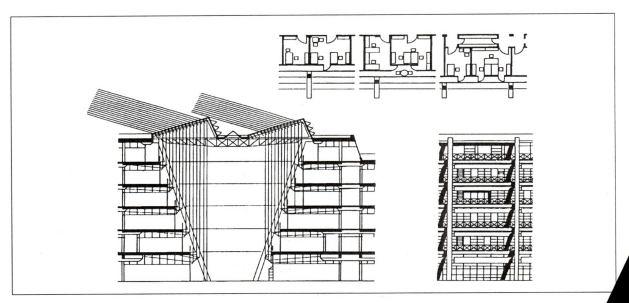

Figure 45 - Atrium design for an office building (Sozialamt der Bundespost, Germany, Architect: G. Willbold-Lohr, 1982): mirrored louvres are used to reflect sunlight to adjacent spaces.

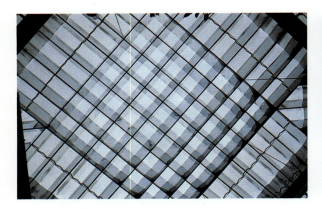 

Figure 46 - Challenger Building, Paris, France (Architect: Kevin Roche): trellised atrium roof allows diffuse light penetration from zenith whereas angled sunlight will be redirected.

Diffusing the sunlight can be achieved mainly by using an indirect daylighting system design such as light wells or fabrics; white finishes have little impact on the colour composition of daylight. Fabrics are usually used as interior shading devices. If sunlight strikes their surfaces the luminance of these architectural elements increases and may cause discomfort for the occupants. A trellised roof structure can act as solar baffles, allowing zenithal light through, but blocking or reflecting sunlight at certain angles (Figure 46). With such trellises, the resultant light entering the atrium has a downward direction, allowing no direct solar gain to atrium-adjacent spaces, which can be exploited by the use of lightshelves to redirect light deep into these spaces (Figure 47).

The control of sunlight in atria is determined by:
- orientation of the building
- fenestration of the roofing area
- roofing construction
- light-directing elements in the roofing area and facades
- glazing material
- shading devices (fixed, movable).

Design Recommendations

The following paragraphs give design recommendations for enclosed atria (four-sided) designed for regions with mainly overcast sky conditions:

Purpose and use of the atrium. In the very early design stage the decision has to be made whether the atrium is to be used mainly for:
- growing plants, or
- improving natural illumination in spaces facing the atrium.

Plant growth in the atrium.
- The species of plants intended to occupy the atrium determine the geometrical proportions of the atrium by their light requirements.
- If however the specific geometry and the geometrical proportions of the atrium are given, and the available daylight level is known, the species of plants must be chosen accordingly.
- Another important selection criterion for plants is whether they require sunshine for growing or blooming.
- Plants need careful maintenance to survive in enclosed spaces, including the cleaning of dust from the leaves.

Daylighting in atria.
- Square atria provide the highest contribution of natural illumination (for a given light-admitting area).
- Atria higher than their width are not suitable for daylighting because the contribution of direct light coming from the sky decreases rapidly with height.
- Atrium façades with highly reflective materials yield the highest natural illumination.

...ses for natural shading are ...a lightshelves for a complete ...anlighting system (7).

- Specular reflective materials perform better than diffuse, but there is potential discomfort due to glare.
- Atrium façades with high reflectivity in the upper storeys maximise the downward reflection of light.
- The size of the windows should be adjusted according to the available light level.
- Set-back of atrium façades (opening towards the sky) increases illumination levels.

Roofing of atria.
- Minimum reduction of the light-admitting area is an important criterion
- Glazing materials should have a high transmittance.
- Diffusing glazing materials reduce the light transmittance for diffuse skylight.
- Fixed shading devices reduce the light-admitting area and therefore are not recommended (movable shading is preferred).
- The construction for the atrium roof should sit on top of the adjacent building or be fixed in front of the atrium walls at a certain distance from them so that light can wash the walls.

Shading.
- Movable shading is recommended (fabrics).
- Shading can also be provided by designing large highly reflective vertical baffles (trellises) in the roofing area which exclude the sun but minimise the reduction of the light-admitting area (problem: weight).

Illumination of spaces facing atria.
- The size of the windows should be adjusted according to the available light level.
- Transmittance of the separating glazing should be as high as possible; if building codes and fire regulations permit, single-pane glazing would suit best.
- Minimisation of the framing in the window area is necessary.
- Rooms with a high contribution of reflected light from the atrium floor must have highly reflective ceilings.
- Light-directing elements increase the quantity and quality of light in spaces next to atria with dark surfaces. Light-directing elements can improve the quality of light where atria have highly reflective walls.
- Light-directing elements in the roofing area for sunlighting do not suit northern climates.

CONTROL ELEMENTS

Introduction

The window as a light-source element (pass-through component) is not necessarily an efficient component in the building structure. Being a transparent part of the building envelope, it establishes contact with the exterior world but it also produces glare and thermal problems. Shading devices and low transmittance, heat-absorbing or heat-reflecting glasses are used to restrict solar heat gain and to reduce the glare component of sky brightness; however, reflective and solar-control glasses also reduce the transmittance of light.

Elements for glare reduction combined with deeper light penetration within the space have long been under investigation. A very interesting approach found in the literature was the Luxfer-prism, published in a German architects' magazine in June 1902 (8). In the 1950s, Hopkinson started research on lightshelves at the Building Research Station (9). The idea of increasing the daylight level in a space and at the same time reducing glare from the sky is well established, but for some reason this is not part of the current architectural language.

Function of control elements. In a conventional side-lit space the daylight distribution falls off rapidly with distance from the window wall. Some form of supplementary (artificial or natural) lighting system is necessary therefore if the depth of such a room exceeds 2.5 times the window height, not only to provide an adequate illumination level but also to provide an optimum visual environment.

The aim is to provide a system of illumination to light deep spaces economically and simply, and to conserve energy. In order to achieve this goal, four major issues need to be addressed:
- increasing the daylight illuminance level
- improving the illumination distribution
- reducing glare due to sky brightness
- control of direct sunlight.

To study the problems of deeper light penetration, beam sunlighting is being investigated. It is possible to use sunlight and to control it directly, by reflecting, refracting or diffracting its rays with a light-directing system onto a ceiling of high diffuse reflectance to provide better illumination deep inside a room.

Beyond the energy-related issues of daylighting, there are important qualitative issues to be addressed. The primary purpose of lighting is to enhance visual performance while providing comfort. Deep spaces lit by conventional window systems can produce intolerable glare due to contrast between sky brightness seen through the window and the relatively dark areas of the interior space. Additional reflected sunlight will increase the ambient light level by lighting the ceiling. This will reduce contrast and therefore reduce discomfort due to glare.

The daylighting effect of control elements can be based on different physical phenomena: reflecting (lightshelf), refracting (prism) or diffracting (holographic elements) daylight from its natural path. Within this analysis, reflecting and diffracting elements are compared by scale model testing.

Need for control elements. Direct sunlight at workplaces causes discomfort; shading devices are therefore needed. However, shading devices, especially fixed devices, reduce the quantity of light in the space and thus increase the electric lighting load. Energy-saving issues require a careful evaluation of energy input. If properly designed, control elements offer the possibility of reducing the use of electric lighting by redirecting natural light.

The use of VDUs in office spaces is becoming more and more common. In rooms with large side windows it can be difficult to orient VDUs to avoid both unwanted reflections on the screen and uncomfortable contrasts between the dark screen and a bright window. Control elements such as lightshelves, prismatic or holographic optical elements can reduce these problems by directing most of the incoming light onto the ceiling.

Thermal aspect. As regards the thermal aspect of the use of sunlight for illumination, Ruck (10) and Rosenfeld and Selkowitz (11) present a calculation based on the luminous efficacy of sunlight and daylight.

If a control element using sunlight as a light source is to be adopted, then the heat gain from sunlight must be compared with the heat gain from the artificial lighting (fluorescent or incandescent lamps) that may be offset. Skylight and sunlight have a higher luminous efficacy than most lamps, consequently the energy input of light-directing systems for sidelit spaces is lower (ie, less heat per lumen) than most electric alternatives.

Lightshelves

Lightshelves are flat or curved elements in the window façade aperture above eye level. They redirect the incoming light by reflection (Figures 48 and 49). At the same time they protect occupants from direct sun penetration (Figure 50).

Function of lightshelves. Lightshelf configurations can generally be classified as interior, exterior or combined lightshelves. An interior lightshelf extends from the plane of the aperture into the space. An exterior lightshelf extends from the plane of the aperture away from the space and is normally outside the building. A combined lightshelf is an interior and exterior lightshelf in one unit.

The lightshelf divides the façade aperture into a view-window, below the shelf, and a clerestory window above the shelf. This separation allows the design of these windows to suit their purpose: the view-window provides mainly for contact by the occupants with the exterior world and the clerestory window is used mainly as a light-window.

The light distribution pattern of the view-window in the space is similar to that of a window with an overhang, potentially taking advantage of high ground reflectance. The light-window above the lightshelf gives a similar light distribution pattern to that of a clerestory window.

Lightshelves were studied in the 1950s by Hopkinson (9) in relation to their control and distribution of diffuse light from the sky, and for glare reduction in hospitals and schools. The re-awakened interest in lightshelves has come about because of their ability to beam sunlight into the space. The original functions of the lightshelf were to act as:
- sunshading device
- control of diffuse light.

These have been extended by:
- redirecting sunlight into the space.

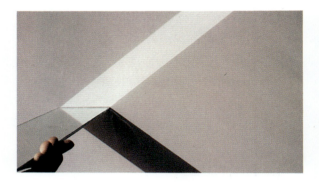

Figure 48 - Reflection of a beam of light.

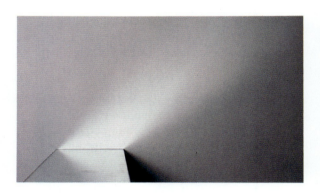

Figure 49 - Reflection of diffuse light.

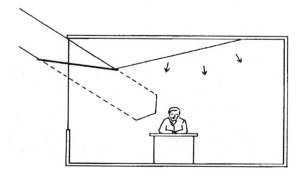

Figure 50 - Function of a lightshelf in design.

Each part of the lightshelf system (interior / exterior / combined) contributes in a specific way to one of the functions mentioned above.

The object of this study is to analyse the effects of each part of the lightshelf system on the illumination in the space and it is therefore split into three divisions:
- effect of lightshelves on glare
- daylighting studies for diffuse skylight
- sunlighting studies for direct sunlight.

Outline of the model testing. Daylighting and sunlighting studies have been carried out with the aid of scale models, whereas the study on glare issues was performed using the simulation program DIM.

A scale model (1:20) was used for parametric testing of various lightshelf configurations. Daylighting studies were carried out in the artificial sky, sunlighting studies under real sky conditions.

In order to define the impact of the lightshelf system on the Sky Component and the Internally Reflected Component in the room, two identical models are needed: one model with completely black interiors to determine the Sky Component, and the other model with the following interior reflectances - walls 0.6, ceiling 0.7 and floor 0.25. Both models are equipped with photometric sensors and measured simultaneously.

For sunlighting studies these two models are tilted and rotated for given solar altitudes and measured in two runs: first with sunlight on the façade aperture and sunlight on the exterior illuminance meters (global illuminances); in the second run, immediately after the first, the façade aperture of the models has to be shaded as well as the exterior illumination meters (Figure 51). The second run provides data of the illumination in the space due to the skylight alone, whereas the first run results in data based on the global illumination. From these two runs, the Sunlight Factor SB can be derived, by subtracting the second from the first.

Various lightshelf systems have been tested with two fenestration configurations:
- a clerestory window (9x1 m) below the ceiling with the lightshelf system at sill height
- a band window opening (9x2 m) with the lightshelf elements in the middle of the window area.

The distance of the lightshelf elements from the ceiling remained the same in both test conditions.

Effect of lightshelves on glare. Glare can be caused by a direct view of the bright sky from the interior of a building. This glare can cause more or less serious discomfort.

Chauvel et al. (12) find in their study that discomfort glare resulting from direct view through a single window (window area greater than 2% of the

Figure 51 - Model setup for sunlight studies with lightshelves.

floor area of the room) is independent of the size of the window and the position and distance of the observer, but is critically dependent on the sky luminance. The Glare Index (Table II) used to assess such conditions therefore depends mainly on the luminance of the sky seen through the window and to a lesser extent on the reflectances of the room surfaces. Glare therefore can be reduced by cutting down the brightness of the visible patch of the sky and by increasing the interior brightness by a rational use of surface areas of high reflectance to reduce the relative brightness of the visible sky.

TABLE II - Classification of daylight glare index (12)

Glare Criterion	Daylight Gare Index
Just imperceptible	16 - 18
Just acceptable	20 - 22
Just uncomfortable	24 - 26
Just intolerable	above 28

Results of glare investigations.
- glare can cause discomfort in summer although the use of lightshelves improves the situation (Table III).

For the 2 m high window:
- the lowest glare index is achieved with the combination of internal and external parts of the lightshelves; this is true for all tested sky conditions; the glare effect becomes almost imperceptible
- the effect on reduction of glare is slightly better than the effect of the lightshelf with the 1 m window.

TABLE III - Glare index for the middle of the test room for various lightshelf configurations

	Summer Clear	Summer Overcast	Winter Clear	Winter Overcast
Clerestorey window 9 x 1 m, lightshelf below				
No shelf	12.9	10.6	14.4	7.2
External	11.0	7.9	13.3	4.8
Internal	11.4	8.1	13.8	5.1
Combined	9.9	5.8	12.2	3.8
Band window 9 x 2 m, lightshelf at mid-height				
No shelf	15.9	12.7	18.8	9.6
External	14.3	10.8	17.2	7.7
Internal	11.8	8.4	15.0	5.6
Combined	10.1	6.6	7.0	3.5

Lightshelves are modelled as perfectly diffusing surfaces situated horizontally and perpendicular to the façade. The calculation is made for a single measurement point near the centre of the room.

Conclusions of glare investigations. It is found that, for a lightshelf situated within a window area, the combination of external and internal parts of the shelf results in a good glare performance. For narrow horizontal windows near the ceiling, only the exterior part of the lightshelf improves the glare situation.

Results of daylighting studies. For different lightshelf parameters (Figure 52) the overall efficiency has been investigated, taking into consideration both the reduction of light entering the space due to the obstruction of the lightshelf in the window frame and the positive effect of light reflection deeper into the space (Figures 53 and 54).

The efficiency is highest for systems causing low obstruction in the window area and having specular reflective surfaces. This is true for all external lightshelves with mirrored surfaces. The internal part of any lightshelf acts mainly as obstruction and therefore reduces the amount of daylight entering the space. To summarise:
- the external part of the lightshelf has a positive effect on the distribution of light inside the space, independently of the size of the window
- the use of a mirrored finish on the exterior part of the shelf, and the slope of the external plate, are of interest as they contribute to a higher exploitation of daylight in the space (Figure 55). For a window height of 1 m above the shelf, a shelf depth from 1:1 to 1:1.5 of window height above the shelf provides the highest contribution of daylight for a slope of 30° (Figure 56)
- the reflectivity of the ceiling is of importance for the efficiency of the system: a mirrored lightshelf in combination with a white ceiling performs better than a white lightshelf with a mirrored ceiling. Therefore it is recommended that the lightshelf is designed in combination with a highly reflective ceiling surface.

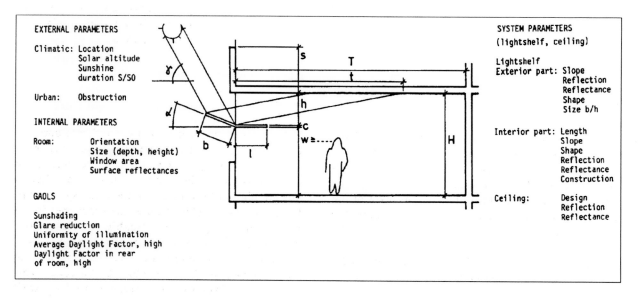

Figure 52 - Design parameters for lightshelves.

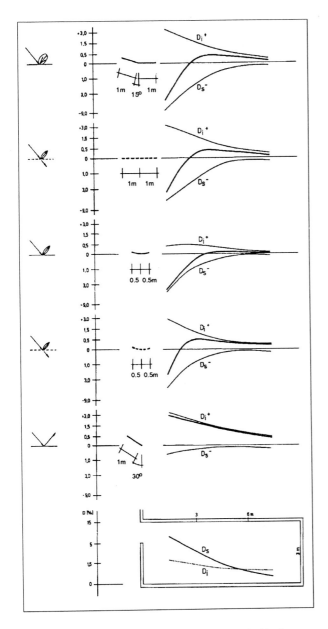

Figure 53 - Efficiency of various lightshelf shape and refectance types located at the bottom of a 1 m high window

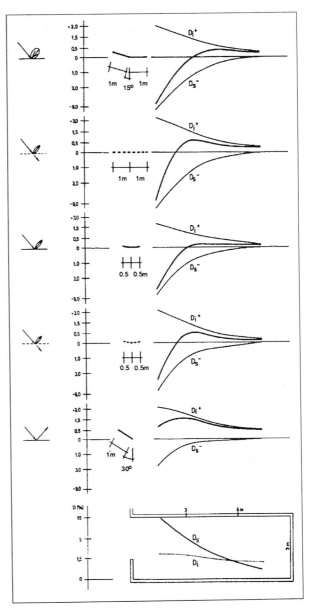

Figure 54 - Efficiency of various lightshelf shape and refectance types located at mid-height of a 2 m high window

Conclusions of daylighting studies. Lightshelves can be used to improve the quantity and the quality of light in the space, for diffuse light as well as direct light. Only the exterior part of the lightshelf contributes to a higher light level in the space, whereas the interior part mainly reduces the sky component and therefore contributes to the quality of light in the space.

It is important to combine the lightshelf with a highly reflective ceiling in order to achieve the best possible efficiency.

It has also been found that the efficiency of a lightshelf system depends on the relative importance of the internally reflected component: the smaller the IRC in the space the greater is the quantitative contribution that a lightshelf can make. If the IRC is already high in the space then the same system will be less efficient. This leads to the recommendation that lightshelves are practicable for improving daylighting situations in relatively poor daylit environments, although a sufficient window size would allow enough daylight to enter the space. This situation exists in all offices with low wall reflectances due to dark furniture or paint.

5.54 Daylighting in Architecture

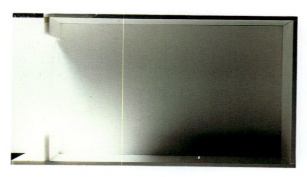

Sidelit room, diffuse light.

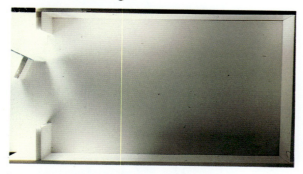

Mirrored exterior lightshelf, declination 20°, reflected light on to the front part of the ceiling.

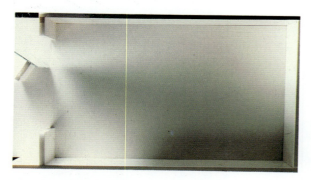

Mirrored exterior lightshelf, declination 30°, reflected light illuminates the rear of the room.

Figure 55 - Two-dimensional performance studies with tilted mirror lightshelves for diffuse light.

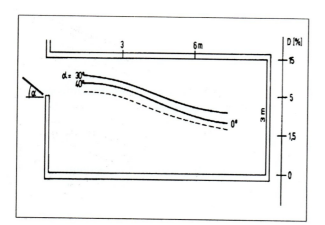

Figure 56 - Mirrored exterior lightshelf: daylight distribution relative to the tilt of the lightshelf.

Results of sunlighting studies. The use of lightshelves as fixed control elements for sunshading is not practicable for northern climates with very low sun angles. When used for sunlighting, these control elements have to be adjusted to the sun angle to provide the best contribution of sunlight.

Sunlight Factors (horizontal and vertical) have been calculated and are presented in graphs (Figure 57). The sunlight factor graphs are to be seen as comparative information and do not provide information on the annual performance of the system unless the factors given are converted into illumination values with the aid of daylight availability data.

- Most systems admit direct sunlight into the space at low sun angles.
- The performance of the lightshelves for high sun angles depends on the size and shape of the lightshelf. A convex-shaped lightshelf rejects most of the incoming light from high sun angles while admitting light from low sun angles.
- Compared to the reference case, almost all tested lightshelf configurations admit direct light for low and even for high sun angles.

Conclusions of sunlighting studies. Fixed lightshelf systems do not prevent direct sun-penetration in the space for low sun angles; additional movable sunshading devices are required. Furthermore, the efficiency of a fixed lightshelf in terms of daylighting for diffuse light interferes with the possible effect of sunlighting. If there is no need for glare reduction, fixed lightshelves have a greater negative effect on the amount of daylight than their positive contribution. Finally, lightshelves for sunlighting are not applicable for east and west oriented façades, where due to the low sun angles the glare of direct sunlight cannot be avoided.

The angle of tilt of a lightshelf, as a function of clerestory height and room depth, can be determined from the graphs in Figures 58 and 59. Figure 60 demonstrates lightshelf angles for various locations.

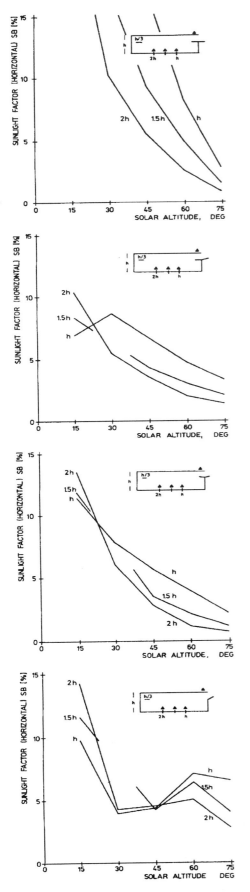

Figure 57 - Contribution due to sunlight (Sunlight Factor) for different lightshelves.

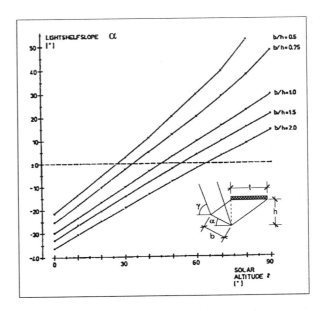

Figure 58 - Determination of the tilt of a lightshelf for a given clerestory height.

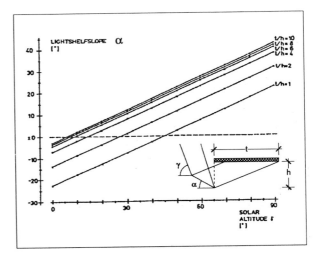

Figure 59 - Determination of the tilt of a lightshelf for a given ratio of room depth to clerestory height.

5.56 Daylighting in Architecture

Sunlighting with lightshelf elements

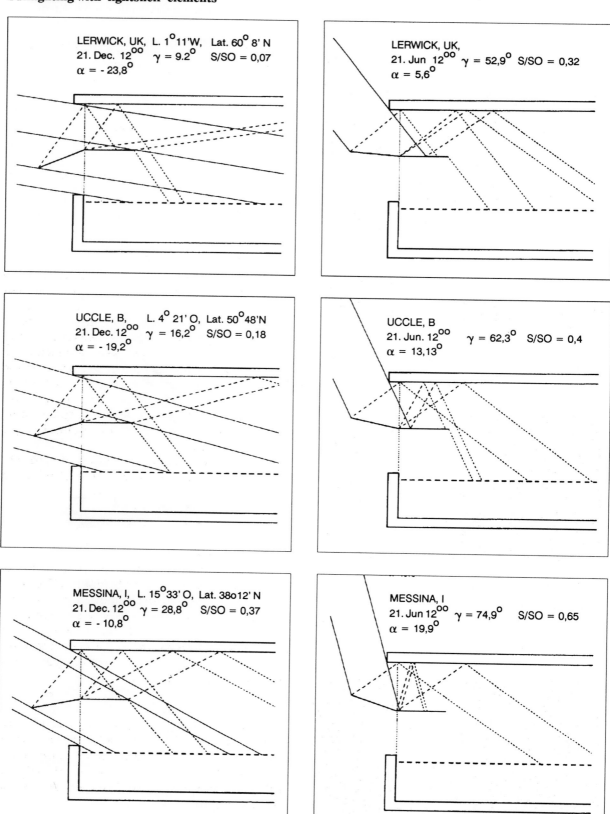

Figure 60 - Lightshelf slope angles catching the maximum amount of sunlight for different European locations on June 21 and December 21. Low sun angles result in a downward slope, which is good for shading but not for redirecting light.

PRISMATIC SYSTEMS

Function

Prismatic panels control transmitted light by refraction. The direction of the incoming daylight is altered by passing through a prism or a triangular wedge of glass (Figures 61 to 64). Normally a prismatic system which will refract the light to the ceiling consists of two sheets of prismatic panels, with the prismatic faces placed internally for dust protection.

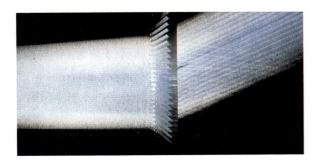

Figure 61 - Light control of diffuse daylight towards the ceiling using prismatic glass (13).

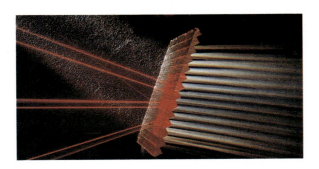

Figure 62 - Prismatic panels, set at a fixed angle so that the specular surface on one side ensures that all possible sunlight angles are within the panel's cut-off range (13).

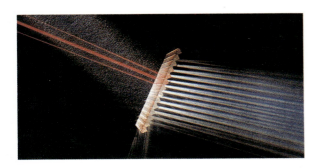

Figure 63 - The moveable prismatic panels do not have a specular surface and move in line with the position of the sun. The cut-off range is +/- 4.5° in relation to the perpendicular (13).

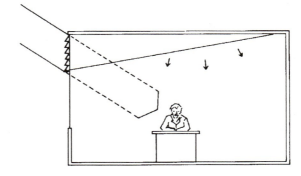

Figure 64 - Function of a prismatic system in design.

Introduction

Prismatic elements for daylighting were in use in Berlin at the beginning of this century (8). Poor quality electric lighting at that time encouraged the investigation of deeper daylight penetration techniques, especially in densely built-up cities like Berlin. These Luxfer-prisms were designed to be used in heavily obstructed rooms to redirect diffuse light from the sky zenith towards the back of the room, which would otherwise receive no direct sky light (Figure 65). They were applied in spaces facing narrow streets, basements, industrial buildings, hospitals and greenhouses.

Further investigation of the idea of prismatic systems has brought about two forms of prismatic glazing:
- sunlight directing prisms
- sunlight excluding prisms.

Sunlight Directing Prisms

Such systems were investigated in Australia during the 1970s by Ruck and Smith (10). The initial approach was a fixed prismatic panel (enclosed by two panes of glass) above eye level; however this application tended to deflect light downwards at high solar altitudes. To overcome this problem a movable tilted prismatic panel with seasonal adjustment was suggested by the research team.

Sunlight Excluding Prisms

These prismatic systems, investigated during the 1980s by C. Bartenbach in Austria (15, 16), aim to reject direct sunlight while admitting near-zenithal sky light. These systems work on the principles of both reflection and refraction. The main functions of such systems are:
- sun-screening
- daylight control and distribution.

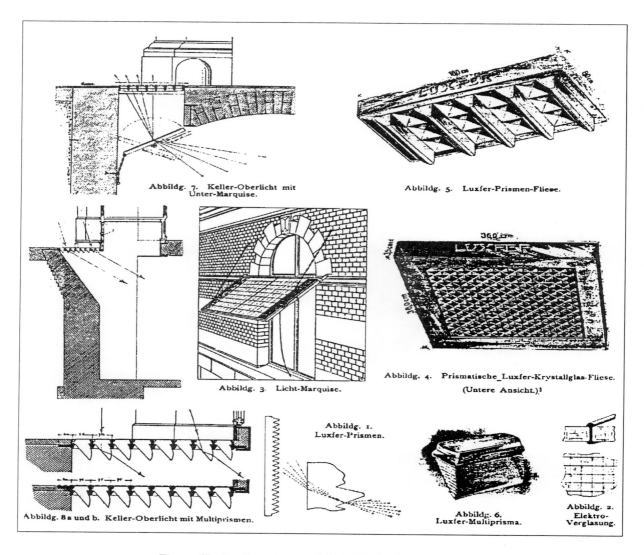

Figure 65 - Luxfer-prisms and their applications in 1902 (8).

Prismatic panels (typically 206 mm x 206 mm in size) are designed to be used also within double-glazed units and are available in 11 different types, each designed to perform a different task. Four different angles of prism, used according to their location in the building, are produced either unsilvered or with a part-silvered prism for solar rejection (13).

Systems of this type have been installed in several buildings in Europe (for example see Figure 66).

The diagrams in Figure 67 demonstrate various design options for sidelighting using prismatic elements.

Advantages of Prismatic Systems
- Prismatic elements are translucent; the sky cannot be seen through these elements, but it remains perceptible. Thus, this control element does not disturb the overall appearance of a window seen from the inside.
- Glare due to the sky brightness seen through a conventional window is reduced to a high degree: sun-excluding prisms have a luminance, viewed from below, of only 100-300 cd/m^2 compared with 2000-6000 cd/m^2 for overcast sky viewed through a conventional window; therefore it becomes an interesting daylighting system for rooms with VDU use.

Disadvantages of Prismatic Systems
- With fixed prism systems the view out is permanently obscured.
- Constructional aspects can be problematic: the single prism requires a 20 mm space between panes of glass, increasing to 48 mm for the three prism version.
- The cost can be considerably higher than conventional systems: however, some of this cost may be offset by savings in building cost (no sunshading devices, reduced HVAC system, reduced installation of luminaires) and also by potential energy savings (lighting and air conditioning running cost).

Daylighting Components 5.59

Figure 66 - Landeszentralbank in Cologne, Germany: Daylighting design: C Bartenbach

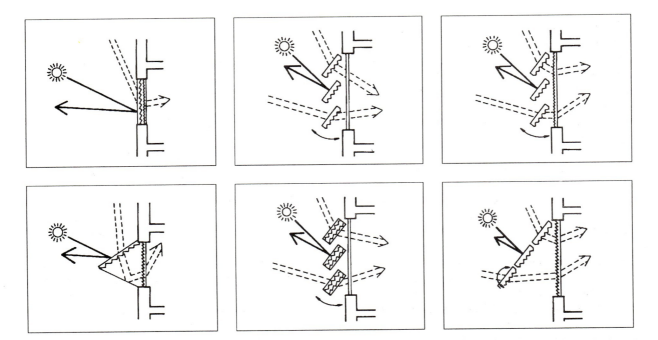

Figure 67 - The use of prismatic elements for sidelighting (13).

HOLOGRAPHIC OPTICAL ELEMENTS

Holographic optical elements (HOEs) bend the light by diffraction (Figure 68).

Function of HOE

The principle of this technology is that windows are coated with a transparent coating in which an invisible diffraction pattern is "printed" by a holographic technique. The window will then deflect transmitted direct and diffuse solar radiation over a well defined angle (which is defined by the diffraction grid characteristics) deep into the building. Similar grids can also be used to reflect away solar light which impinges on the window from well defined angles (Figure 69).

One of the main advantages of HOEs is the fact that, unlike conventional optical elements, their function is essentially independent of substrate geometry. Another advantage is the possibility of spatially overlapping elements, since several holograms can be recorded in the same layer (Figure 70). Volume holograms, recorded in dichromated gelatin, can take up to four different images with specific information.

Design of HOEs

The diffraction pattern of transmittance and reflection holograms has to be defined before recording. Light then emerges at set exit angles, but only if the angle of incidence and wavelength of the incident light satisfy a particular relationship. It is possible to obtain a diffraction efficiency of 90%, which means that 90% of the incoming light can be redirected to a specific point in the space.

The diffraction of light waves - the way in which they bend when they pass an obstruction - is a well known phenomenon. Light waves resemble ripples on water: when one encounters another, interference patterns are created. Such an interference pattern, refined by the coherence (single wavelength) of laser light, is frozen on a recording material to create a hologram. The recording process is similar to the recording procedure of display holograms, but without the use of an object. A split laser beam is aimed at the recording material to create the interference pattern.

Recording Material for HOEs

The ideal recording material for HOEs should have a spectral sensitivity well matched to available laser wavelengths, linear transfer characteristics, high resolution and low interference.

Several materials do more or less meet these requirements:
- Silver halide photographic emulsions are the most widely used recording material for holography. After recording, these holograms need to be wet processed and dried using a

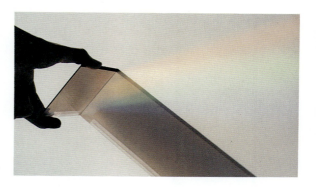

Figure 68 - Diffraction of light using HOE.

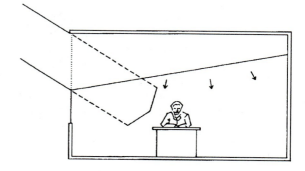

Figure 69 - Function of HOE in design.

Figure 70 - Two-dimensional performance of top-lighting using two layers of HOE.

technology similar to that for normal photographic materials. The optical properties do not change during exposure; this advantage makes it possible to record several holograms in the same photographic emulsion without interaction between them.
- Dichromated gelatin. In some respects this material represents the ideal recording material for phase holograms; it provides large refractive-index-modulation capabilities, high resolution, low absorption and scattering. The gelatin layer has to be prepared in the laboratory, fixed on the glazing material, exposed and dried after exposure. The process involves a sequence of chemical and physical reactions. The final volume hologram needs to be protected against humidity. One of the most attractive aspects of using dichromatic gelatin is the high resolution of the material in which the interference patterns can be created with a high density of lines. This results in a non-disturbed view through the hologram perpendicular to the glazing in building applications.
- Other recording materials such as photoresists (light-sensitive organic films), photopolymers, photochromics, photothermoplastics and photorefractive crystals can be used.
- Embossed holograms, a technology used for reflective holograms, need to be developed at a large scale for transmittance holograms. As these holograms are printed on a polymer material, the potential for low-cost production is very good. However these holograms present several disadvantages: only one hologram per layer, reduced diffraction efficiency, reduced transmittance, and disturbed view out.

Criteria for the Use of Holograms in Building Applications
- To redirect daylight or sunlight which strikes the window surface over a large area of the interior of the room, HOEs need to be developed that allow for a 40° (northern Europe) - 45° (southern Europe) difference in summer and winter noon altitudes.
- White light is needed in the space: single layer holograms disperse the incoming light wavelength and therefore are not appropriate, unless this effect is wanted.
- A clear view out is required: no distribution in the direct field of view is acceptable.
- High optical transmittance for diffuse light is aimed at, because for regions with mainly overcast sky conditions, the amount of daylight in the space should not be reduced.
- UV resistance

Model Set-up
Within the CEC project BUILDING 2000 one participant designed a building using HOEs for daylighting. Several small prototype plates of HOEs recorded by the group were tested using scale models (17).

Specifications of the samples used for testing. Samples of these HOEs could be used within the frame of this analysis on selected components to introduce the third physical phenomenon to alter the path of light, namely diffraction.

The holographic prints of these samples are recorded in a photographic emulsion fixed on a single pane of glass. The dimension of one holographic element is about 15 x 15 cm. The exit angles are 18°, 30°, 45° and 60°, each plate with a different exit angle. The HOEs are recorded with a diffraction efficiency of about 50%; about 50% of the incident light travels straight through the HOE. The transmittance for diffuse light is very low; the range differs between 50 and 60% due to the uneven thickness of the photographic emulsion over the plate.

Two-dimensional performance studies. Studies were first carried out in a photographic laboratory using a projector as light source. Different design options for the use of the available HOEs samples for sidelighting and top-lighting for different sun angles have been studied and documented (Figure 71).

Quantitative photometric measurements. Measurements were taken from the scale model of the basic office space (Figure 72). Due to the specifications of the samples and the fact that these samples are prototypes, the results provide only an indication of the potential of uses of HOEs in architectural design.

The basic office model (scale 1:20) with the clerestory window was used with two different application options of HOE (exit angle 45°):
- HOE as vertical window aperture above eye level, clerestory band window, height 1 m
- HOE as tilted window (same window opening) in combination with a sloped mirror, which also could be a reflective hologram (declination 30°)

Results of Sunlighting Studies
HOE vertical in façade aperture
- Due to the specifications of the tested HOE with an exit angle of 45° and a range of incidence angles of +/- 15°, the sun at high altitudes is automatically blocked out (sunshading): winter sun penetrates the space whereas summer sun is shaded (Figure 73).

5.62 Daylighting in Architecture

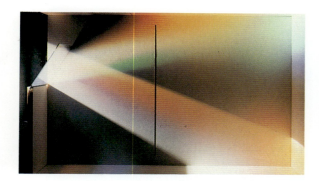

HOE, exit angle 45°, solar altitude 30°

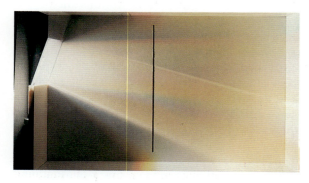

Combination of two HOE layers in order to mix the diffracted light to white light, solar altitude 45°.

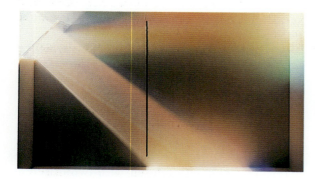

Combination of four layers of HOE, two in window aperture, two tilted to redirect transmitted light.

Combination of two transmission HOEs in window aperture with one reflection HOE, simulated by a mirror.

Figure 71 - Two-dimensional performance studies with different holographic elements.

Figure 72 - Model set up for quantitative analysis of HOEs, in sunlight (top) and artificial sky (bottom).

HOE tilted in combination with a tilted mirror

- The specification of the HOE supports the penetration of sunlight at higher solar angles since the most effective angle to strike the HOE window is normal incidence.
- The effect of the tilted mirror is to increase the contribution of sunlight in the room for high sun altitudes (Figure 73).
- Low sun angles are not controlled by the sloped mirror.
- Reflections on the ceiling near the window can cause glare at high sun altitudes.
- The use of HOEs in combination with the mirror reduces the glare potential near the window but still admits a high contribution of sunlight for low sun altitudes.

Results of Daylighting Studies
HOE vertical in façade aperture

- due to the low transmittance of the available HOE for diffuse light, the illumination level in the space is decreased without any significant effect on the light distribution (Figure 74).

HOEs sloped in combination with a sloped mirror
- The use of a sloped mirror increases the daylight level in the space by a factor of 2.
- The use of HOEs in this combination enhances the uniformity of the illumination in the space and also indicates a higher contribution of daylight in the middle part of the room.

Conclusions

The performance of the HOEs tested for diffuse light was very poor. No significant effect is visible because light waves of diffuse light are coming from various directions; if there was any specific light-directing effect it was so small that it could not be detected.

When using the HOE for sunlighting there is a high potential for application in design. HOEs can be designed as sunshading and light-directing devices in combination, without altering the distribution of the window area. However, at certain viewing angles rainbow effects are created.

There is a very high potential for the application of HOEs in buildings for improving the luminous environment. Among the advantages are:
- the conventional appearance of a building does not need to be changed when using HOEs: thus there is a high potential for its application in buildings to be retrofitted
- HOE elements can fulfil several functions of the building aperture in one element: light-admitting element, sunshading, passive control of energy consumption
- HOE elements can also be combined with various functions of the glazing material: insulating glazing, security glazing, acoustic glazing, etc
- for properly designed HOEs there is no need for sun-tracking
- the design of the HOE integrates the requirements of the user and the architecture.

Systems are currently under development (18).

GENERAL DESIGN RECOMMENDATIONS FOR CONTROL ELEMENTS (19)

First of all, the need to install light-directing elements has to be established. The fields of application can be where:
- requirements for visual comfort are demanding (ie office space with VDUs, classrooms)
- large external obstructions with low reflectance are located outside the room (ie urban space, street, atrium)
- the availability of sunshine is high (south-facing window)

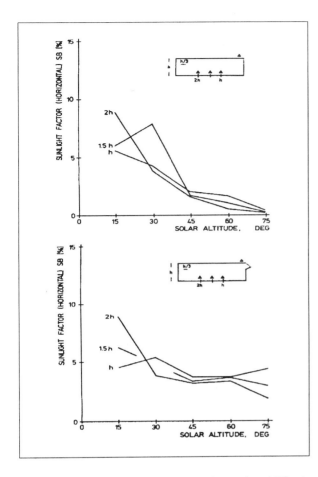

Figure 73 - Sunlight Factors for HOE with a 45°° exit angle: a vertical window aperture (top) and a tilted aperture with mirrored lightshelf (bottom).

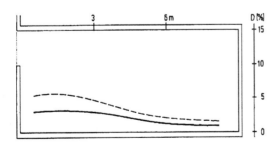

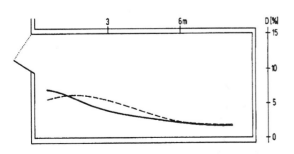

Figure 74 - Daylight distribution using HOE (45°) as glazing material (dotted line represents the daylight distribution in the reference case without glazing).

adequate uniformity of natural illumination cannot be met with a side window in a large office space. Minimum uniformity ratios are given in DIN 5034 (20) for side-lit spaces: $DF_{min} / DF_{max} > 0.16$. The CIBSE *Window Design Guide* (21) gives the following criteria for acceptably uniform diffuse daylighting:

$$l/w + l/h < 2/(l-R_B)$$

where:
l = depth of the room
w = width of the room
h = window head height above floor level, and
R_B = average reflectance of the surfaces at the rear half of the room.

GENERAL CHECKLIST FOR DESIGN (19)

Quantity of Light

Enough diffuse light must be allowed to enter the window. Light-directing elements partially obstruct the window area for diffuse daylighting. A check is recommended of the Daylight Factor (average Daylight Factor, min./max. Daylight Factor) for the conventional window without control elements. If the calculated Daylight Factor is less than required, the reduction of daylight access caused by any device in the window area is not acceptable.

The CIBSE *Window Design Guide* (21) provides a formula for calculating the average Daylight Factor for a side-lit space:

$$D_{av} = TW\Phi / (A(1-R^2))$$

where:
T = transmittance of the glazing material
W = net glazed window area
Φ = angle in degrees of sky visible from the centre of the window, measured in the vertical plane through the window
A = total area of interior surfaces
R = area-weighted average reflectance of interior surfaces, including window

Control System for Sunlighting or Daylighting

Does the environment of the building allow access to sunlight all year and all day long? If sunlight is available, the system can be designed to use sunlight in a controlled or diffused way. Therefore the position of the sun has to be reviewed, with respect to façades, using a sunpath diagram appropriate to the latitude of the site. If the system is designed to use sunlight, the system should cope with sunlight for all possible sun positions: solar tracking systems are most effective, but also expensive. For fixed systems the sun position of March 21 at noon is appropriate to represent the whole year. The design should not allow low winter sun penetration through the upper part of the system because of possible glare problems in the rear part of the room.

If the system is designed for daylighting it should give a high average daylight factor and a low uniformity ratio value.

Ceiling is Part of the Control Element

The ceiling is an important design element. The ceiling acts as a secondary diffuser of the incoming light; therefore the ceiling must have a very high reflectance and the ceiling area should be plain to avoid shadows. Although metal ceilings (ie aluminium) have very high reflectances, they reflect the surroundings and all movements, and the noise in the space may be increased and cause discomfort to the occupants.

Control of Electric Lighting

Good control of electric lighting is essential. For sunlighting systems the controls have to be designed very carefully because of the rapid changes in illumination level of the sun. Instead of controlling broad zones, it seems appropriate to have sensors at the workplaces, that is if the electric lighting is not designed for task lighting.

Maintenance Aspects

Any kind of system which is protected by glazing can be recommended as it minimises the loss of efficiency due to dust and dirt; however the gradual reduction of transmittance has to be taken into account. Mirrors and highly reflective specular surfaces can be damaged by cleaning; scratches and mat areas in those surfaces will be seen as shadows or dark areas on a brightly lit ceiling. The possibility of cleaning the façade from the inside as well as from the outside should be considered during the design process.

Capital Costs

Any kind of sun-tracking element requires a sophisticated computer-controlled system and therefore becomes expensive. Fixed elements are cheaper, but their efficiency is lower because they are not optimal for all possible sun positions.

The feasibility of sun-tracking elements is dependent on the location of the site - it is not very useful to design sun-tracking systems for regions with a relative sunshine duration of less than 0.4. This aspect has to be solved in the very early design stages with a design tool which gives an annual energy balance.

REFERENCES

(1) Saxon, R., *Atrium Buildings*, Development and Design, The Architectural Press, London, 1986.

(2) IES, RP-5, *Recommended Practice of Daylighting*, Illuminating Engineering Society of North America, New York, 1978.

(3) Willbold-Lohr, G., *Increasing Daylighting of the Court Building Oranienstrasse 9 in Berlin*, Daylighting Study within the CEC project BUILDING 2000, 1989 (not published).

(4) Navvab, M., Selkowitz, S., "Daylighting Data for Atrium Design", *Proc. 9th National Passive Solar Conference*, Columbus, Ohio, 1984.

(5) AAMA, *Skylight Handbook*, Design Guidelines, American Architectural Manufacturers Association, Illinois, 1987.

(6) Lam, W., *Sunlighting as Formgivers for Architecture*, Van Nostrand Reinhold Company, New York, 1986.

(7) IES., *Lighting Handbook: Application Volume*, Illuminating Engineering Society of North America, 1987.

(8) *"Über Luxfer-Prismen und deren Anwendung im Bauwesen"*, *Deutsche Bauzeitung*, July 1902, pp. 374-376.

(9) Hopkinson, R.G., "Lighting: daylighting a hospital ward", *Architects Journal*, Vol. 115 (2973), 1952.

(10) Ruck, N., "Daylight in Practice", *Building Services*, Vol. 7, 4/1985.

(11) Rosenfeld, A., Selkowitz, S., "Beam Daylighting: an Alternative Illumination Technique", *Energy and Building*, 1, 1977.

(12) Chauvel, P., Collins, R., Dogniaux, R., Longmore, J., "Glare from windows: current view of the problem", *Lighting Research and Technology*, Vol. 14, No. 1, 1982.

(13) Daylight System, *Siemens Product Information*, Traunreut, Germany, 1987.

(14) Ruck, N., Smith, S.C.J., "Solar beam lighting using a prismatic panel", *Proc Windows in Building Design and Maintenance*, Göteborg, Sweden, 1984.

(15) Bartenbach, C., Klingler, M., *"Lenken und Spiegeln"*, *Werk, Bauen und Wohnen*, Nr.12, 1987.

(16) Bartenbach, C., *"Neue Tageslichtkonzepte"*, *Technik am Bau*, No. 4, 1986.

(17) Willbold-Lohr, G., *Sunlighting with HOE in a Community Centre in Papenburg, Germany*, Daylighting Study within the CEC Project BUILDING 2000, 1989 (not published).

(18) Ian, R., King, E., "Holographic glazing materials for managing sunlight in buildings", *The Journal of Architectural and Planning Research*, No. 1, 1987.

(19) Littlefair, P., "Innovative Daylighting Systems - A critical review", *Proc. (CIBSE) National Lighting Conference*, Cambridge, 1988.

(20) DIN 5034, *Innenraumbeleuchtung mit Tageslicht,* German Standard.

(21) *CIBSE Window Design Guide*, The Chartered Institution of Building Services Engineers, London, 1987.

Chapter 6
ELECTRIC LIGHTING

INTRODUCTION

Electric lighting plays an essential role in determining the environmental quality of building interiors. Furthermore, the choice of artificial lighting systems and the relationship to daylighting strategies is central to the achievement of energy efficiency. In the design of electric lighting systems there are three levels of initial decisions to be made relating to:
- the lighting strategy
- the type of lamp
- the type of luminaire.

In terms of a lighting strategy for commercial situations the choice is typically between providing:
- general lighting (uniform illuminance, regular layout, monotonous effect, independent of furniture layout)
- localised lighting (provides illuminance where it is required, more energy efficient, relocatable systems are preferable to achieve flexible)
- task lighting (with minimal background lighting, typically one third of working illuminance, efficient when high illuminances are required, flexible, permits individual control).

The choice of lamp will be determined by its characteristics, such as colour rendering properties, efficacy, run-up time when switched on, control, cost, life and maintenance.

In determining the choice of luminaire, the following are issues that should be considered: light distribution, utilisation factor (which determines the installed efficacy of a system), safety, robustness, reliability and life.

Once these strategic decisions have been determined, more detail is required at a tactical level of the following equipment for electric lighting installations:
- lamps
- control gear for lamps
- luminaires
- luminaire mounting systems
- lighting control gear.

Of this, the lighting control gear will be described in Chapter 7, "Control Systems".

LAMPS

Of the many lamp types available, only two main groups find universal use in interior lighting:
- incandescent lamps
- low-pressure mercury fluorescent lamps.

A third group, high-pressure discharge lamps, is at present used in a few specialised applications only, but this situation may well change in the future.

Both classes, incandescent lamps and fluorescent lamps, can be further divided. The following survey gives this subdivision, in which only the lamp types relevant to this publication are considered.

Standard Incandescent Lamps

The first practical carbon-filament electric lamps date from 1879. In the period from 1902 to 1913 various types of metal-filament lamps came on the market, and by 1933 the tungsten-filament lamp had been developed to its present form. Since then, no significant improvements have been added to the technology of the standard incandescent lamp.

Essentially a standard incandescent lamp consists of a single or double-coiled tungsten wire, or filament, which is mounted in a glass bulb filled with an argon/nitrogen gas mixture (Figure 1). The bulb of such a general lighting service (GLS) lamp can be clear, satin-finished ("frosted"), white-coated ("opalised") or colour-coated. Frosting gives no significant light loss; opalising results in a light loss of about ten per cent.

The wire, or filament, is brought to incandescence by passing a current through it, the wattage dissipated in the lamp being principally determined by the supply voltage and the length-to-diameter ratio of the filament. With some restrictions, the tungsten filament can be regarded as a black-body radiator. This means that an increase of the filament temperature will not only result in an increase of the total radiation, but also of the proportion that is radiated in the shorter wavebands. Thus, with increasing temperature, the ratio of visible radiation to infrared radiation, or radiant heat, shifts in favour of the visual waveband.

6.2 Daylighting in Architecture

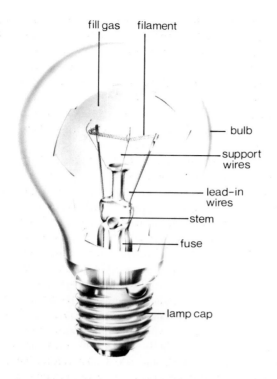

Figure 1 - Standard incandescent lamp.

There is therefore every reason to make the filament temperature as high as possible, but a limit is set by the melting point of the tungsten (3680°K). Even before this point is reached the tungsten will start to evaporate, which is in fact the chief restricting factor for lamp life. So a compromise has to be made between luminous efficacy and lamp life (Figure 2).

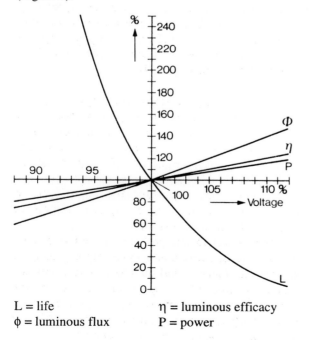

L = life
φ = luminous flux
η = luminous efficacy
P = power

Figure 2 - Effects of voltage variations on lamp performance.

By international agreement, the lamp industry has opted for an operating life of 1000 hours for GLS lamps. This corresponds to a filament operating temperature of approximately 2700° K, and a luminous efficacy of about 15 lm/W for a 100 W lamp.

The depreciation of the light output after 1000 hours is of the order of 20 per cent, which is almost entirely due to bulb blackening (Figure 3).

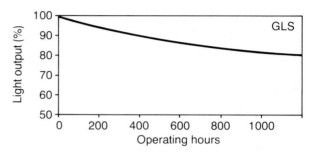

Figure 3 - Depreciation curve of incandescent lamps.

Reflector Lamps

Reflector lamps have an internal aluminium mirror coating over part of the bulb surface. The purpose is to obtain a directional light output. There are two main types of reflector lamps, with the reflector behind or in front of the filament. Those with a parabolic mirror behind the filament produce a more or less concentrated light beam. The shape of the mirror and the size and position of the filament determine the degree of divergence of the beam. If this beam spread is less than 20°, the lamp is designated "spot", if between 20° and 40° it is said to be "flood", and if more than 40° it is "wide flood".

Constructionally there are two types of parabolic reflector lamps, one with a blown bulb, as a GLS lamp, the other with a bulb of moulded glass. The latter is called a PAR-lamp (parabolic aluminised reflector). This type offers tighter beam control and has a higher mechanical and thermal strength than a blown-bulb lamp, which makes it more suitable for outdoor use (Figure 4).

Figure 4 - Reflector lamps: blown bulb (left) and pressed glass PAR bulb (right).

Some PAR lamps employ dichroic reflectors, instead of aluminium mirrors. Such a reflector has selective transmission and reflection properties, which allow visible radiation to be reflected, but heat to be transmitted behind the lamp. These "cool-beam" lamps are popular for the illumination of heat-sensitive goods, but need a well-ventilated luminaire, as the heat is radiated backwards (Figure 5).

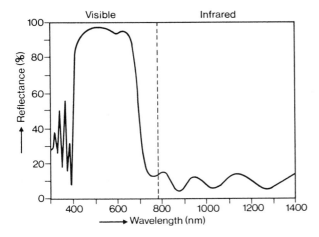

Figure 5 - Reflectance curve of a typical "cool-beam" reflector.

Lamps with a spherical mirror in front of the filament are called bowl-reflector lamps. The purpose is to screen the filament from direct view. The light is directed backwards and either diffusely reflected from a wall or other light surface, or from a parabolic mirror behind the lamp, which produces a sharply defined beam of light. Because the mirror is spherical, bowl-reflector lamps generally have the same bulb shape as normal GLS lamps (Figure 6).

 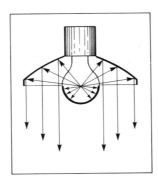

Figure 6 - Bowl-reflector lamp (left), and the same placed in a parabolic reflector (right).

Mains-voltage Tungsten Halogen Lamps

Halogen lamps are incandescent lamps in which special measures have been taken to counteract bulb blackening with age, as a result of evaporation of the filament. It had long been known that lamp life could be considerably extended - or the filament temperature, and thus the luminous efficacy, increased - if the gas pressure inside the bulb could be increased. This, however, was only feasible with much reduced bulb dimensions, but the reduction would lead to an unacceptable degree of blackening because of the smaller surface area.

In 1959 it was discovered that bulb blackening could be eliminated altogether by adding a halogen (normally a bromine compound) to the fill gas. The vapourised tungsten forms a volatile compound with the free bromine, and thus stays gaseous instead of settling on the relatively cool bulb wall. Near the hot filament, the tungsten bromide decomposes, whereby the tungsten is deposited on the filament (but not necessarily on the place from which it was evaporated, so tungsten halogen lamps do not have eternal life!) and the bromine becomes available again for a new cycle. In order to maintain the halogen cycle, the bulb temperature must be fairly high, between 200°C and 300°C. Therefore, halogen lamps must be made of hard-glass or pure quartz.

Halogen lamps for operation on the mains voltage in standard applications have a filament temperature of 2800°K to 3000°K and a luminous efficacy of 16 to 18 lm/W. Wattage ratings range from 75 W to 2000 W. The average life is 1000 to 2000 hours. The following main varieties exist:
- tubular double-ended
- tubular single-ended
- double-envelope lamps
- reflector lamps.

Double-envelope lamps are provided with a separate outer envelope and a standard screw or bayonet cap. They are intended as direct replacements for GLS-type standard lamps. Reflector halogen lamps are single or double-ended capsules sealed in a PAR-type outer envelope (Figure 7).

Low-voltage Tungsten Halogen Lamps

An attractive feature of tungsten halogen lamps is their small size compared with standard lamps, which allows for a great freedom in luminaire design. It also makes them ideal for use in narrow-beam spotlights, but in that case the filament must also be as compact as possible, because only then is tight beam control with an optical system of restricted dimensions possible.

The answer to this problem is the use of low-voltage lamps, because at a lower supply voltage a shorter and thicker - and therefore more compact - filament is needed to dissipate the same power. The consequence of using low-voltage lamps is that, with mains supply, a transformer is needed, which is usually accommodated inside the luminaire.

Low-voltage lamps exist as a plain bulb, for use with a separate reflector, or are sealed in an external

6.4 Daylighting in Architecture

Figure 7 - Double envelope (left) and PAR-type reflector (right) mains voltage halogen lamps.

parabolic mirror reflector. The latter is usually provided with a front glass. Coloured lamps have a dichroic filter in front of the lamp. The operating voltage of low-voltage lamps is generally 6 V, 12 V or 24 V. Wattage ratings range from about 15 W to 100 W. The luminous efficacy is up to 25 lm/W.

Tubular Fluorescent Lamps

A tubular fluorescent lamp is a low-pressure discharge lamp. It basically consists of a glass tube with a metal conductor, called an "electrode", at either end. The tube is filled with an inert gas and mercury vapour at low pressure. If an electric current is passed between the electrodes through the gas mixture, the mercury vapour atoms will start to emit electromagnetic radiation. The actual light-emitting agent is a thin coating of a fluorescent powder (also called "phosphor"), or a mixture of fluorescent powders, applied to the inside of the tube. This coating absorbs the short-wave ultraviolet radiation (184 nm and 254 nm) from the low-pressure mercury discharge and re-emits it as visible radiation (Figure 8).

The composition of the phosphor(s) chiefly determines the colour characteristics (colour appearance and colour rendering), the luminous efficacy and also the price of the fluorescent lamp. There is a global rule that any attempt to obtain a very high average colour rendering index (R_a) results in a lower luminous efficacy, and, conversely, that a very high luminous efficacy can be obtained if an R_a of no higher than 80 is considered acceptable.

The survey of currently available fluorescent colours and their characteristics shown in Table I is based upon the Philips classification, but the performance of other major brands is comparable.

The luminous flux, and to a certain degree also the luminous efficacy, depend on the ambient temperature. The values given in Table I refer to an ambient temperature of 25°C in still air. With higher or lower temperatures the luminous flux drops considerably (Figure 9).

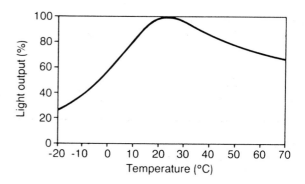

Figure 9 - Luminous flux of a fluorescent lamp as a function of ambient temperature.

Fluorescent lamps first appeared on the market in 1938, but their widespread introduction in Europe had to wait until the end of World War II. Although other types of low-pressure discharge lamps, such as the Moore and neon tubes, had already existed for a long time, the introduction of commercially successful fluorescent lamps only became possible after the discovery of efficient phosphors with good colour characteristics, and the development of oxide-coated electrodes, allowing operation on the standard mains voltage.

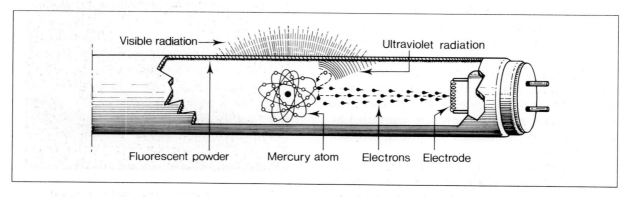

Figure 8 - Working principle of a fluorescent lamp.

TABLE I - Characteristics of tubular fluorescent lamps

Colour Code	Colour Appearance	Colour Temperature (°K)	Colour Rendering	Luminous Flux (lm)[1]	Luminous Efficacy (lm/W)[2]	Price Index
29	Warm-white	2900	51	3000	83	1
82	Warm-white	2700	81	3250	90	2
83	Warm-white	2900	82	3450	96	2
92	Warm-white	2700	94	2250	63	2, 3
93	Warm-white	3000	95	2300	64	2, 3
94	White	3800	96	2350	65	2, 3
33	Cool-white	4100	63	3000	83	1
25	Cool-white	4100	70	2500	62	1
84	Cool-white	4000	80	3450	96	2
85	Cool-white	5000	80	3300	92	2
95	Cool-white	5000	98	2350	65	2, 3
54	Daylight	6200	72	2500	69	1
86	Daylight	6300	77	3250	90	2
57	Daylight	7300	94	1800	45	1, 7

Notes:
1 The luminous flux is given for the 36 W / 40 W version
2 Not including ballast losses

In common with practically all discharge lamps, tubular fluorescents need a current-limiting device, called "ballast". In practice this is either an inductive coil ("choke ballast"), or an electronic circuit ("electronic ballast"). Furthermore, most fluorescent lamps need a "starter" or "igniter" to initiate the discharge. This is either an automatic bimetallic switch, which sends a high voltage pulse generated by the choke ballast through the lamp ("glow-discharge starter"), or, again, an electronic circuit ("electronic starter"). The operation of a glow-discharge starter is easily recognisable because often several attempts are needed to ignite the lamp.

The use of inductive ballasts induces a phase shift between voltage and current, resulting in a low power factor. For larger installations, electricity supply companies require this to be compensated for, which is done by inserting a series or parallel capacitor in fifty per cent of the lamp circuits (Figure 10).

The most sophisticated type of ballast is an electronic circuit that converts the 50 Hz mains into a regulated high-frequency current ("HF electronic ballast"). This expensive solution is justified by providing a higher luminous efficacy of the lamp and longer lamp life.

To facilitate starting, the electrodes of most fluorescent lamps are preheated for a few seconds before ignition, the preheating current being switched on and off by the starter. As soon as the lamp is operating, the discharge takes over the heating of the electrodes.

Smaller lamps (of maximum 1.2 m length, 40 W) can be used in circuits without a separate starter. To facilitate ignition, these lamps mostly employ some form of external conductor along the discharge path, for example a conductive coating or a conductive strip on the tube wall. The commercial designation of these lamps is "rapid-start" lamps. Because there is no device to switch off the heating current after ignition (the glow-discharge starter), the current is maintained during operation of the lamp, but at a lower level. An advantage is that lamps with continuously heated electrodes are the most suitable for dimming installations.

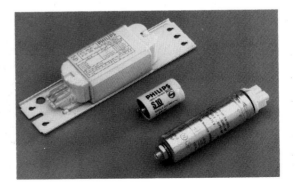

Figure 10 - Ballast, starter and compensating capacitor for fluorescent lamps.

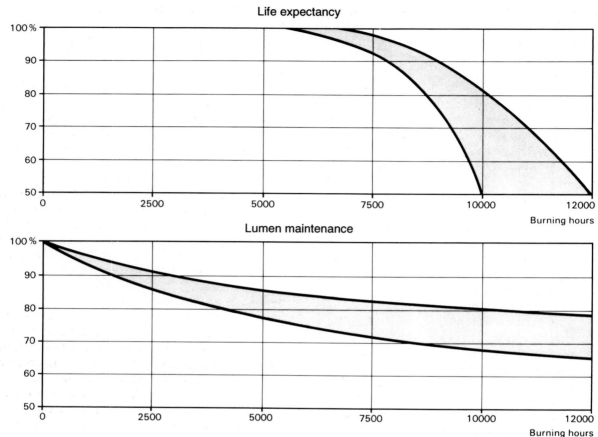

Figure 11 - Life expectancy and light output depreciation curves of fluorescent lamps.

Finally, there are fluorescent lamps without preheating of the electrodes, so-called cold-start or "instant-start" lamps. These either use an extra large choke ballast, which gives a very high voltage pulse when the lamp is switched on, or are provided with an auxiliary electrode, in the form of an internal conductive strip, to facilitate starting. The strip is connected to one of the electrodes, and reaches close to the other one.

Fluorescent lamps are far more standardised than incandescents are. In Europe just three tube diameters are in use:
- 38 mm for rapid-start and older lamps
- 26 mm (1 in) for present-day lamps
- 16 mm (0.5 in) for miniature lamps.

Lamp length is related to lamp wattage, although within rather wide margins. In practice, standard dimensions of luminaires and ceiling systems are more critical factors with respect to lamp length. Extremes vary between 150 mm and 2400 mm, but the most frequently used lengths for general applications are:
- 590 mm, corresponding to 20 W (38 mm Ø), 18 W (26 mm Ø) or 16 W (26 mm Ø HF)
- 1200 mm, corresponding to 40 W (38 mm Ø), 36 W (26 mm Ø) or 32 W (26 mm Ø HF)
- 1500 mm, corresponding to 65 W (38 mm Ø), 58 W (26 mm Ø) or 50 W (26 mm Ø HF).

Apart from straight lamps, there are lamps bent into a curve, a circle, a U-form or a W-form. There also exist fluorescent lamps with an internal reflecting layer over part of the circumference, giving the lamp a directional aspect.

The end of life of a fluorescent lamp is normally determined by the condition of the electrodes. If these have lost most or all of their emitter material, the lamp will refuse to start or begin to flicker. Lamp life is influenced by external factors such as the switching frequency and the type of ballast and/or starter used. Standard lamps operated on an inductive ballast have an average rated life of about 12000 hours. With a HF electronic ballast an average value of 15000 hours is easily achieved. At the end of life, the light output has dropped considerably (Figure 11).

Compact Fluorescent Lamps

Compact fluorescent lamps have been developed for use in those applications that were traditionally reserved for incandescent lamps. Basically, a compact fluorescent lamp is a standard or narrow-bore fluorescent tube, bent into a compact form and sometimes provided with an integral ballast and/or starter. At present the following four basic forms of compact fluorescent lamps can be distinguished (Figure 12):

1. A standard-diameter (26 mm) fluorescent tube, bent into a circle, fitted with an integrated ballast and starter, and given a standard screw or bayonet cap.
2. A single small-bore fluorescent tube folded into a flat compact form, or two or more parallel small-bore tubes interconnected at or near the ends in such a way as to offer a continuous pathway for the electric discharge. The lamp is provided with a starter and either a conventional or an electronic ballast, and a screw or bayonet cap. Sometimes the lamp proper is detachable from the unit containing the gear and cap.
3. A small-bore fluorescent tube folded into a three-dimensional compact form and placed in a glass or plastic outer envelope. Integrated with the lamp are a starter and a conventional or electronic ballast, and the whole is fitted with a screw or bayonet cap.
4. Two or more parallel small-bore tubes, interconnected at or near the ends. The lamp is fitted at one end with a special cap. This sometimes contains a starter, but the lamp operates on a separate ballast.

Wattage ratings of compact fluorescent lamps range between 5 W and 40 W. The stated wattage and luminous efficacy of lamps with integrated gear (types 1 - 3) always includes the ballast losses. This results in a seemingly lower luminous efficacy, which can be quite considerable, as up to 25 per cent of the total power is dissipated in the small integrated ballasts of these low-wattage lamps. Table II gives the electrical and light characteristics of compact-fluorescent lamp types and wattages marketed by Philips.

The number of light colours available is generally more restricted than with normal fluorescent lamps. Warm-white (2700°K or 3000°K) prevails in lamps that are intended to replace incandescents. Most types, however, are also available in cool-white (4000°K). The colour rendering index is generally about 80.

Because of the higher load on the tube wall, caused by the small dimensions of the discharge tube, and the complicated construction, the economic life of compact fluorescent lamps is generally shorter than that of standard fluorescents. Typical values are between 6000 and 10000 operating hours.

High-pressure Discharge Lamps

Although widely used in outdoor applications, such as street and sports lighting, and also in some specialised indoor applications such as factory halls, the application of high-pressure discharge lamps in interior lighting is still comparatively rare, and until recently virtually non-existent. The reasons were the moderate colour characteristics of the light (colour appearance and colour rendering) and the large lumen packages.

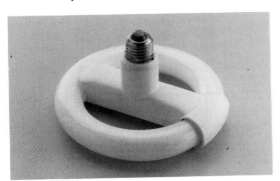

Type 1

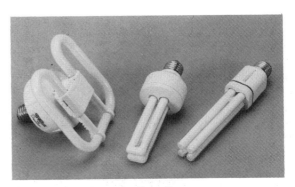

Type 2

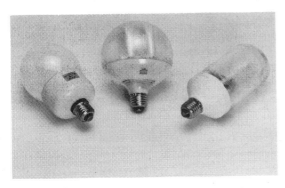

Type 3

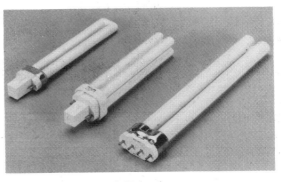

Type 4

Figure 12 - Compact fluorescent lamp types.

TABLE II - Characteristics of compact fluorescent lamps

Designation and Wattage	Integrated Ballast [1]	Integrated Starter [2]	Cap [3]	Outer Envelope [4]	Luminous Flux (lm)	Luminous Efficacy (lm/W) [5]
PLC*E 7 W	yes E	yes E	E27/B22	no	400	57
PLC*E 11 W	yes E	yes E	E27/B22	no	600	55
PLC*E 15 W	yes E	yes E	E27/B22	no	900	60
PLC*E 20 W	yes E	yes E	E27/B22	no	1200	60
SL* 9 W	yes I	yes G	E27/B22	yes C	450	50
SL* 9 W	yes I	yes G	E27/B22	yes O	400	44
SL* 13 W	yes I	yes G	E27/B22	yes C	650	50
SL* 13 W	yes I	yes G	E27/B22	yes O	600	46
SL* 18 W	yes I	yes G	E27/B22	yes C	900	50
SL* 18 W	yes I	yes G	E27/B22	yes O	800	44
SL* 25 W	yes I	yes G	E27/B22	yes C	1200	48
SL* 25 W	yes I	yes G	E27/B22	yes O	1050	42
SL*D 9 W	yes I	yes G	E27/B22	yes O	400	44
SL*D 13 W	yes I	yes G	E27/B22	yes O	600	46
SL*D 18 W	yes I	yes G	E27/B22	yes O	850	47
SL*DE 11 W	yes E	yes E	E27/B22	yes O	550	50
SL*DE 15 W	yes E	yes E	E27/B22	yes O	850	57
SL*DE 20 W	yes E	yes E	E27/B22	yes O	1200	60
PL-S 5 W	no	yes G	spec. 2p	no	250	50
PL-S 5 W	no	no	spec. 4p	no	235	47
PL-S 7 W	no	yes G	spec. 2p	no	400	57
PL-S 7 W	no	no	spec. 4p	no	390	56
PL-S 9 W	no	yes G	spec. 2p	no	600	67
PL-S 9 W	no	no	spec. 4p	no	610	68
PL-S 11 W	no	yes G	spec. 2p	no	900	82
PL-S 11 W	no	no	spec. 4p	no	990	90
PL-C 10 W	no	yes G	spec. 2p	no	600	60
PL-C 10 W	no	no	spec. 4p	no	560	56
PL-C 13 W	no	yes G	spec. 2p	no	900	69
PL-C 13 W	no	no	spec. 4p	no	920	71
PL-C 18 W	no	yes G	spec. 2p	no	1200	67
PL-C 18 W	no	no	spec. 4p	no	1250	69
PL-C 26 W	no	yes G	spec. 2p	no	1800	69
PL-C 26 W	no	no	spec. 4p	no	1800	69
PL-L 18 W	no	no	spec. 4p	no	1200	67
PL-L 24 W	no	no	spec. 4p	no	1800	75
PL-L 36 W	no	no	spec. 4p	no	2900	81

Notes:
[1] I = inductive; E = electronic. [2] G = glow discharge; E = electronic. [3] E27/B22 = standard screw or bayonet; spec. 2p/4p = special two-pin or four-pin. [4] C = clear; O = opalised (white). [5] For lamps without integrated ballast these figures may vary, according to the type of gear used.

In recent years, compact low-wattage lamps with good to very good colour characteristics have entered the market. These are used to an increasing extent in office and shop lighting and, especially, in spot and accent lighting. Two types will be described here:
- metal halide lamps
- high-pressure sodium lamps with an increased sodium vapour pressure ("White SON") lamps.

These high-pressure discharge lamps generally consist of a small discharge tube of a chemically highly resistant material (eg quartz or sintered aluminium oxide), which is surrounded by an outer envelope of glass. The outer bulb is either fitted with a single bi-pin cap, or provided with an electrical contact at both ends (double-ended lamps) (Figure 13).

The filling of the discharge tube consists of an inert gas (usually xenon), mercury, and the actual light-emitting material. For high-pressure sodium lamps this is sodium, and for metal halide lamps it is a mixture of the iodides of various elements such as dysprosium, thulium and holmium, or tin and sodium. Each type has a different spectral distribution (Figures 14, 15 and 16),

The operating pressure of high-pressure discharge lamps ranges, depending on type and wattage, between one and ten atmospheres, compared with a few hundredths of an atmosphere for low-pressure (e.g. fluorescent) lamps. The result is a much higher gas temperature in the discharge, and a much higher luminous flux in relation to the dimensions of the discharge tube. This makes these lamps very compact, which is of special importance if accurate beaming is required, for example in spotlights.

Basic characteristics of typical high-pressure sodium and metal halide lamps are summarised in Tables III and IV.

As with all discharge lamps, "White SON" and metal halide lamps must be connected to the mains via a current-limiting device (ballast). They also need an electronic starter, and sometimes a voltage stabiliser to avoid unacceptable shifts in the light colour as the result of mains voltage fluctuations.

The lamps require a run-up time, after switching on, of five to eight minutes before full light output and stable colour characteristics are obtained. Also, after a voltage interruption, a cooling-down period of some minutes is necessary before the lamp can be re-ignited.

The average life of these lamp types is 5000 hours or more.

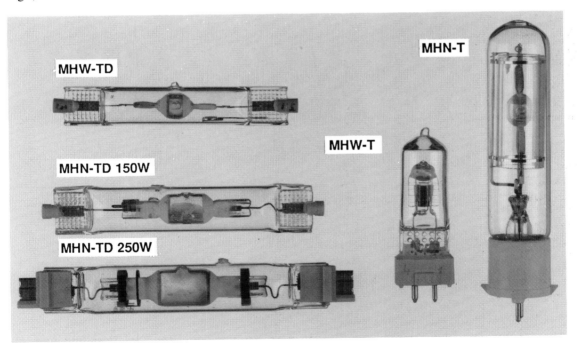

Figure 13 - Double-ended and single-ended metal halide lamps for indoor lighting use.

TABLE III - Characteristics of high-pressure sodium (White SON) lamps

Designation and Wattage	Cap[1]	Luminous Flux (lm)	Luminous Efficacy (lm/W)	Colour Temperature °K	Colour Rendering Index (R_a)
SDW-T 35 W	PG12 s.e.	1300	40	2500	80
SDW-T 50 W	PG12 s.e.	2300	44	2500	80
SDW-T 100W	PG12 s.e.	4800	49	2500	80

[1] PG12 s.e. - bi-pin, prefocus, single-ended.

TABLE IV - Characteristics of metal halide lamps

Designation[1] and Wattage[2]	Cap[3]	Luminous Flux (lm)	Luminous Efficacy (lm/W)	Colour Temperature °K	Colour Rendering Index (R_a)
MHN-T 70 W	PG12 s.e.	5100	68	4200	80
MHN-T 150 W	PG12 s.e.	11000	73	4000	85
MHW-TD 70 W	R7s d.e.	5000	67	3000	74
MHN-TD 70 W	R7s d.e.	5500	73	4200	80
MHN-TD 150 W	R7s d.e.	11200	75	4300	86
MHN-TD 250 W	Fc2 d.e.	20000	80	4300	90

[1] Halide filling: MHW - tin and sodium; MHN - rare earths.
[2] The power dissipated in the 70 W lamps is actually about 75 W.
[3] PG12 s.e. - Bi-pin, single-ended;
 R7s d.e. - Recessed, single contact, double-ended;
 Fc2 d.e. - Special pin, single contact, double-ended.

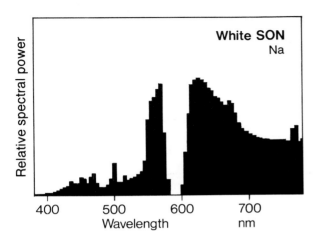

Figure 14 - Spectral distribution of a high-pressure sodium lamp.

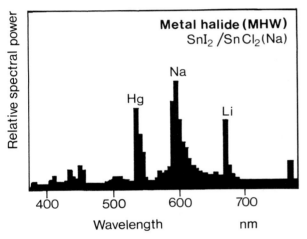

Figure 15 - Spectral distribution of a metal halide lamp with tin/sodium filling.

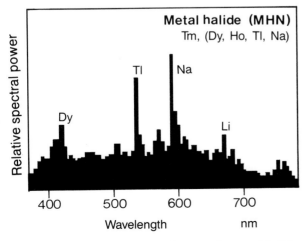

Figure 16 - Spectral distribution of a metal halide lamp with rare-earth metals filling.

CONTROL GEAR FOR LAMPS

Of the lamp types described in the previous section, only standard incandescent and mains-voltage halogen incandescent lamps can be directly connected to the public mains. All others need some form of control gear to regulate the supply voltage and/or current, which often also incorporates a device that provides the high voltage peak necessary to start discharge lamps. In the preceding section, the subject of control gear has already been touched upon. The overview in Table V gives the more common gear combinations for the various lamp types, when operated on 220/240 V mains supply. Switching and dimming equipment will be described in Chapter 7, "Control Systems".

If not actually forming part of the lamp - as is the case with many compact fluorescents - gear for lamps for indoor lighting is nearly always incorporated in the luminaire.

LUMINAIRES

The *Commission Internationale de l'Eclairage* (CIE) defines a luminaire as follows: *"An apparatus which distributes, filters or transforms the light transmitted from one or more lamps and which includes, except the lamps themselves, all the parts for fixing and protecting the lamps and, where necessary, circuit auxiliaries together with the means for connecting them to the electric supply".*

This definition divides the functions of a luminaire into three groups:
- mechanical - to support, fix and protect the lamp(s) and auxiliary equipment
- optical - to distribute, filter or transform the light
- electrical - to connect the lamp(s) to the electrical supply and control the electrical performance of the lamp(s).

The first group of functions is performed by the luminaire housing, the second by the optical system, and the third by electrical components such as lamp holder, cabling and control gear.

In addition to these purely technical aspects, luminaires also form part of the interior design, and from this point of view have to fulfil an aesthetic function as well, which under certain circumstances may be more important than the illumination of a task.

Mechanical Functions

The chief mechanical functions that a luminaire should perform are:
- providing support to the lamp holder(s) and other electrical parts
- protecting the lamp(s) against hostile environments
- providing mounting facilities.

TABLE V - Control gear combinations with lamp type

Lamp Type	Gear
Standard incandescent (GLS)	none
Standard incandescent (reflector)	none
Mains-voltage halogen	none
Low-voltage halogen	step-down transformer
Switch-start fluorescent	inductive ballast + glow-discharge starter (+ compensating capacitor) inductive ballast + electronic starter (+ compensating capacitor) HF electronic ballast with built-in starter
Rapid-start fluorescent	inductive ballast + heating transformer (+ compensating capacitor) inductive ballast with extra heating windings (+ compensating capacitor) semi-resonant ballast + capacitor
Cold-start fluorescent	ballast (+ compensating capacitor)
Compact fluorescent	as switch-start fluorescent, but the entire control gear - or a glow-discharge starter only - may be incorporated in the lamp
High-pressure sodium ('White SON')	inductive ballast + control unit (combined voltage stabiliser and electronic starter) (+ compensating capacitor)
Metal halide	inductive ballast + electronic starter (+ compensating capacitor)

TABLE VI - Classification system of dust and moisture protection characteristics for luminaires

Dust protection	Moisture protection
1 Not applicable to luminaires	1 Drip-proof (vertical falling drops)
2 Objects greater than 12 mm (fingers)	2 Drip-proof up to 15° from vertical
3 Objects greater than 2,5 mm	3 Rain-proof up to 60° from vertical
4 Objects greater than 1 mm	4 Splash-proof from any direction
5 Dust-proof, against harmful deposit of dust	5 Jet-proof, against water jets
6 Dust-tight, against ingress of dust	6 Heavy sea-proof

The environmental protection that a luminaire should provide may include:
- protection against shock, vibration, impact, etc
- protection against ingress of dust
- protection against ingress of water
- protection against chemical aggression
- protection against explosion risk
- protection against very low temperatures.

The International Electrotechnical Commission (IEC) has laid down a classification for protection against the ingress of solid particles (dust) and water. It consists of the letters IP (International Protection) followed by two figures, the first indicating the degree of protection against dust, and the second indicating the degree of protection against moisture, as shown in Table VI.

General-purpose indoor luminaires are typically classified as IP20.

Protection against chemical attack, explosion and very low temperatures is provided by specially constructed luminaires, which will not be described here.

Optical Functions

The optical characteristics of a luminaire may govern one or more of the following functions:
- spatial distribution of the light
- directional characteristics of the light
- colour of the light
- glare control
- luminaire efficiency.

The spatial distribution of the light is in the first place determined by the mechanical construction of the luminaire. The *Commission Internationale de l'Eclairage* (CIE) distinguishes six main classes of general-lighting luminaires, according to the percentage of the total light output (light output ratio, LOR) that is emitted downward to the horizontal plane (ie towards the visual task), and upwards (Figure 17).

A further classification can be made according to the directional characteristics of the light output, which can be more or less diffuse, or focused into a well-defined beam. In the latter case, different types of beam can be distinguished according to the beam spread, which is defined as the angle over which the luminous intensity drops to 50 per cent of its peak value. A practical distinction often made is between:
- narrow beams - beam width < 20°
- medium beams - beam width from 20° to 40°
- wide beams - beam width > 40°.

Directional control of the light is mostly achieved by optical components, for example diffuse or mirror reflectors of various geometrical shapes (flat, spherical, elliptical, parabolic or a combination), lenses, prismatic refractor panels, parabolic louvers, diaphragms or directional shields (Figures 18, 19 and 20).

Other optical functions to be performed by the luminaire may include:
- influencing the spectral composition of the light by means of filters or coloured reflectors
- glare control, obtained by screening or shielding the lamp(s) from direct view by means of diffusing screens, baffles or louvres
- achieving a favourable light output ratio, or luminaire efficiency, which is the ratio of the luminous flux produced by the luminaire to the luminous flux(es) of the lamp(s) inside. This should, however, not be achieved at the expense of other optical parameters, as it is of primary importance that the light arrives at the place where it is actually needed.

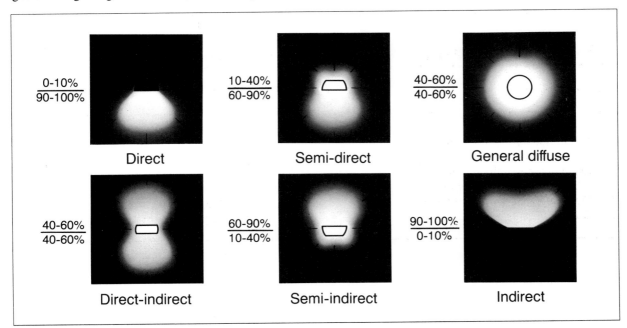

Figure 17 - The CIE classification of luminaires with respect to spatial light distribution.

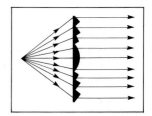

Figure 18 - Directional control by means of a Fresnel lens.

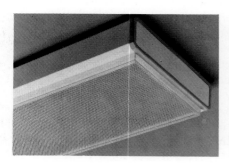

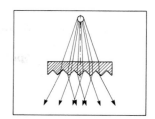

Figure 19 - Directional control by means of a prismatic refractor panel.

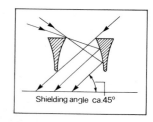

Figure 20 - Directional control by means of a parabolic louvre.

For most lamp-luminaire combinations used in professional lighting installations, the optical characteristics are measured by the manufacturer and published in the form of so-called photometric data sheets. These include for example:
- the luminous intensity distribution in one or more planes through the luminaire (often the planes of symmetry)
- glare-control angles for different directions of view
- light output ratio under various angles
- tables required for illuminance calculations.

Electrical Functions

The electrical function of a luminaire is to provide the correct voltage and current for the proper operation of the lamp(s), in such a way as to ensure the electrical safety of the luminaire.

The electrical parts in a luminaire may include:
- lamp holder(s)
- control gear
- electrical wiring
- internal connectors, switches, fuses, etc
- mains connection.

Lamp holders exist in a wide variety of types and sizes, the most important being screw (or Edison), bayonet and single and double pin-and-socket types. Standardization of lamp holders is regulated by the International Electrotechnical Commission (IEC).

The various types of control gear described above in the section "Control Gear for Lamps" are often - but not always - located in the luminaire housing. In that case care must be taken that the heat generated by, for example, the ballast can do no harm to other components.

Internal wiring of luminaires must satisfy certain minimum requirements with respect to the heat resistance of the insulation and current-handling capacity. The IEC has laid down a minimum cross-section of the copper core of 0.5 mm^2 (0.4 mm^2 if space is restricted) for live and neutral wires and 2.5 mm^2 for earthing wires.

The international colour coding of electric wiring (also specified by IEC) is:
- brown - live
- blue - neutral
- yellow/green - earth.

If, for aesthetic reasons, uniformly coloured wiring is used, the connection blocks must be properly labelled.

Luminaires are either permanently connected to the mains, or provided with a flexible mains cable and plug for connection to a wall socket. In both cases the cable entry must be suitably sealed against the ingress of dust and water, as specified for the rest of the luminaire (see IP classification). Also a cable clamp must be inserted between the cable entry and the internal connection block for mechanical strain relief. Luminaires used for end-to-end mounting, eg in trunking systems, are often not individually connected to the mains, but through-wired, ie the supply cable is run through one luminaire onto the next one. This has the advantage of quicker installation and savings on cabling.

Electrical Safety

In nearly all countries, luminaires, like other electrical appliances, have to comply with local and/or international safety requirements. Approval marks of the official testing authorities are found on the type plate of the luminaire, as well as on all relevant electrical parts (eg lamp holders).

The International Electrotechnical Commission (IEC) has drawn up the following four electrical safety classes for luminaires, according to the construction (Figure 21):
- Class 0 - Functional insulation only; no provision for earthing
- Class I - Functional insulation only; provision for earthing of all exposed metal parts
- Class II - Double or reinforced insulation; no provision for earthing
- Class III - Luminaires supplied at an extra-low, safe voltage (max. 42 V ac).

Class 0

Class II

Class I

Class III

Figure 21 - Typical examples of electrical classification of luminaires (Class 0, I, II, III).

LUMINAIRE MOUNTING SYSTEMS

Luminaires in lighting installations are often combined to form integrated systems. This may apply to mechanical functions, electrical functions or both.

Mechanical Systems

Ceiling systems. The best known of these is the suspended integrated ceiling. This consists of a framework of aluminium or steel profiles, which are suspended at a certain distance from the structural ceiling by adjustable rods. The framework is filled in with panels or equipment, both of which perform functions required by the interior environment. These functions are mainly related to:
- lighting
- interior climate
- acoustics.

The lighting requirements are generally met by luminaires for fluorescent lamps that are suitable for recessed mounting in the ceiling suspension system. Ideally, the dimensions of the luminaires are keyed to the grid of the suspension system, otherwise the remaining openings must be filled in by small panels (Figure 22). A recessed power track and special electric functions (eg remote control of the lighting) may also be incorporated in the suspended ceiling.

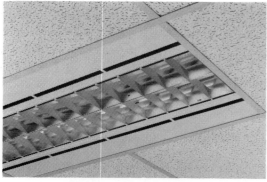

Figure 22 - Luminaire for flush-mounting in an integrated ceiling, seen from above and from below. At both sides of the luminaire are slots for air supply.

The space between the false and the structural ceiling - called the plenum - is in many cases used to accommodate the air-handling ducts. Often there are two ducts, one for supplying fresh air, the other for removing the return air. In other cases there is only a single duct, serving either for fresh-air supply or for the return-air exhaust. The plenum itself then takes the role of the other duct, as it is kept at a lower or higher pressure than the room underneath. Often air conditioning and lighting equipment are combined in such a way that the return air is exhausted through the luminaires. This is done primarily in order to:
- reduce the heat load from the lamps and luminaires on the room
- ensure that the temperature of the air surrounding the fluorescent lamps is maintained at about 25°C, at which temperature the highest luminous flux and efficacy are obtained
- making use of a single ceiling element for both lighting and air-handling (for aesthetic and cost reasons).

Such luminaires are provided with duct connection or, in the case of plenum exhaust, slots through which the return air can pass (Figure 23).

Although electro-acoustic components, such as loudspeakers, may be incorporated in false ceilings, what is primarily meant by the acoustic function of a ceiling system is how it influences, or more specifically reduces, the noise level in the room. Effective noise reduction can be achieved using sound-absorbing materials in the false ceiling, either by using panels of a porous composite material or using mineral or glass wool blankets on top of a perforated metal suspended ceiling.

Trunking system. Where aesthetic considerations are not of first importance, trunking systems can be employed to achieve lighting systems that are quick and easy to install, and which at the same time offer a high degree of flexibility.

A trunking system consists of steel support rails - generally inverted U-sections - which are mounted against the ceiling or roof support structure. On these rails special fluorescent lamp luminaires can be fastened by means of a quick-fit system. Remaining empty sections are closed by cover strips, so that the tracking also acts as a cable duct (Figure 24). Special attachments can be used to mount non-standard luminaires and other special equipment to the support rail.

Electrical Systems

Prewired trunking systems. The trunking systems described above are also available already provided with electrical wiring and connectors, in such a way that the electrical connection to the luminaire is made at the moment that it is mechanically fastened to the rail. The cabling is usually of the flat type, laid out for three-phase supply, with end-on connectors for through wiring

Electric Lighting 6.17

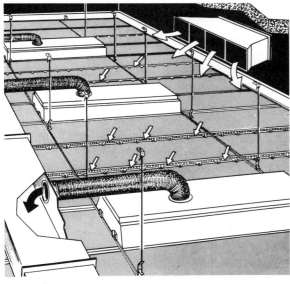

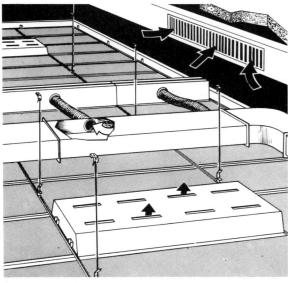

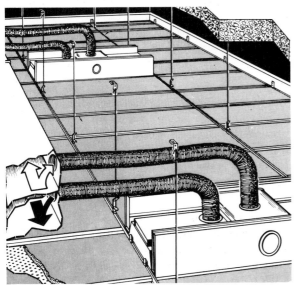

Figure 23 - Air-handling systems: (1) Positive pressure plenum (supply), (2). negative pressure plenum (extract), and (3) two-duct air handling.

and three-pole socket connectors for the electrical connection of the luminaire. Phase-selector switches allow equal distribution of the electric load over the three phases of the mains supply. Sometimes, for more decorative applications, various modular lighting systems consisting of through-wired sections are employed. These sections can be interconnected into almost any geometric arrangement and allow for a great flexibility in the choice of light sources and accessories to be attached, for example luminaires for fluorescent and compact fluorescent lamps, incandescent or halogen incandescent spots, downlighters, lengths of power track, low-voltage equipment, etc.

Power track. A power track consists of extruded sections, in which electrical conductors are incorporated in such a way as to provide at any desired place along its length both mechanical support and electrical connection for luminaires and other electrical equipment provided with special attachments. The track is generally suitable for mounting at or near ceiling or wall surfaces, but some types are also available for recessed mounting in suspended ceiling systems (Figure 25).

There is little standardisation of the mechanical and electrical construction of power tracks, so that with a certain brand or type only luminaires that are fitted with certain adaptors that match the type of track can be used. The following electrical arrangements can be distinguished:

- single, two, three or four-circuit track, provided with one neutral and respectively one, two, three or four live conductors
- low-voltage track: a single-circuit track with conductors of a large cross-section for the connection of low-voltage spots without built-in transformers.

Most types of track are provided with a separate earth conductor (except low-voltage tracks). Typical maximum loads are 16 A or 3500 W (all circuits combined) for mains voltage track, or 25 A or 300 W for 12 V low-voltage track.

The track adaptor mounted to the luminaire contains, apart from a device for making the mechanical and electrical contact, often a circuit-selector switch, and sometimes an on-off switch and/or a fuse. Apart from the standard components, a choice of special connecting pieces and attachments is usually available.

6.18 Daylighting in Architecture

Figure 24 - Trunking system used in an aircraft hangar.

Figure 25 - Power-track mounted spotlights in a fashion shop.

Chapter 7
CONTROL SYSTEMS

INTRODUCTION

In order to take full advantage of the energy saving potential obtainable from daylight, it is essential to provide the building with control systems capable of adjusting electric light output to the available daylight. If such a system is not provided, electric lights will tend to stay on, as can be seen in many office buildings. Frequently the artificial lighting system is switched on in the morning and off in the evening, regardless of available daylight. This is particularly the case in open plan spaces where there is a central switch. In cellular building designs the occupant is more in control of and feels responsible for light switching, resulting in an improved lighting performance.

The graphs in Figure 1 illustrate the impact of different control systems on energy consumption. They show daylight levels rising from a minimum at the start of the working day, reaching a peak and then reducing towards evening. This natural light trend is followed by the artificial lighting control systems, ranging from simple manual control to the most complex continuous automatic systems.

The technologies and systems used to control lighting and environmental plant are of great importance in the process of design and construction in accordance with the energy saving and natural light use criteria described in other chapters of this book. The selection and/or the practicability of the control system can permit optimum use of the design decisions, but it can equally make them ineffective.

In addition, good control systems can also provide appreciable economic benefits in existing buildings not designed in accordance with the criteria described in other chapters of this book.

This chapter discusses the hardware and software needed to acquire data on natural and artificial illumination, as well as other aspects of the environment, in order to determine comfort parameters. It also examines the integration of various parameters at the local level and of building and system parameters. It deals with the definition of local and general management policies and their implementation through the use of controls on lighting, environmental and sun-screening systems.

The observations in this chapter are discussed with reference to a system designed in accordance with optimisation criteria in which the methods of calculation and the control components are both available and used. The optimum control system is therefore applied to a building which has already been optimised in its architectural and plant aspects in terms of the comfort-to-construction cost ratio. The control system optimises operation in terms of the comfort-to-running cost ratio.

In approaching the problem it is first necessary to identify the maximum predicted number of input variables to the control system, and also the maximum number of output variables. Thereafter, the upper and lower size limits for the field to be controlled must be determined. Within such limits, different levels of integration will be determined which allow fields of small size to be grouped until the maximum field size is reached, while allowing the small fields appropriate independence.

Finally, specific control policies appropriate to every level of sophistication must be formulated. These must combine maximum flexibility and adaptability to the field with the widest possibilities for integration at higher levels. It is clear that what has been said in other chapters on materials, comfort parameters and economic aspects is of great importance in defining all the above.

None of the systems constructed so far has gone much beyond the concept of supplementary lighting, which essentially consists of measuring the level of daylight and supplementing it with artificial light. A system based on this concept makes it possible to keep the level of illumination in the room at the design level as daylight varies, and the coefficient of uniformity at pre-selected values. It also permits a considerable saving of electricity, longer life for lighting equipment, and simple management of reserve lights. A system based on this concept may incorporate some of the following:

- Continuously controlled lighting units
- Computerised panels for on/off switching of a large number of lighting units
- Central computers to manage a number of computerised panels
- Timer-controlled switching systems
- Telephone actuated controls for individual lighting units
- Systems for detecting the presence of people in a room.

It is then not difficult to move on to programmed maintenance of lighting units based on hours of

7.2 Daylighting in Architecture

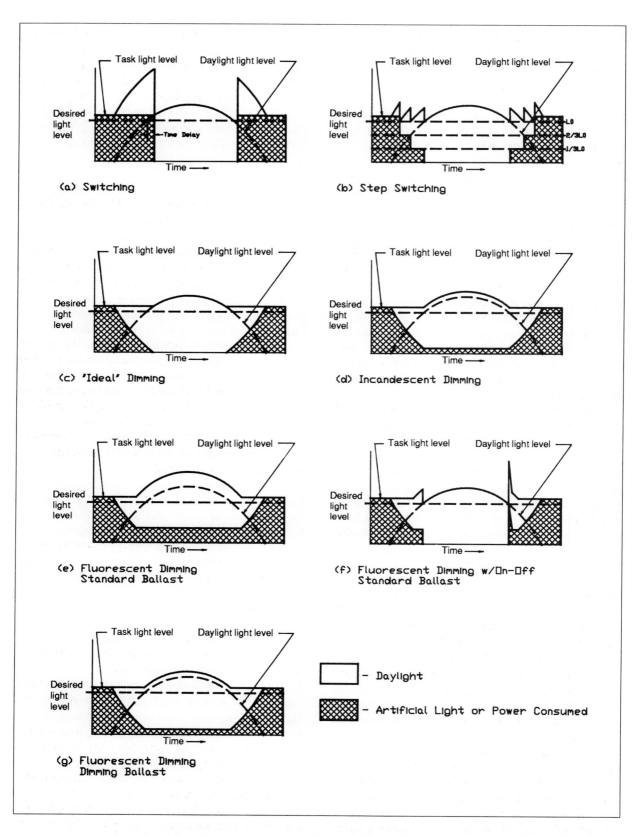

Figure 1 - Energy consumption over a day for different lighting control systems.

operation as memorised directly by the central computer, or to measurement of electricity consumption, which may also be shown in relation to other energy consumption in the building.

The current practical approach to the problem is limited both by the number of input and output variables taken into account and in the control policy adopted. The sensors used are always photometric and measure light intensity, or at most, uniformity of light intensity. Actuators are almost always on/off controls on lighting units. Energy management programmes are mainly based on presumed rather than real time data.

It thus seems that the principal need is to supply the control system with data from the field which it is controlling in real time, data not only on light intensity but also on all other comfort factors (light contrast, uniformity, temperature, humidity and glare). The control system would then be able to compare these data with comfort standards and instruct the various devices concerned (sun blinds, lighting units) to take the corrective action necessary if a comfort parameter goes beyond its limits of acceptability.

The difficulty of creating specific sensors and devices, or at least of creating them at reasonable cost, means that not all comfort variables can be included at the simplest levels of control. The same will hold true for energy savings and lighting maintenance variables. The simpler control units must be able to partially modify their cut-in parameters on the basis of information received from higher level control units and so on up to a central unit.

The control system must be modular and hierarchical, but direct ring connections between units to by-pass the hierarchical system may also be required for safety or other reasons. The possibility of action by human operators to cut-out limiting parameters and force the system to non-standard operation for unforeseen reasons should also not be forgotten. It is clear that it should be possible to vary reaction times and control policies according to the field and the variables to be controlled. This means incorporating great flexibility in both hardware and software.

The calculation programmes developed in other chapters of this book are also important here. They can be used for continuous recalculation of optimum regulation on the basis of real data arriving from the field. However, the results obtained in this chapter can be applied quickly to buildings of non-optimised design and result in useful savings.

A general theoretical layout for a control system is shown in Figure 2.

What has been said above defines a very wide but nonetheless circumscribed field for this chapter, covering the regulation for a single variable or a combination of several variables.

This chapter covers not only compact devices for immediate application, but also complex laboratory methods capable of simplification or adaption for application. Furthermore, existing control systems, their flexibility and the possibilities for using them as part of a larger and more complex whole are part of the topic. Façade components, their influence on comfort parameters and their adaptability to control systems are of particular importance.

CONTROLS FOR ARTIFICIAL LIGHTING SYSTEMS

Before starting to design a control system, one must first be aware of the most common design techniques and the products already available on the market.

In approaching the problem, one must not forget that properly applied lighting control systems cannot ignore other comfort parameters which may not always be obviously closely related. This section will therefore provide a review not only of sensors and equipment for artificial lighting equipment, but also of other devices that can be used to regulate artificial lighting systems and optimise energy use.

Catalogues published by the following European companies provide valuable information on the state-of-the-art in lighting technology:
- Philips
- Siemens
- Osram
- Bruel & Kjaer

as do those of the following American companies:
- General Electric
- Conservolite Inc
- Novitas Inc
- Multipoint Control Inc
- Lutron
- XO Industries Inc

The following overall conclusions can be drawn from examining the documentation:
- There is no difficulty in finding temperature and humidity sensors of various levels of quality and cost.
- Instruments for measuring photometric quantities are sufficiently easy to find (as discussed in Appendix C).
- It is not easy to find sensors for photometric quantities for widespread use at low cost.

Since these latter devices are specifically designed control systems, they have been included in the data sheets in Appendix E and only a general description of them is provided in this chapter.

This chapter does, however, include information on sensors to detect the presence of people, since use of these devices in offices can result in significant savings in lighting energy use as well as air conditioning or heating energy.

7.4 Daylighting in Architecture

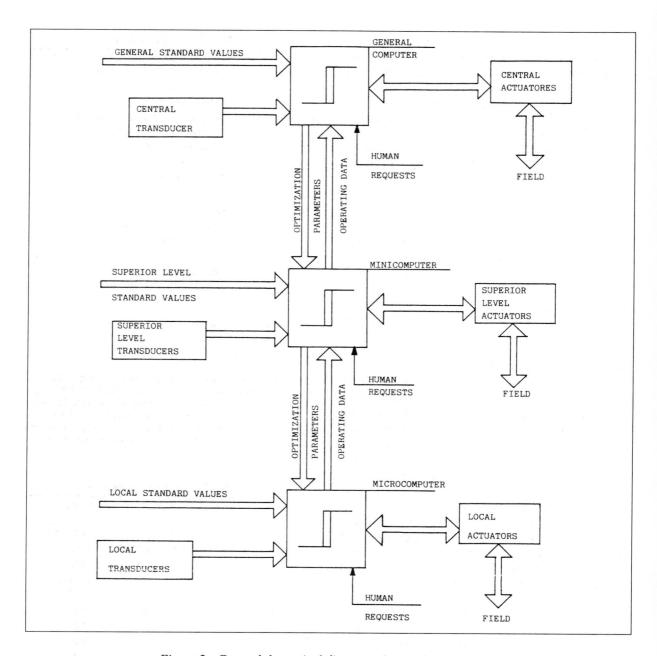

Figure 2 - General theoretical diagram of a regulation system..

Lighting Sensors

All automatic artificial light control systems are based on sensors. The most commonly used versions are basically photoelectric cells used to measure the light level.

Table I lists some of the most common lighting sensors used in control systems.

In work environments where VDUs are used, aspects of glare and light distribution are of great importance in optimising plant to meet the more stringent requirements.

It would be useful, in practical applications, to have low cost sensors or instruments available to measure the total radiant energy, including light, passing through glazed surfaces. This would make it possible to envisage systems to integrate natural and artificial light, with motorised blinds to reduce entry of solar radiation, and to optimise air conditioning under computer control.

From a practical point of view it is interesting to study the optimum location of lighting sensors. For example, in an office with people seated at desks, the theoretically ideal placement of sensors would be on the desks, where the lighting level is particularly important. However, in real life situations, the need to prevent interference due to accidental causes (eg the sensors becoming buried under papers, etc) suggests that the locations used will not be the optimum from the point of view of the measurement to be made, but rather should be those which are least affected by external interference (eg the ceiling or the walls).

With respect to the number of sensors to be installed and the area they relate to, this is not determined by the sensor but rather by the building form and the distribution of natural daylight in spaces within the building. The various use patterns of the rooms (timetable, number of rooms used at the same time and so on) will also have a role to play. It is these latter elements in relation to the cost of the sensors and the comfort level desired that effectively define the number of, the positioning of and the area covered by the sensors.

People Detectors

These are sensors which can detect whether there are people in a room. Their typical operating method is to detect movement of a mass by sensing reflected beams (infrared, radar, ultrasound). The most common type uses infrared rays. Technical and construction specifications can be found in the Data Sheets in Appendix E.

Temperature and Humidity Sensors

These sensors have not been listed here since they are widely available at very reasonable prices, being commonly used in air conditioning systems (1). They have an important role in the overall energy management of buildings. There is a very large number of firms manufacturing highly standardised temperature and humidity controls.

Actuators for Electric Lighting Systems

In artificial lighting, the concepts of "actuation" and "control" are often closely interrelated in practice. The only way of controlling artificial lighting from lamps or light units is by on-off switching or regulating their brightness. "Actuators" are therefore systems for either switching groups of lamps or for regulating their brightness.

The systems commonly used fall into three main categories: on-off step regulation of lamp numbers, continuous regulation of lamp brightness, and a third system which is a combination of the two, avoiding both the crudeness of the former and the manufacturing complications of the latter. The main systems examined and their manufacturers are listed in Table II.

Step regulation. Step regulation controls the light level simply by switching lights on or off, and it is used at the most primitive level in small rooms by switching off lights as the amount of natural light increases. Systems for large areas work on the same principle by switching off individual lights in multi-light fixtures until ultimately all the lights are off. Clearly, switching off will always begin at the light fixtures closest to the windows where natural light levels are greatest.

A step regulation system does not require large changes to existing systems, since this method uses power sources and accessories which are in common

TABLE I - Lighting sensors

Firm	Model	Notes
Philips	LRF100 photoelectric cell	1
Siemens	LDR photoelectric cell	1
Multipoint	Photo-conductive sensor 0-50 cd/m^2	
	Photo-diode sensor 0-1000 cd/m^2	
Conservolite	Universal mount	
	T-Bar mount	
	Surface mount	
General Electric	System sensors	
	Remote control	
	Smart remote control	
	Programmable light control	
Lutron	DACD	1
	WLCSD	2
Bruel & Kjaer	1100	3
	1101	
	1105	

Notes:
1. Ceiling mounted
2. Wall mounted
3. These sensors are extremely precise measuring instruments and therefore are used in research laboratories or for sample surveys or to calibrate other measuring instruments. This same company also manufactures precision sensors and measuring instruments to determine glare and to measure light contrast. For this reason, they do not seem to be suitable for general use as system control sensors.

TABLE II - Control systems

No.	Type of Control	Firm	Model
1	Step on/off	Siemens	Altoswitch
2	Continuous (CV)	Siemens	Altomat
3	Combination	Siemens	Partial regulation of light values
4	Step	Philips	Integrated Function System
5	Continuous (M/A)	Philips	System HF dimming
6	On/Off	Conservolite	Series IV
7	Continuous (M/A)	Conservolite	Eclipse
8	On/Off	Novitas	Light-o-matic
9	On/Off	Multipoint	Mark series
10	Continuous	Lutron	Paesar
11	Continuous	XO Ind.	Auto tuning controls
12	Step	XO Ind.	
13	On/Off	Gen. Elec.	Remote control
14	On/Off	Gen. Elec.	Smart remote control
15	On/Off	Gen. Elec.	Programmable lighting control

CV: Constant value
M/A: Manual /Auto

use. More refined systems can be fitted with delay circuits to prevent the system from cutting-in inappropriately due to temporary changes in the natural light.

The controls listed in Table II numbered 1, 4, 6, 8, 9, 12, 13, 14 and 15 belong to the step regulation with time delay category. Control types 13, 14 and 15, though they use low voltage electrodynamic relay switches to control the lighting level (not the most advanced in modern equipment), have the additional advantage of being programmable.

Control type 14 has a simple computer, which can store a time map of the light requirements of the room during the course of the day and in relation to whether people are present or not.

The control type numbered 15 is controlled by software which can run an entire system. It can be programmed and controlled by telephone and can also control remote units in the same way. It uses photo-sensors to compare the natural light with the level of artificial lighting. It can detect when people are in the room and will switch off the lights in the area (after the set delay time) once all the workers have left. There is another delay circuit to prevent incorrect operation due to temporary or unusual natural lighting conditions.

Control type 8 is based on sensors which can detect people walking in front of the control switch. It will switch off the lights in the area when it has been vacated. The delay can be programmed from 30 seconds to 12 minutes.

Continuous lighting control. These systems regulate the total light output of the lighting units within a 25-100% range. Light from fluorescent lamps cannot be dimmed below this 25% threshold, except by turning the unit off. While there are no great problems for incandescent lamps, fluorescent tubes require a series of modifications to enable this kind of control. Accessories have to be altered, which increases both design and system implementation costs (2).

The control systems available on the market go from simple manual dimming switches, which use rheostats, to automatic systems, which use sensors to determine the level of natural light and thus the required supplementary artificial light (3, 4). The more sophisticated of these use software programmes to control and run the systems and can even achieve different lighting levels in various zones within a single room.

Some manufacturers produce special accessories for fluorescent tubes to work with lighting control systems, as well as tubes specially designed for electronic control systems (3).

An example of manufacturers' control systems is the dimming system from Philips (No.5 in Table II). This offers manual (rheostat) or automatic (photocell) control of new type tubes fed from a high frequency electronic converter. This modification to the power supply and stabilisation systems is required to obtain reductions in light output below 50% of the maximum without malfunctions, but it also results in a significant reduction in the amount of power lost as heat.

Another good example of this kind of system is "Altomat" (No. 2 in Table II, manufactured by Siemens), which regulates brightness continuously from 100% to 25%. This system again requires tubes fed from an electronic converter. It could suitably be combined with the new Quicktronic system using Lumilux tubes from Osram.

Numbers 7, 10 and 11 in Table II belong to this same category. The first and third are continuous control systems, and they can be manual or automatic using a light level detecting sensor. They use special electronic circuits to control the tubes. Usually these types of control systems require a switch and a sensor for each lighting fixture.

System No.10 is compatible with a large number of accessories and can increase the level of the control system from a simple on/off system to graduated zone regulation.

Mixed control systems. Mixed control systems are based mainly on the use of standard tubes without complicated electronic control units on the tube itself. Provided suitable starters are used, the light produced by fluorescent tubes can be reduced by up to 30% without having to use auxiliary cathode preheating circuits. Any further level of reduction generates instability and reduces tube life drastically. Thus to obtain energy savings when the natural light permits, the lighting is left at 70% for a temporary period and then the tubes are turned off one by one with an ordinary step on/off control. The tubes are brought back to 100% lighting when the daylight falls below the set threshold. Item No. 3 in Table II, manufactured by Siemens, is a good example of this kind of system.

Starters

The most important recent changes in the technology of starters relates to fluorescent tube stabilising inductors. New concept starters work by means of special high frequency (around 30kHz) electronic circuits. The result is a 10-15% reduction in the amount of electric power dissipated as heat generated in the tube-inductor unit. This reduction would increase the amount of light produced for the same amount of power, though usually the choice made is to decrease the specified tube rating to produce the same amount of light. This results in power savings. The total energy saving from a high frequency power source tube is 20-25% for the same light output.

Another important aspect is that where less power is used less heat is introduced into the room. This can allow, in air conditioned spaces, the cooling load to be reduced or the plant rating to be reduced.

High frequency starters can be installed on normal fluorescent tubes but, as mentioned previously, a number of firms are also manufacturing this new type of tube. Table III shows a list of firms and starters.

TABLE III - Starters

No.	Type of Starter	Firm	Model
1	High frequency actuator	Philips	ITF Dimming Series
2	HF fluorescent tube	Philips	TLD Super 80
3	HF actuator	Siemens	RELF
4	HF power supply actuator	Osram	Quicktronic
5	Fluorescent tube	Osram	Lumilux Plus/20
6	HF power supply actuator	Conservolite	Eclipse
7	HF power supply actuator	Conservolite	
8	HF power supply actuator	XO Ind.	Smart Power

Façade Actuators

The concept of an actuator particularly applies to ways of controlling natural light.

There is a large variety of façade actuators available (sun-blinds, Venetian blinds, etc) and they are often equipped with their own motor drives and various types of control methods. All such systems, though with different constructional features, allow a room to be partially (in some cases entirely) screened.

Electronic control systems use electric motors to move the screen elements into position, using the sun or other control variables from the room to determine their position.

Thermal actuators are not covered in this book since, like thermal sensors, they are well-known, widely used and comparatively cheap.

MANAGEMENT STRATEGIES

Other chapters of this book describe ways in which buildings can be designed to optimise energy use and in particular to exploit daylight to the full. Such buildings, which have already been optimised for natural conditions, need additional artificial systems to achieve the required comfort conditions at the right times.

A number of systems to control artificial light levels are available for this purpose, as well as air conditioning and heating systems to optimise room temperature and humidity levels (8).

There are, also, façade screening systems which allow integrated control of the radiative environment. These go from the traditional forms, such as moving window screens, mobile curtains etc, to more sophisticated versions.

The designer of lighting control systems has to face the problem of working on a building constructed to energy optimisation parameters on the basis of annual weather statistics. The aim is to adjust, dynamically and in real time, and allow for any divergence from the set comfort parameters with respect to the radiation data used for the design.

The current state-of-the-art offers three different approaches to this problem:
1. Control systems applied only to the lighting system.
2. Combination control systems (also called building control systems) of the heating or air conditioning systems (8) plus the lighting system.
3. General remote controls to be used entirely for the lighting systems or the HVAC systems.

The first approach is the most commonly used and most of the lighting equipment manufacturers devote their research efforts to it.

The various systems are quite dissimilar both in the input data used and in the management procedures adopted. The more evolved systems can be summarised under the following typologies (Figure 3):
- room photometric sensors
- room control units (usually switches) for light fixtures
- transmission lines (9)
- centralised control units for the entire system or sub-system, and this can also be integrated with a more complex general system (10, 11).

The management policy is simply to reduce energy consumption by reducing the amount of artificial light, with consequent savings in the amount of electricity use by the lights. This type of control can be step (by cutting out sets of lighting units) or continuous (by reducing the amount of light put out by each unit) but both systems will use the control devices and actuators described in the previous pages: lighting sensors usually located inside the rooms to be regulated by, and in communication with, the central control unit which receives data from the field.

In practical terms, these systems do not use models or calculate the effect of daylight inside the spaces as a function of the external variables. In fact, they merely "measure" ambient conditions in real time.

The various control systems adapt to varying needs in different ways depending on the type of plant and the set comfort levels. American standards call for high lighting levels, whereas the general European level and the present trends tend to concentrate on other parameters (contrast, uniformity, colour, etc). The purpose is to obtain good comfort levels with lower lighting levels. Much

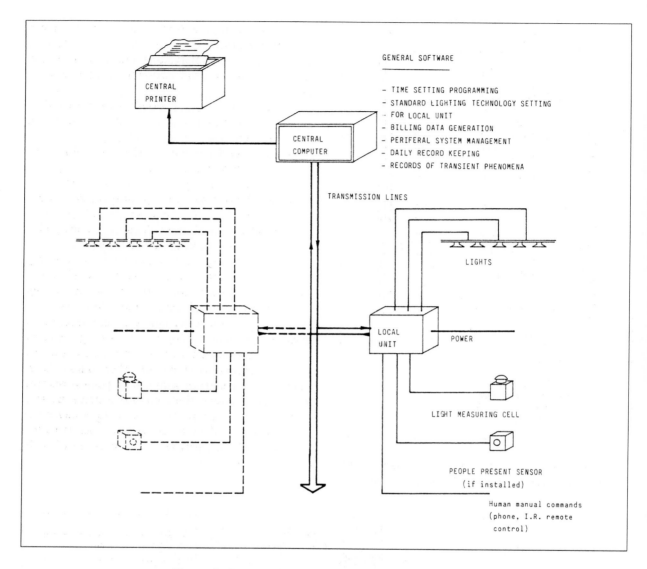

Figure 3 - Layout of control system for artificial lighting.

attention must, therefore, be given to the selection of the control system to prevent sudden changes in artificial lighting which could negatively affect the overall comfort status. For example, a step control system on entire blocks of lighting units could, if the rows are fairly distant one from the other, cause an excessive instantaneous, and therefore unacceptable, change in the lighting level.

In defining the minimum parameters for the control units, the following factors should be kept in mind:
- uniformity in relation to the daylight factor
- current and future use of the spaces
- possible need to build-in local human adjustment to the parameters
- different visual needs and consequent different lighting levels, uniformity and so on.

Once the minimum control levels have been defined on the basis of these parameters and needs, the sensors will be positioned in the light of these needs. Their precise location will be chosen so as to shield the sensors, as far as possible, from interference, which the software is unable to eliminate.

Additional information on layouts for this method of control can be found in Figure 4 and also in Appendix E, where a number of Fact Sheets, summarising the specifications of actual systems with the main management strategies employed, can be found.

The second approach, which integrates lighting control with room comfort control, marks an important step forward towards energy use optimisation in a building. Such systems use a single centralised control unit to automate all the building's safety and service equipment subsystems. These systems use both control systems and software modules for all the temperature and humidity control systems, plus the electric systems, by means of actuators on the various systems. Figure 5 illustrates the layout for this kind of system manufactured by an Italian firm, Italtel - Telesis.

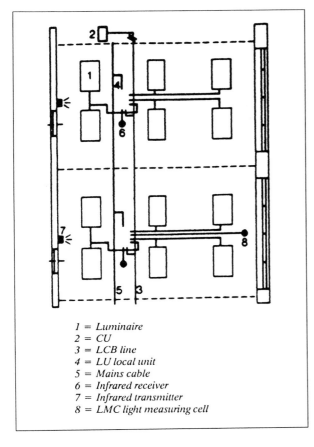

1 = Luminaire
2 = CU
3 = LCB line
4 = LU local unit
5 = Mains cable
6 = Infrared receiver
7 = Infrared transmitter
8 = LMC light measuring cell

Figure 4 - Example of lighting control system layout (programmable on/off type - Philips IFS System).

The third approach mentioned does not use specifically designed control systems but applies general remote control systems already widely used in other fields. The main features of these systems are:
- wide availability and highly flexible local or central hardware units plus transmission equipment
- large number of inputs and outputs for every type of sensor and actuator
- high speed data acquisition and response time
- user-friendly software packages are widely available to design customised control systems
- the end user must set up the management policy for the system on a case-by-case basis.

The criteria underlying these systems are probably the most interesting when it comes to overall energy integration of the lighting subsystem and the heating and humidity control subsystems. This approach allows these two subsystems to be connected to a third very important subsystem for daylight control needs, i.e. the actuators controlling the sun-screens for the building windows, or other natural light controls (12). However, it should be pointed out that in such applications, these control systems are still rarely used or only partly used. Undoubtedly the main obstacle to their wider use continues to be the cost, not only of hardware but especially of their customised, software.

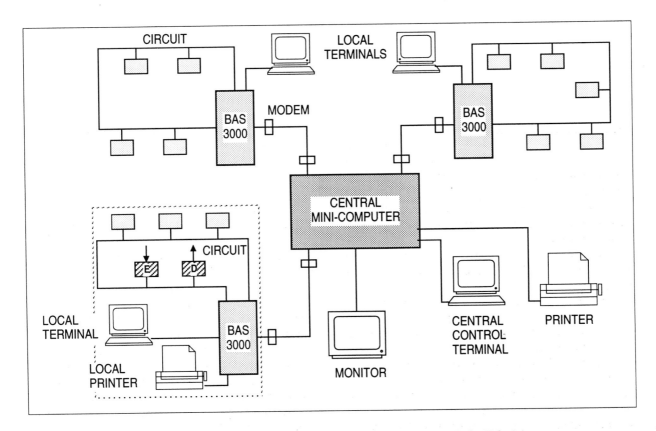

Figure 5 - Building energy management system layout (Italtel - Telesis)

As an example, it would be possible to use standard components to construct an energy control system for a building with motorised façade units as its final actuator. A management strategy of this type would be of limited value in winter, since the cost of energy for lighting is about three times the cost of that for heating. On the other hand it could be economical in southern Europe in summer, when solar radiation entering through the windows increases air conditioning costs by much more than the cost of increased artificial lighting. A very simplified management strategy could therefore be as follows:

- During the winter: the façade actuators are always open, or they follow the sun's position to prevent direct radiation, or any other control strategy is implemented.
- During the summer: software calculates the energy costs per unit of time, it changes the position of the façade actuators and recalculates the energy costs. It continues to adjust the actuators in an attempt to achieve minimum costs and thus compensates for changes in weather conditions. The control units automatically adjust the lighting and air conditioning levels. The time constants built into the system prevent unstable room conditions.

Naturally, building cost parameters are calculated on the basis of all the investment and management data. Furthermore, these parameters can be made more or less flexible by designing the software to take into account stock levels, possible price increases, seasonal supply contracts, etc.

While these higher level programmes are operative, other standard procedures will continue to control the individual lighting and air conditioning units. Figures 6 and 7 summarise the hardware and software structures for this simple application.

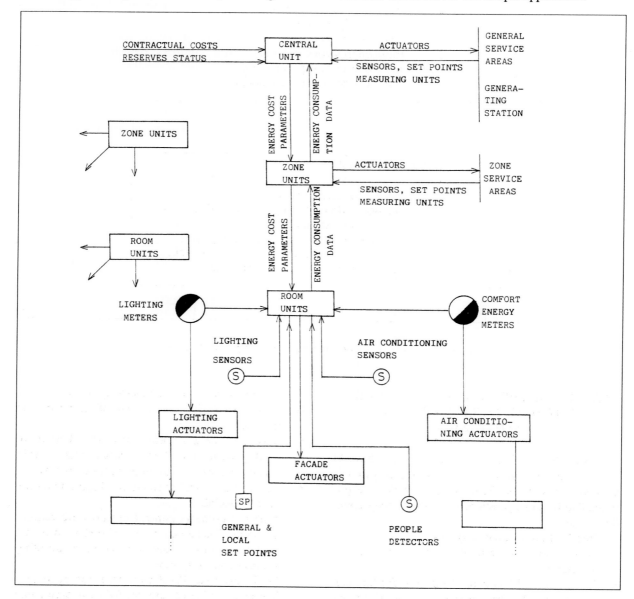

Figure 6 - Hardware system for building energy management.

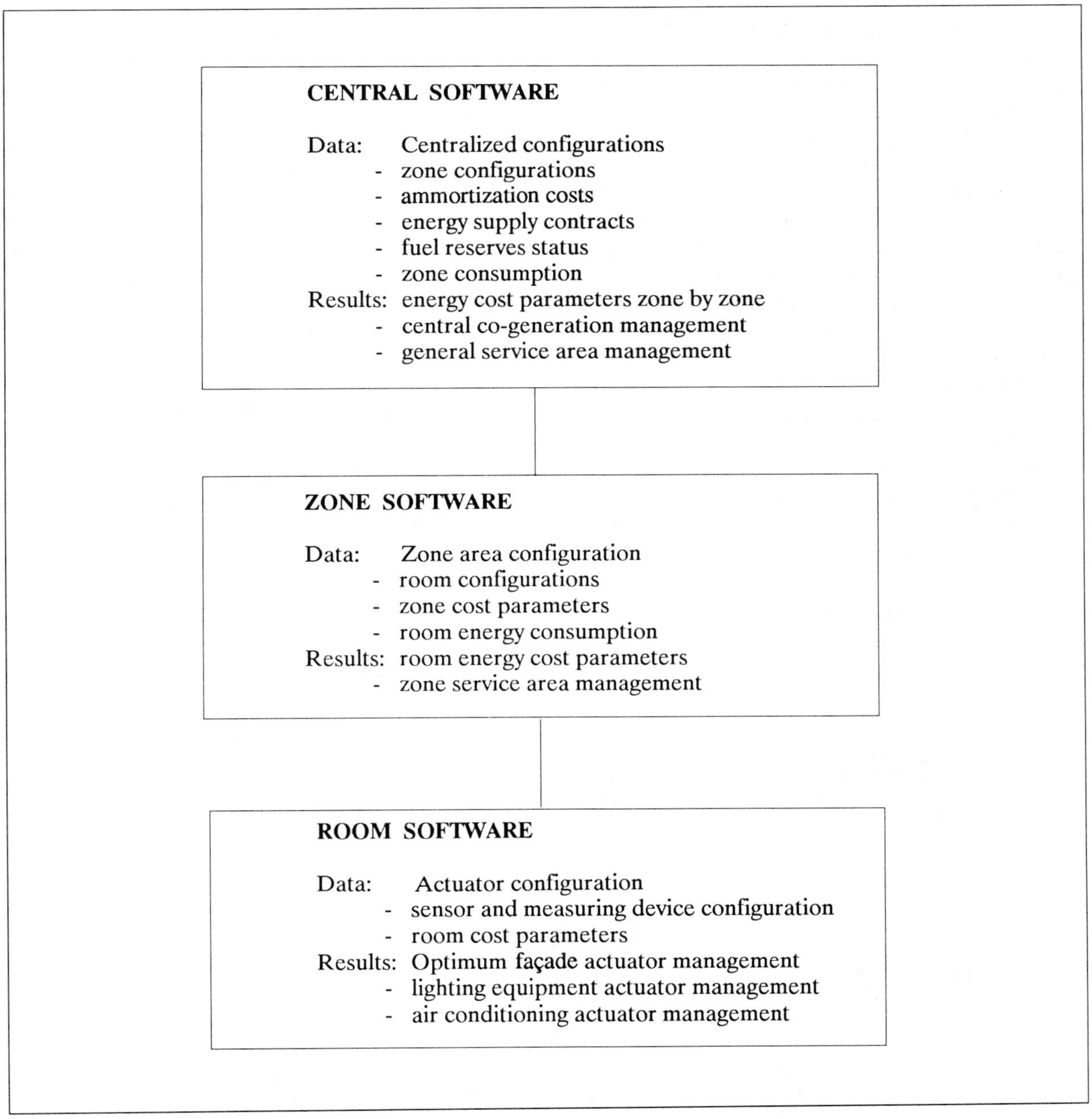

```
┌─────────────────────────────────────────────────────┐
│              CENTRAL SOFTWARE                        │
│                                                      │
│   Data:    Centralized configurations                │
│            - zone configurations                     │
│            - ammortization costs                     │
│            - energy supply contracts                 │
│            - fuel reserves status                    │
│            - zone consumption                        │
│   Results: energy cost parameters zone by zone       │
│            - central co-generation management        │
│            - general service area management         │
└─────────────────────────────────────────────────────┘
                          │
┌─────────────────────────────────────────────────────┐
│              ZONE SOFTWARE                           │
│                                                      │
│   Data:    Zone area configuration                   │
│            - room configurations                     │
│            - zone cost parameters                    │
│            - room energy consumption                 │
│   Results: room energy cost parameters               │
│            - zone service area management            │
└─────────────────────────────────────────────────────┘
                          │
┌─────────────────────────────────────────────────────┐
│              ROOM SOFTWARE                           │
│                                                      │
│   Data:    Actuator configuration                    │
│            - sensor and measuring device configuration│
│            - room cost parameters                    │
│   Results: Optimum façade actuator management        │
│            - lighting equipment actuator management  │
│            - air conditioning actuator management    │
└─────────────────────────────────────────────────────┘
```

Figure 7 - Software layout for building energy management.

EXAMPLES OF CONTROL SYSTEMS FOR OPTIMISED BUILDING ENERGY MANAGEMENT

Examples of what the standard products have to offer illustrate some of the most important achievements in the field, which can also be put immediately into operation. Essentially, they cover four fields:
- interior lighting control systems
- solar radiation control systems with façade actuators
- integrated building control systems (for lighting, unauthorised entry prevention, overall energy control, etc)
- general purpose remote control systems.

In general terms, the present state-of-the-art can be summarised as follows:
- Specific control systems for artificial light are readily available. These range from simple on/off systems to continuous control systems.
- Automatic control systems for façade units are readily available.
- Ambient comfort control systems are readily available. These can be associated with on/off artificial lighting control systems or building security control systems.
- In all the cases illustrated here, the necessary interfaces and hardware needed to integrate one system with others are available.

- Software for the various applications is also available.
- Software packages to integrate all or many of the aspects covered can be found only with difficulty or not at all.

The cost-effectiveness and usefulness of the different systems may vary considerably due to a large number of parameters: geographical position, available energy sources, radiation, thermal conditions and lighting levels.

A cost-effectiveness analysis should be made before the definitive choice is made of one or other of the possible projects.

Below is a list summarising the management potential achieved by various building automation systems:

Security Systems
Fire:
- zone or precise location detection
- building evacuation management, fire-doors, lifts
- fire fighting control

Unauthorised entry:
- control of unauthorised entry sensors
- alarm management

Access control:
- badge readers and secret code input keys
- control of revolving doors
- management of time bands

Energy and Services Management Systems.
Energy use management:
- electricity
- heating.

Service system management
- lifts
- level detectors
- motor / technical equipment
- analogue measurement and remote controls

Video equipment remote control:
- control of cameras and video recorders.

A building management system can handle centralised supervisory and control tasks for:
- electricity, water and gas distribution networks
- land reclamation and irrigation plants
- water treatment plants
- railway networks
- telecommunications networks
- technological plants
- meteorological and territory data collection
- security systems.

To summarise, the system is made up of the following:
- Telecontrol stations (both remote and central)
- Communication units (one or more operator stations, mimic board, etc)
- Central computer (often duplicated). In some small systems the computer may not be present, due to the fact that Selimatic central stations can perform all the essential telecontrol functions. The Digital Equipment Computer (micro VAX series) is generally used. IBM PCs are provided in smaller systems.
- Software for central computer. A standard software package is available that performs all the general telecontrol functions. This package can be integrated with packages oriented to specific applications.

Solar Control Systems.

There are many examples of venetian blinds and sunshades which can be adapted for use with automatic control systems with power packs and control sensors. Typical specifications for a sun-blind may include the following:
- weather-resistant and robust, with reflective or translucent louvres or screen to control diffuse light and sunshine.
- adjustability to allow for complete obstruction of direct sunshine.
- automatic stop when the screen reaches its limit.
- manual operation (eg a clutch winder unit) or operation by an electric motor which is acoustically and visually unobtrusive, with optional centralised control.
- multi-unit screening systems with centralised control.

Various types of programmable control units are available designed for curtains venetian blinds, vertical strip sun screens, roller blinds, cinema screens and canopies units. For example, Griesser are currently manufacturing a series of control panels, and in particular, the Multitronic (model MZ20) version has the following specifications:
- 20 control motors connected
- 8-stage timer programme with programmable raise/lower switching
- impulse powered step-by-step strip positioning
- direct connection to the drive motor
- up to six motors with centralised control
- solar power pack, anemometer for automatic screen position changing.

Another building façade element which can be centrally controlled and regulated is the vertical strip sun-blind. This is in wide use and available from a large number of manufacturers. Automated control presents no problems since the directional gearings can easily be motorised. A special feature which "tracks" the sun can be added to a standard automatic control system. Most types do not have this feature factory supplied. Such a type of solid state electronic control ensures the best luminous intensity inside the room without direct sunlight. A revolving sensor that follows the sun's path is installed on the top of the building and sends the equipment data which will direct the sun-screen blades to the proper position to prevent direct sunlight penetration, or open them if solar intensity is low. The automatic device may also be disconnected for different internal requirements and manually operated in a different way in the different rooms (for instance, total closure due to storms, during the night, due to strikes, etc).

EXAMPLES OF INTEGRATED BUILDING ENERGY MANAGEMENT SYSTEMS

Described below are three examples of integrated building energy management systems applied to building designs.

Example 1: Canton Hospital, Lucerne (Switzerland)

The project for central control and supervision of a large hospital complex included the general hospital itself, various special clinics, a personnel building, as well as administration, technical and support facilities. In total there are approximately 1000 beds. It is essential that the staff be fully supported by automation devices in their wide range of exacting duties, which include:
- central supervision of systems and facilities
- immediate indication of any deviations from the normal state of patients' health
- optimal employment of the technical personnel
- maximal personal comfort for patients and staff
- ideal environmental conditions for specialised equipment
- central control of energy use.

On the basis of the requirements summarised above, Landis & Gyr adopted the Visonik 4000 building energy management system. The main areas supervised by Visonik 4000 are:
- the central heating plant and the heat recovery system
- more than 75 air conditioning plants
- the supply plants (electricity, gas, water, steam)
- the security systems (fire protection, alarm)
- the transport installations (lifts, escalators).

The best results that can be obtained with this kind of system are summarised below:
- building and plants are well controlled, with more than 2200 information points and 1000 process or time reactions
- no time-consuming daily routine tasks needed to control environmental conditions
- high operational reliability of the expensive technical equipment is assured
- regular information due to system and maintenance records
- energy costs are noticeably reduced.

Example 2: Samda Assurance, Paris (France)

The project is for the central control and supervision of a large administration building. The Samda headquarters is planned as two large linked buildings and is designed to accommodate up to 800 people. The administrative centres contain large and complex installations and the energy demand is high.

The main needs are:
- temperature and humidity levels which provide agreeable conditions for the personnel and also an optimal environment for the specialised equipment and installations
- automatic switching of various operating systems
- optimum use of energy
- central supervision and control of system and facilities
- indication of deviations from the normal state and related alarm functions.

The system designed by Landis & Gyr included a Visonik 4000 system to handle building energy control management automation, a Polygyr system to handle the air conditioning plant and a Sygmagyr-Eco system to control and optimise heating. The following main areas are controlled and supervised:
- the heating, ventilation and air conditioning plants
- the security plants (fire protection, access control)
- the supply plants (electricity, water, etc) and the lifts.

About 2500 information points and 700 process or time reactions are presently supervised by the building management system.

An analysis of the system actually installed (on the basis of references supplied by Landis & Gyr) confirms the following points:
- environmental conditions in the building are agreeable
- energy costs are held at a minimum level
- the modular concept of the Visonik and the straightforward software has allowed extension of the system to cover expanding requirements
- the system is user-friendly and reliable, with simple person/machine dialogue and an efficient operating centre
- buildings and plant are well controlled, the security requirements are fulfilled.

Example 3: Office Building

Figure 8 represents an existing building project using an energy management program for the building's main technical systems. The building is a management and administrative office tower complex with eleven floors planned at the present time.

The systems proposed control factors influenced by outside environmental conditions such as heating, air conditioning and lighting. The types of systems are determined by building specifications, by the activity to be carried out, and the degree of comfort required.

This study deals with the main construction features. The exterior walls are completely constructed in glass and the interior walls are movable with built-in office equipment, so the office layout can be changed frequently to provide maximum flexibility. The company's various activities (management, administration, business, etc) at different times require this flexibility, which will influence the approach to energy management.

The air conditioning systems are centrally regulated with respect to outside weather conditions and the building's location. Sensors will determine the local energy management for each office according to the office's specific thermo-hygrometric conditions and also in accordance with the needs of those working in the office.

Lighting is provided by a fluorescent ceiling system that can be turned on manually in each office. The continuous energy management system is equipped with sensors to reduce energy consumption by regulating the amount of light and limiting consumption to what is absolutely essential to maintain the minimum lighting level.

The building has only one system to supervise the security and technical systems for the entire building including fire warning, trespassing, etc.

The equipment used for environmental regulation comes from Staefa Control Systems (SCS), the humidity and temperature sensors being of types FT3OH90 and PUT30. The diffusers are Carrier type 37 AG, with air flow modulation.

For continuous light regulation, the Philips dimming system described above is used. The lighting level sensors are Model LRF100. The lighting fixture units are controlled by Model LRA110 signal amplifiers.

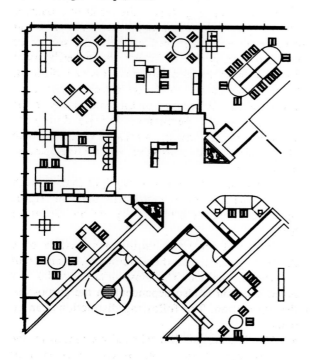

Figure 8 - Partial plan of an office building designed with an energy management system. (Project by CIAM, Modena, Italy)

REFERENCES

(1) Junker, B., *La regolazione automatica negli impianti di climatizzazione*, PEG Editrice, Milano, Italy, 1978.

(2) IES, *Lighting Handbook: Application Volume*, Illuminating Engineering Society of North America, New York, 1981.

(3) Maraschin, G., Van Westering, S.P., "*Nuovo sistema IFS (Integrated Function System) per una razionale gestione degli impianti di illuminazione*", AIDI Conference, Milan, Italy, 1983; published in *Luce*, May-June, 1983.

(4) Corsi, P., Preti, P., "*Il contributo al risparmio energetico attraverso la regolazione del flusso luminoso delle lampade fluorescenti in ambienti interni*", AIDI Conference, Milan, Italy, 1983; published in *Luce*, May-June 1983.

(5) ASHRAE, *Handbook: Fundamentals*, American Society of Heating Refrigerating and Air Conditioning Engineering, New York, 1977.

(6) ASHRAE, *Handbook: Systems*, American Society of Heating Refrigerating and Air Conditioning Engineering, New York, 1980.

(7) Mialich, R., Rossi, G.G., *Applicazioni di elettronica industriale - Elettronica industriale sistemi e automazione*, Calderini, Bologna, Italy, 1982.

(8) *Air Conditioned and Computers International Congress Acta*, AICARR, REHVA, Milan, Italy, 24-27 February, 1982.

(9) Macchi, C., Guilbert, J.F., "*Telematica - Trasporto e trattamento dell'informazione*", Tecniche Nuove, Milan, Italy, 1983.

(10) Coppelli, M., Stortoni, B., *Microprocessori: teoria e progetto*, La Sovrana Editrice, Fermo, Italy, 1983.

(11) Baranzini, R., Dugnani, G., *Microprocessori e microcomputers*, Hoepli, Milano, Italy, 1983.

(12) Pintus, M., "*Progettazione di alberghi e di complessi turistici in rapporto all'illuminazione naturale e artificiale*", AEI/AIDI Conference, Cagliari, Italy, 18 February, 1986; published in *Luce*, November-December, 1986.

Chapter 8
LIGHT TRANSFER MODELS

INTRODUCTION

The distribution of luminances in the built environment is the result of the complicated transfer of light from sources to different surfaces. A physical model of this process will require data on light sources and on the optical properties of surfaces and materials, which are often not available.

Many models have been developed and differ in terms of complexity and accuracy. An important distinction can be made between approximate manual calculation methods and more accurate computer-based models (see Chapter 9).

A detailed solution of a daylighting problem within an acceptable calculation time is particularly difficult when the geometry of the transmitting and reflecting surfaces is complex. To obtain realistic results a fine level of detail and analysis is needed. This is only possible on computers with vast memories, resulting in long computation times. The differences between sophisticated models can be regarded as the differences in methods adopted to solve the same governing equations. Simplified models are often capable of analysing only simple (often rectangular) geometry and use crude data on light sources and optical properties of materials.

The light falling on a window can be separated into the three components of direct sunlight, (diffuse) skylight and light reflected by external surfaces such as the ground and other buildings. This last contribution is generally small compared to the former two components. Hence the modelling of exterior reflections and interreflections can afford to be less accurate than the modelling of the interior reflections. Often different models for exterior and interior are used.

The illuminance of a surface in a room without the contribution of the interior interreflections is called the direct illuminance. The illuminance of any room surface is proportional to the luminance level of the sky. This relation gave rise to the concept of a daylight factor. The daylight factor is defined as the illuminance received at a point indoors from a sky of an assumed luminance distribution and without direct sunlight, expressed as a percentage of the horizontal illuminance outdoors from an unobstructed hemisphere of the same sky. The daylight factor depends for a given horizontal illuminance on the actual luminance distribution of the sky (1) and the geometry and material properties of the surrounding surfaces.

The daylight factor has been standardised by the CIE for the horizontal working plane and completely overcast sky conditions. The advantage of using a standard luminance distribution representing minimal daylight conditions is that then the daylight factor is only design-dependent. It provides a simple way to compare possible designs.

For many other practical daylighting problems however, such as energy calculations, daylight factors for the different sky luminance distributions are required.

Another approach to account for the daylight variations resulting from different luminance distributions of the sky is proposed by Tregenza (2) with the concept of the "daylight coefficient". The daylight coefficient is defined as the illuminance of an interior surface caused by the light from a small part (within a defined solid angle) of the sky divided by the normal illuminance from that part of sky.

The spectral composition of daylight is continuously varying. This is the result of a number of atmospheric processes (absorption, refraction and diffraction) and (inter) reflections at the surface of the earth and by clouds. To take the actual spectral composition of daylight into account is very difficult. In theory, a separate calculation is necessary for each wavelength, but because of the complexity, most calculations are performed independently of wavelength, giving an average value for the illuminances. This is also the approach adopted in this chapter.

Other physical details of daylight transfer such as the effects of the polarisation of light, phosphorescence and luminosity are not discussed here.

It is not only the required accuracy of a daylighting calculation that determines the choice of a particular method. The most suitable model in a given situation is also dependent on the purpose of the calculation. The following distinctions can be made:

- Window design: The calculations needed for the design of windows are based on average or minimal standardised daylight conditions, combined with estimates of the percentage of the working hours that the required levels are exceeded. Regulations used for window design

are often based on manual methods for calculation of the daylight factor. There is a growing demand for visualisation techniques for daylit spaces, and advanced methods such as 'ray-tracing' are used to create a realistic image of the room.
- Energy behaviour: The calculations to determine the energy performance of a building are often made on an hourly basis. As daylighting calculations on an hourly basis are time-consuming, the relevant daylighting parameters (such as daylight factors or daylight coefficients) are often pre-calculated for the hourly energy simulations on computers.
- Control strategies: To determine the effectiveness of daylighting or artificial lighting controls, calculations have to be made on the basis of detailed daylight data. As daylight levels change continuously, depending on the sky conditions, statistical methods, combined with the methods used in energy calculations, are an appropriate solution.

DIRECT ILLUMINATION

Direct Illumination by Sunlight

The sun is the primary source of daylight. The direct beam of light coming from the sun is attenuated by scattering, absorption and reflection processes in the atmosphere. A part of the scattered light reaches the surface of the earth in the form of diffuse sky light.

The illumination of an interior surface by the direct solar radiation gives rise to high illuminances. The variability of direct sunlight, due to clouds and the motion of the sun across the sky, leads to an illumination that is difficult to control. In daylighting calculations therefore the light coming directly from the sun is traditionally excluded from the calculations. It is assumed that in practical situations shading is used to control the amount of direct sunlight. This situation has changed, and direct sunlight can now be used effectively by the adoption of diffusing devices, light-guides and sophisticated artificial lighting control (see Chapters 5 and 7).

As direct light from the sun has an almost parallel beam, shadowing calculations or other predictions involving the direct sunlight can be achieved by simple parallel projections.

If the illuminance from the sun on a plane perpendicular to the direction of the sun's rays is known, then the illuminance of a plane of arbitrary orientation is given by:

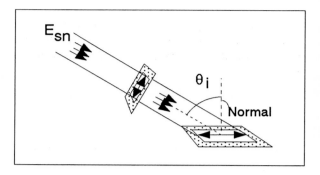

Figure 1 - Horizontal illumination by direct light.

$$E = E_{sn} \cos \theta_i$$

where:
E_{sn} = normal solar illuminance
θ_i = angle of incidence
(See Figure 1).

To determine whether or not a surface in a room is illuminated by direct sun through a non-diffusing aperture four methods can be distinguished:
- Determination of the intersection of a line from a point on the considered surface towards the sun and the interior and exterior surfaces: This method is used in ray-tracing programs (3).
- Projection methods such as the projection from the direction of the sun of external and window surfaces on the internal surfaces: If the geometry is simple this can be much quicker than the above method.
- Analytical methods: For simple geometries, such as overhangs perpendicular and parallel to the window plane, one can find formulae in the literature to determine the shadowed part of the window. In general this approach is not very useful for lighting calculations as only the amount of light entering the room is calculated using such algorithms and not the surface it falls upon.
- Graphical methods: In these methods the amount of direct daylight is determined by drawing a projection of the window on solar curves. Examples of this method have been developed and are available (4).

Direct Illumination by Diffuse Skylight

If the surface on which the light falls has optical properties which are dependent on the angle of incidence, the diffuse light has to be modelled in a number of discrete beams. This is the way light is considered in ray-tracing programs. In simple models this aspect is often too detailed and is ignored, leaving only the calculation of illuminances to be performed.

The diffuse sky can be described by a luminance distribution. The illuminance of a surface lit through a small solid angle by the sky is expressed as follows:

$$dE = L(\theta,\phi) \cos\theta_i \sin\theta \, d\phi \, d\theta$$

where:
$L(\theta,\phi)$ = sky luminance distribution as a function of θ and ϕ

Thus, the illuminance from the whole sky on a horizontal unobstructed plane can be defined as:

$$E_h = \int_{2\pi} L(\theta,\phi) \cos\theta \, d\Omega$$

where:
$d\Omega$ = solid angle

For simple sky luminance distributions the horizontal illumination can be calculated. If an overcast sky luminance distribution defined by Moon and Spencer (5) is substituted, then:

$$E_h = L_z \left\{ \frac{2\pi}{1+b} \left(\frac{1}{2} + \frac{b}{3} \right) \right\}$$

where:
L_z = luminance at zenith
$b = 0$ for an overcast sky
$b = 2$ for a CIE overcast sky

The illumination by the sky on a vertical plane for this sky can be determined as:

$$E_v = L_z \left\{ \frac{1}{1+b} \left(\frac{\pi}{2} + \frac{2}{3} b \right) \right\}$$

For a uniform sky ($b = 0$) and the CIE overcast sky ($b = 2$) the ratio of the vertical to the horizontal illuminance becomes:

$E_v / E_h = 0.5$ (uniform sky)
$E_v / E_h = 0.396$ (overcast sky)

From this it is clear that the illuminance of a window pane is very dependent on the luminance distribution model of the sky.

The illumination of rooms is found to be more strongly correlated with the illuminance of the apertures, such as a vertical window, than with the horizontal exterior illuminance.

The illuminance by diffuse light of a point in a room is determined by integration of the solid angles subtended to the window apertures, and their luminances, with respect to the reference point. There are three possible approaches to this:

- Numerical integration: eg the Monte Carlo techniques (discussed below at p.8.11).
- Analytical integration: only possible for simple luminance distributions and simple geometries
- Graphical methods: these are used in manual methods for clear glazing and standardised skies (4).

An example of an analytical solution of the integral is given below. The conditions are:
- The sky luminance distribution is uniform or CIE overcast sky
- Illumination of a horizontal surface is by a rectangular vertical window or horizontal rooflight.

Vertical Window
- If the sky has an uniform luminance:

$$E_p = E_h \frac{(\phi_2 - \phi_1 \cos\theta)}{2\pi}$$

- If the sky has a CIE overcast luminance distribution:

$$E_p = E_h \frac{3}{7\pi} \left\{ \begin{array}{l} \frac{1}{2}(\phi_2 - \phi_1 \cos\theta) + \\ \frac{2}{3}\tan^{-1}(\sin\theta \sin\phi_2) - \frac{1}{3}(\sin 2\theta \sin\phi_1) \end{array} \right\}$$

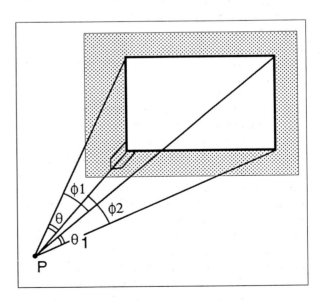

Figure 2 - Illumination of a horizontal plane through a window in a vertical wall.

These results can be used in the condition represented in Figure 2, or the more general situation by using addition and subtraction of four surface areas, as illustrated in Figure 3. In this situation the direct sky illuminance of a horizontal plane through the vertical window (D) can be found by adding the sky illuminance from an imaginary window of size (A+B+C+D), and subtracting "windows" of the sizes (A+B) and (B+C), and adding window (B).

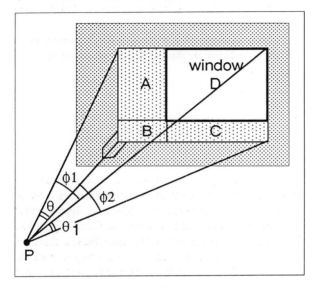

Figure 3 - Imaginary windows for determining horizontal illumination.

Horizontal Rooflight

The direct illuminance by a horizontal rectangular opening under a CIE overcast sky is:

$$E_p = \frac{E_h}{7\pi} \left\{ \begin{array}{l} \frac{3}{2}(\phi_1 \sin\theta + \theta_1 \sin\phi) + \sin 2\phi \, \sin\theta_1 + \\ \sin 2\theta \, \sin\phi_1 + \pi - 2 \sin^{-1}(\cos\alpha \, \cos\theta) \end{array} \right\}$$

where:
$\alpha = \tan^{-1}(\tan\theta / \tan\phi)$

REFLECTION AND TRANSMISSION

Direct light falling on surfaces will be reflected, transmitted or absorbed by those surfaces. Reflection and transmission of light are complex interactions of light with matter. Generally the process is dependent not only on the angle of incidence but also on the wavelength (spectral composition) of the light (seen as "colour of the surface") and on the polarisation of the light. In this chapter spectral distribution and polarisation are not discussed. In Chapter 4 the optical properties of materials were described.

For the mathematical treatment, the concept of bi-directional distribution function will be used which relates the reflected (transmitted) luminance in a given direction to the incident luminance distribution (6).

Bi-directional Reflection Distribution Function

The bi-directional reflection distribution function is defined for a local coordinate system on the surfaces, where the directions of incident and reflected luminance are defined as shown in Figure 4.

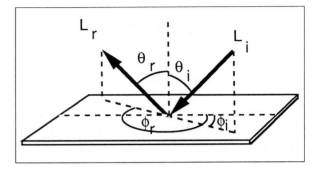

Figure 4 - Bi-directional reflection.

The luminance in a direction θ_r, ϕ_r from a reflecting surface can be written as follows:

$$L_r(\theta_r, \phi_r) = R\, L_i = \int_{2\pi} L_i(\theta_i, \phi_i)\, fr(\theta_i, \phi_i, \theta_r, \phi_r)\, \cos\theta_i \, d\Omega_i$$

where:
R = reflection operator
$fr(\theta_i, \phi_i, \theta_r, \phi_r)$ = bi-directional reflection distribution function

The bi-directional reflectance distribution function is symmetrical with regard to the directions of incidence and reflection:

$$fr(\theta_i, \phi_i, \theta_r, \phi_r) = fr(\theta_r, \phi_r, \theta_i, \phi_i)$$

This property is called the reflectance reciprocity (7). If the surface is isotropic the function is rotation invariant in the plane of the surface, thus:

$$fr = fr(\theta_i, \theta_r, |\phi_r - \phi_i - \pi|)$$

where:
$0 \leq |\phi_r - \phi_i - \pi| \leq \pi$

With the bi-directional reflection function many different types of reflectances can be defined (6), for example, the directional reflectance as follows:

$$\rho(\theta_i, \phi_i) = \int_{2\pi} fr \cos\theta_r \, d\Omega_r$$

The (bi-hemispherical) reflectance can then be defined by:

$$\rho = \frac{\int_{2\pi} \rho(\theta_i, \phi_i) L_i \cos\theta_i \, d\Omega_i}{\int_{2\pi} L_i \cos\theta_i \, d\Omega_i}$$

Reflectance is in general dependent on the directional characteristics of the incident radiation and, for simplicity, is often defined for uniform incident radiation.

Specular Reflection

When the angle of an incoming ray to the normal is identical to the angle of the reflected ray with the normal the reflection is specular. Ideal specular reflection can be described with the aid of a Dirac δ-function as:

$$fr(\theta_i, \phi_i, \theta_r, \phi_r) = \rho_s(\theta_i, \phi_i) \, 2\delta(\sin^2\theta_r - \sin^2\theta_i) \, \delta(\phi_r - \phi_i + \pi)$$

where:

$$\int f(u) \, \delta(u) \, du = f(0)$$

and:
$\delta(u) = 0$ if $u \neq 0$

Thus:

$$L_r(\theta_r, \phi_r) = \rho_s L_i(\theta_r, \phi_r - \pi)$$

The reflectance for a specular surface is often calculated with the Fresnel formula in the case of unpolarised light:

$$\rho(\theta_i) = \frac{1}{2} \frac{\sin^2(\theta_i - \beta)}{\sin^2(\theta_i + \beta)} + \frac{1}{2} \frac{\tan^2(\theta_i - \beta)}{\tan^2(\theta_i + \beta)}$$

where:
$\sin\beta = (\sin\theta_i) / n$
n = index of refraction relative to air

Isotropic Diffuse Reflection

This is a diffuse reflection in which the spatial distribution of the reflected radiation is such that the luminance is the same in all directions. Its distribution function can be described as follows:

$$fr(\theta_i, \phi_i, \theta_r, \phi_r) = \rho_d(\theta_i, \phi_i) / \pi$$

Note that this function is not symmetrical, ie it is a simplification of the reflected luminance distribution.

If the directional reflectance is independent of the incidence angles, the reflected luminance is:

$$L_r = \rho_d E_i / \pi$$

where:
ρ_d is the reflectance of the surface.

Mixed Reflection

Specular reflection and isotropic diffuse reflection are two extreme reflection forms. Most surfaces have a reflectance distribution function that is neither isotropic diffuse nor perfectly specular. The angular dependence of the distribution function relates to the material's optical properties and to the surface conditions, such as: surface roughness and the presence of contaminants.

Models can be found based on either physical optics (8) or geometrical optics. The analytical models based on physical optics are often valid only for perfect materials and do not explain the experimentally observed non-specular peaks. More successful are the models based on geometrical optics. This theory is, strictly speaking, valid only when the surface roughness is large compared to the wave-length. A problem for the application of these models is, however, the different empirical parameters one has to know, such as:
- isotropic diffuse reflection (wave-length dependent)
- parameters for the description of the surface roughness
- optical properties of the material (such as index of refraction and absorption index (wave-length dependent).

At an isotropic rough surface the reflected light is from specular reflection at randomly oriented micro-facets of the surface and from multiple reflections between the surface micro-facets. The latter is considered as isotropic diffuse reflected light. There are no expressions for the isotropic diffuse reflectance; this has to be measured experimentally.

The description of the specular reflectance at a micro-facet requires the introduction of two angles:
- the incident angle on the micro-facet: as the reflection is specular, this angle is half the angle between the incident and reflected radiation, given by the formula:
$\cos 2\theta_s = \cos\theta_i \cos\theta_r + \sin\theta_i \sin\theta_r \cos(\phi_i - \phi_r - \pi)$
- the angle between the surface normal and the normal on the micro-facet: this angle is given by the formula:
$\cos\alpha = (\cos\theta_i + \cos\theta_r) / 2\cos\theta_s$

The facets may be more or less randomly oriented and randomly inclined. This characteristic is described by the micro-facet distribution function $P(\alpha)$:

$P(\alpha)\, d\Omega\alpha\, dA$ = the surface of the micro-facets whose normals are within the solid angle $d\Omega\alpha$.

Another issue is the shadowing and masking of the facets at large incident or reflection angles. For this the geometric attenuation factor G is introduced (9). The analytical derivation of this factor is rather complicated as more information is required on the geometry of the surface roughness than is given by the micro-facet distribution.

The bi-directional distribution function for the non-specular peak (only one reflection) can be described as:

$f_{rs}(\theta_i, \phi_i, \theta_r, \phi_r) = F(\theta_{s,n,k}) [(P(\alpha)G)/(4\cos\theta_i \cos\theta_r)]$

where:
F = the Fresnel reflectance

Many formulae for $P(\alpha)$ can be found in the literature (8, 9, 10). One simple version is given here (11):

$P(\alpha) = e^2 / [\pi (e^2 \cos^2\alpha + \sin^2\alpha)^2]$

where:
e = an empirical parameter

If e decreases the surface is less rough and there will be more reflection in the specular direction. The bi-directional distribution function can be written as:

$fr = [\rho_d(\theta_i)/\pi] + f_{rs}$

where:
ρ_d = the (directional) isotropic diffuse reflectance
f_{rs} = the bi-directional distribution function of the non-specular peak.

Bi-directional Transmission Distribution Function

Light coming from the exterior enters the room through transparent and translucent surfaces. Such surfaces in most practical situations are vertical or horizontal daylight openings, but daylight may also enter through other more complicated light guiding systems. In general, a window is a compound system consisting of one or more layers of (coated) glass, often combined with a shading or diffusing device. In this section the discussion will be limited to window systems.

For transmission through a thin plate, or shell, where transmitted rays can be practically considered to emerge from the point of incidence, the concept of bi-directional transmittance distribution function may be applied empirically. The function is defined as follows (Figure 5):

$L_t(\theta_t, \phi_t) = T L_i =$

$\int_{2\pi} L_i(\theta_i, \phi_i)\, f_t(\theta_i, \phi_i, \theta_t, \phi_t)\, \cos\theta_i\, d\Omega_i$

where:
T = transmission operator
$f_t(\theta_i, \phi_i, \theta_t, \phi_t)$ = bi-directional transmission function

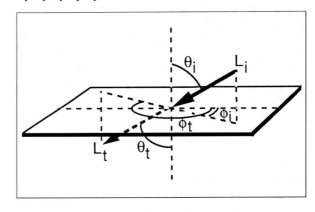

Figure 5 - Bi-directional transmission.

Analogous to the expressions for directional reflectance and (hemispherical) reflectance, expressions for the directional transmittance and (bi-hemispherical) transmittance can be given by the bi-directional distribution function.

Direct Transmission

Direct transmission is that process by which incident flux passes through a surface or medium without scattering. For an element of a transmitting surface, there is a one-to-one correspondence between incident and transmitted rays. The angle of the transmitted ray is governed by Snell's law of refraction. For clear glazing the transmission can be considered as direct. The bi-directional distribution function can be written analogous to specular reflection, as:

$$f_t(\theta_i, \phi_i, \theta_t, \phi_t) = \tau_c(\theta_i, \phi_i) \, 2\delta(\sin^2\theta_t - \sin^2\theta_i) \, \delta(\phi_t - \phi_i + \pi)$$

Thus:

$$L_t(\theta_t, \phi_t) = \tau_c \, L_i(\theta_t, \phi_t - \pi)$$

where:

τ_c = transmittance for clear glazing

In crude light transfer models, glazing (single or double) is described by a transmittance that is independent of the angle of incidence (eg in the Building Research Establishment split-flux method, a single glazing transmittance of 0.85 is assumed for all angles of incidence; for double glazing the value is 0.77). More complex models use a transmittance depending on the angle of incidence, such as the following formula:

$$\tau_c = 1.018 \, \tau_o \, (\cos\theta_i + \sin^3\theta_i \, \cos\theta_i)$$

where:

θ_i = angle of incidence
τ_c = transmittance at the given angle of incidence
τ_o = transmittance at normal incidence

Isotropic Diffuse Transmission

In the case of perfectly diffusing glazing (eg opal glass) the reflection analogue is the Lambert's Law for diffuse reflection, so:

$$f_t(\theta_i, \phi_i, \theta_t, \phi_t) = \tau_d(\theta_i, \phi_i) / \pi$$

where:

τ_d = diffuse transmittance

If the (directional) transmittance is independent of the incidence angles, the transmitted luminance is:

$$L_t = E_i \, \tau_d / \pi$$

Mixed Transmission

If a window system contains a layer of a translucent medium, the transmission will be neither direct nor isotropic diffuse.

The analytical treatment of a transmission model is beyond the scope of this chapter. It is very complicated and an experimental determination of the bi-directional properties is more appropriate. A linear combination of direct (regular) and isotropic diffuse transmission can be useful for sensitivity studies and for systems with (not fully closed) venetian blinds.

If the bi-directional properties of each layer in a window system are known, the overall bi-directional functions can be derived (see below).

Transmittance of Multi-layer Fenestration Systems

As an application of the reflection and transmission theory, a system consisting of plane parallel panes (eg triple glazing) will be described. It is assumed that the distances between the glass-panes are small compared to the surface dimensions of the panes, furthermore the thicknesses of the panes are negligibly small. In the case of three panes it is convenient to denote the luminances as in Figure 6.

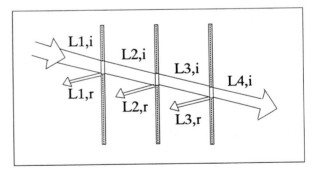

Figure 6 - Transmission of light through triple glazing.

The luminance $L_{1,r}$ is the result of reflection of $L_{1,i}$ and the transmission $L_{2,r}$. In the same way $L_{2,i}$ is the result of transmission of $L_{1,i}$ and reflection of $L_{2,r}$.

The relations between these luminances can be given by a number of equations. For the first pane:

$$L_{2,i} = T_{1,f} L_{1,i} + R_{1,b} L_{2,r}$$
$$L_{1,r} = T_{1,b} L_{2,r} + R_{1,f} L_{1,i}$$

where:
T = the bi-angular transmission operator
R = the bi-angular reflection operator
f = front side
b = back side

The equations can also be written in a matrix form:

$$\begin{pmatrix} L_{1,i} \\ L_{1,r} \end{pmatrix} = \frac{1}{T_{1,f}} \begin{pmatrix} 1 & -R_{1,b} \\ R_{1,f} & T_{1,b}T_{1,f}-R_{1,b}R_{1,f} \end{pmatrix} \begin{pmatrix} L_{2,i} \\ L_{2,r} \end{pmatrix} = \frac{1}{T_{1,f}} [A_1] \begin{pmatrix} L_{2,i} \\ L_{2,r} \end{pmatrix}$$

For the light falling through the next two panes we can derive similar equations. So for three panes the light transfer is:

$$\begin{pmatrix} L_{1,i} \\ L_{1,r} \end{pmatrix} = \frac{1}{T_{1,f} T_{2,f} T_{3,f}} [A_1][A_2][A_3] \begin{pmatrix} L_{4,i} \\ 0 \end{pmatrix}$$

This can easily be extended to more layers. The above equation offers only a formal solution. In some simple cases this equation can be worked out analytically. Assuming specular reflection and direct transmission for all surfaces, the solution for two panes (double glazing) is:

$$\begin{pmatrix} L_{1,i} \\ L_{1,r} \end{pmatrix} = \frac{1}{\tau_1 \tau_2} \begin{bmatrix} 1 & -\rho_1 \\ \rho_1 & \tau_1^2-\rho_1^2 \end{bmatrix} \begin{bmatrix} 1 & -\rho_2 \\ \rho_2 & \tau_2^2-\rho_2^2 \end{bmatrix} \begin{pmatrix} L_{3,i} \\ 0 \end{pmatrix}$$

So:

$$L_{3,1}/L_{1,i} = \tau_{tot} = \tau_1 \tau_2 / (1 - \rho_1 \rho_2)$$

The advantage of this method is its applicability in computer calculations. Matrix calculations can easily be done by computers. The luminance can be described as a ray-vector, whose components are the luminances within a certain small solid angle. The reflection and transmission operators are turned into matrices. The solution can then be found by computer (instead of luminances one can also use luminous fluxes from different directions) (12).

For window systems with venetian blinds the situation is more complicated. Often venetian blinds are treated as a layer with averaged optical properties. In this case the properties of the whole fenestration system can be derived with the method described above.

Descriptions of venetian blinds are given in references (13), (14). In ray-tracing models there is no direct need for a different approach as light rays can be traced also within the window system.

Inter-reflections Between Surfaces

The luminance of reflected light inside a room is governed by the following equation:

$$L = R L + R L_s$$

where:
$L = L(\underline{x}, \theta, \phi)$ the luminance of the reflected light at a point $\underline{x}$ of a surface
R = reflection operator working on the luminances of the surrounding surfaces seen from $\underline{x}$ (excluding direct light from the sources)
L_s = the luminance of the light source(s) at $\underline{x}$

The formal solution of this equation is given by:

$$L = R L_s / (1-R)$$

Only by rather crude approximations can the inverse operator be found. The most frequently used solution in computer programs is by successive iteration:

$$L = \Sigma L(k)$$

where:
$L^{(1)} = R L_s$
$L^{(n)} = R L^{(n-1)}$

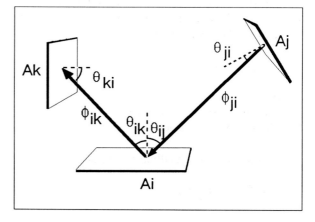

Figure 7 - The exchange of luminous flux between small surfaces.

The physical interpretation of this equation is straightforward: it is a summation of the successive reflections in the room.

This equation can be fragmented in several ways. Suppose that all surfaces are fragmented into a large number of small sub-surfaces with a homogeneous luminance. The luminous flux leaving sub-surface i and arriving at k is (Figure 7):

$$\Phi_{ik} = \int_{A_i} L_{ik} \, dA_i \int_{\Omega_{ik}} \cos\theta_{ik} \, d\Omega_{ik}$$

where:
A_i = area of surface i
Ω_{ik} = solid angle by which A_k is seen from A_i (dependent on the position on A_i)

The fragmented equation is:

$$\Phi_{ik}^{(n)} = \int_{\Omega_k} \cos\theta_{ik} \, d\Omega_{ik} \sum_j f_{rijk} \Phi_{ji}^{(n-1)}$$

where:
Σ = summation over all surfaces seen by A_i (the problem of surfaces partially masking other surfaces will not be treated here)
f_{rijk} = bi-conical reflection function

If the surface elements and the solid angles are small, f_{rijk} can be approximated by the value of the distribution functions for angles determined by one point on each surface.

The equation above can also be interpreted without referring to surface elements but to discrete solid angles and ray directions, such as in ray-tracing programs.

View Factors

The transfer equation can be simplified if all surfaces have an isotropic diffuse reflectance which is independent of the incident angles. The luminance distribution is known if the illuminances of the surfaces are known. Subdividing the surfaces into small elements one can write:

$$L_i = \rho_i E_i / \pi$$

where:
L_i = isotropic luminance leaving surface i
E_i = illuminance of surface i

The equation for the illuminances is:

$$E_i = E_{si} + \Sigma F_{ij} \rho_j E_j$$

where:
E_{si} = direct illuminance of surface i
F_{ij} = view factor

Note that view factors are also called form factors, angle factors and configuration factors.

E_{si} and F_{ij} are defined by:

$$E_{si} = \frac{1}{A_i} \int_{A_i} dA_i \int_{2\pi} L_{si} \cos\theta_{si} \, d\Omega_{si}$$

$$F_{ij} = \frac{1}{\pi A_i} \int_{A_i} dA_i \int_{\Omega_{ij}} \cos\theta_{ij} \, d\Omega_{ij}$$

$$= \frac{1}{A_i} \int_{A_i} dA_i \int_{A_j} \frac{\cos\theta_{ij} \cos\theta_{ji}}{\pi r_{ij}^2} \, dA_j$$

where:
r_{ij} = distance between a point on A_i and one on A_j.

View factors have two basic properties:
- reciprocity: $A_i F_{ij} = A_j F_{ji}$
- the sum of the view factors from one surface to all the other surfaces is 1: $\Sigma F_{ji} = 1$

For a convex or plane surface $F_{ii} = 0$ (the surface does not see itself). View factors can be calculated using standard analytical formulae for many different geometrical situations (15, 16). Also a number of approximate calculation methods exists for situations in which the complete calculation of all view factors is too time-consuming or impossible.

Calculating the "surface to surface" view factor is often possible only by numerical integration, or Monte Carlo methods (17). Without the averaging over surface A_i the factor is called the point to surface view factor. For this factor more analytical formulas can be found in the literature.

As examples, the formulae for rectangular planes are given below (Figures 8 and 9):

$$F_{12} = \frac{1}{2\pi} \left[\frac{x}{\sqrt{(1+x^2)}} \tan^{-1}\left(\frac{y}{\sqrt{(1+x^2)}}\right) + \frac{y}{\sqrt{(1+y^2)}} \tan^{-1}\left(\frac{x}{\sqrt{(1+y^2)}}\right) \right]$$

where:
x = a/c
y = b/c

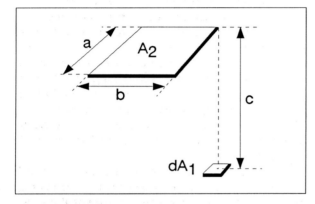

Figure 8 - Plane point source dA_1 and a plane rectangle A_2 parallel to plane dA_1.

$$F_{12} = \frac{1}{2\pi} \left\{ \tan^{-1}\left(\frac{1}{y}\right) + A(x\cos\phi - y)\tan^{-1}A + \frac{\cos\phi}{B}\left[\tan^{-1}\left(\frac{x - y\cos\phi}{B}\right) + \tan^{-1}\left(\frac{y\cos\phi}{B}\right)\right] \right\}$$

where:
x = b / a
y = c / a
A = 1 / $\sqrt{(x^2 + y^2 - 2xy\cos\phi)}$
B = $\sqrt{(1 + y^2 \sin^2\phi)}$

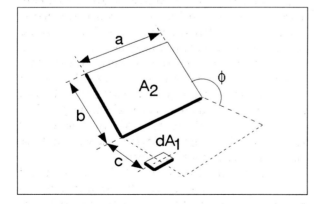

Figure 9 - Plane point source dA_1 and a plane rectangle A_2 in which the planes of dA_1 and A_2 intersect at an angle ϕ ($0<\phi<\pi$).

If the point dA_1 is not situated as indicated in the figures, the method suggested in Figure 3 can be applied.

The last formula for $\phi = \pi/2$ is similar to the formula given for the direct illuminance on a horizontal plane through a vertical window from a uniform sky.

If A/r_{ij}^2 is small, the factor can be approximated with the point to point view factor. Corrections for inevitable inaccuracies in corners can be found (18).

The view factor equation can also be written as:

$$\Sigma \phi_j \left[(F_{ij}/A_i \rho_i) - (\delta_{ij}/A_j) \right] = E_{si}$$

where:
ϕ_j = total luminous flux leaving surface i
δ_{ij} = Kronecker symbol

The matrix consisting of the elements between the brackets is symmetrical and positive. This has advantages for computing the exact solution.

If there are many sub-surfaces, methods with matrix decomposition are much more time-consuming than methods using successive reflections.

There are several programs using the view factor approach (19). Sometimes it is used only for the exterior space to calculate the reflected light from overhangs, buildings, etc. To facilitate the calculations, the building geometries may be shaped as blocks and each side of the block is considered as one sub-surface (18).

CALCULATION MODELS

The integral equation for the luminance distribution in a room cannot be solved exactly. Many approximate methods exist. The next sections describe a number of methods (state-of-the-art and manual methods) to calculate the illuminance of a room with known daylight and artificial light sources.

Ray-tracing and Ray-tracking

Forward ray-tracing. Perhaps the most general method to find the luminance distribution in a room due to one or more light sources is the ray-tracing method. It is based upon the principle of following the light rays along their path through a room. "Light rays" in this context are elementary fluxes coming from a light source. Light rays striking surfaces (walls, furniture, etc) in the room contribute to the luminances of those surfaces. Then a reflection algorithm is applied and new rays are traced. The ray loses its intensity by non-specular reflections or refractions, and by absorption processes. A ray is traced until its intensity has dropped below a certain pre-defined level, when it has become too weak to contribute significantly to the illumination of the room.

An advantage of the ray-tracing method is that there are no restrictions on the geometry of the scene under study. Most other methods can only give solutions for rectangular geometries, in most cases only for simple box-shaped rooms.

Another major advantage of the ray-tracing method is that there are no restrictions on the surface reflection characteristics. Therefore specular reflection, off-specular reflection and other phenomena can be studied in a ray-tracing technique.

It is obvious that if the rays have a very small solid angle it requires much computing time to trace all possible rays of the light from their sources to the final distribution in a scene. For this reason ray-tracing is often combined with a statistical method. This method is called the Monte Carlo method (20).

In this technique a random number generator is used in combination with a pre-defined distribution function to determine the direction in which a ray will be traced. This method can be applied to the whole hemisphere (one ray only) or to a subdivision of the unit hemisphere on a surface element. The last method is called stratified sampling and the number of rays depends on the number of subdivisions (Figure 10).

The accuracy of the calculation result will increase with the number of rays traced, but after only a limited number of rays a first impression of the final result can be obtained.

A model based on this principle is Genelux (21).

Ray-tracing models can be divided into two main categories: programs that trace the light backwards (from the eye towards the sources) and programs that follow the rays in a more natural forward way (from the sources towards the eye).

Following the light rays, forward or backward, involves the same basic problems:
- calculate the direction of the ray
- determine which surface the ray intersects
- calculate the interaction of the light with the surface; this may involve the creation of new rays - those new rays have to be traced also, each of them may interact with a surface giving even more rays, until the ray's intensity is too low
- sometimes the calculation may also account for scattering and absorption of light (e. fog)

Backward ray-tracing. Originally the ray-tracing technique was developed for the visualisation of complex scenes in computer graphics. The aim was to determine which objects are visible from a given point. It was meant as a technique for "hidden line removal". Later this technique found other applications. In lighting calculations, for every pixel of the screen the light is traced back to its sources. The word tracing means "tracing the light sources" (Figure 10).

From the point of the observer, rays are traced through the pixels of the plane of representation (eg the screen of a computer monitor) towards the scene under study. A ray will intersect with one of the surfaces of this scene. Each ray can be thought of as the luminance value that results either directly from an emitting surface, or indirectly from a reflecting surface. If the surface is a reflecting surface the ray can be traced further back, knowing the reflectance distribution of the surface. This is done by resolving the reflectance distribution with the flux arriving at the surface (3). Each new ray has a relative weight depending on the reflection(s) that have occurred, tracing it backwards. This process is repeated until the ray hits a source, or it has become too weak.

For diffuse reflection many rays have to be traced. Therefore the number of rays is often reduced as the cumulative weight decreases. Also, after only a small number of reflections, a constant value is used for the diffuse component. This may be calculated with a simplified method.

A major disadvantage of tracing the light rays backwards is that the results are only valid for one eye point. To visualise from another point of view, the whole calculation has to be performed again. The advantage is that the number of rays that are not relevant for the image is limited and visualisations can be produced with a high quality.

Ray-tracking. Instead of following rays and calculating each intersection with a surface, it is also possible to pre-calculate the ray vectors and trace them afterwards: this method is called ray-tracking. To store all possible rays between surface elements would be unnecessary as there would be a large diversity of solid angles. A method has been developed to reduce the number of rays while keeping the same accuracy. It is explained in the following description of the program DIM (22).

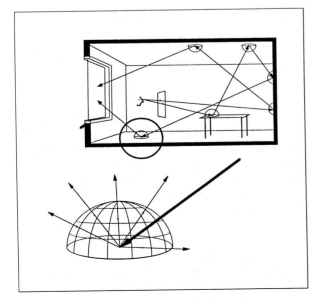

Figure 10 - Ray-tracing.

Description of the DIM system. DIM comprises three program modules. The first program module - DIMdis - accepts user data which describe a zone's topology. This description is conceived as ordered lists of vertices permitting the entry of any three-dimensional zone shape. The objective of DIMdis is to transform this continuous zone model into a discrete numerical equivalent in a manner which maximises numerical accuracy. Conceptually, the technique is described as follows: Each zone surface is subdivided into a number of finite elements. A unit hemisphere is then superimposed on each finite element in turn. This comprises many small patches formed by slicing the unit hemisphere in the horizontal and vertical planes as shown in Figure 11.

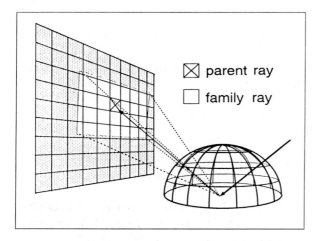

Figure 11 - Ray-tracking.

The projections defined by the lines joining the surface element centre point and the hemisphere patch centre points represent the surface element's radiosity field. Each projection line is traced until an intersection with another surface is encountered. This defines one ray, connecting a source surface element with a sink surface element, along which light may propagate. By repeating the process for all surface elements, a stack of rays isgenerated. Because rays represent cones of light, the greater the distance before intersection the greater is the spread of light. This is handled by next processing all surface sink elements which have not received a hit from the projections associated with one united hemisphere. Lines are formed between each of these sink elements and the source element to determine which hemispherical patch is the associated one. The sink element is then associated with the ray emanating from that patch. In this way, an initially generated ray is able to represent the point view factors between one source and several sink elements. This technique dramatically reduces the ray-tracing task while ensuring high accuracy. The output from DIMdis is a stack of rays representing the total number of possible light flow-paths.

The second program module of DIM is DIMray. This accepts the output from DIMdis and the positioning of light fittings extracted from DIM's fittings database.

Each fitting is conceived as a number of vectors representing the fitting's candle power distribution in three dimensions, with a number of monochromatic wavebands superimposed to represent spectral characteristics. The computational mission of DIMray is to process each vector, for each monochromatic waveband, through the ray stack representing the zone's inter-reflection possibilities. For any given vector, a ray counter is set to zero and the vector is traced to its point of first inter-reflection. A reflection model is then invoked to determine the direction and number of the reflected rays. These are added to the ray counter. The first ray is then traced in the same manner and the rays it generates entered in the stack. The process continues until the initial vector strength has diminished to insignificance. The performance output from DIMray is a database giving, for each surface element, the distribution of reflected light associated with each initial fitting vector and monochromatic waveband.

DIMout is the analysis module which allows the recovery and presentation of these performance data. It has two capabilities. The first allows the performance data to be integrated by fitting and by surface. This allows, for any mix of fittings, the production of surface illumination contours and luminance data for any given eye point. At this stage various lighting control strategies can be investigated by accepting or rejecting fittings against some criteria. The second capability allows the organisation and transfer of data to the VISTA program to permit display realism in terms of surface colour and luminance variation with the possibility of spectral reflections.

Display of images. The output of a program can be such that one can take an arbitrary view point in the room for an image. In that case the output often does not contain the full information of luminance distribution, as this would require a large storage, but only the illuminances of the surfaces. For the image an isotropic luminance distribution is then assumed (21). An exception is DIM where all the information is kept (22).

If only one picture can be made in one run there is no need for the simplification of illuminance data and the pictures can be very accurate (eg backward ray-tracing).

Another problem is the fragmentation. To display a picture, interpolations are made to "smooth" the result. This smoothing is also a source of inaccuracies. Special interpolation techniques have been developed. Nevertheless details can sometimes disappear in this post-processing.

Sometimes the image contains a pattern of points for which the density is proportional to the illuminance. These kinds of pictures are obtained with a scaled random number technique, where all rays have the same luminous flux.

The accommodation of the eye looking at a screen is another source of possible unrealistic images. This might be less of a problem for an expert in this field. Different formulae are used for calculating brightness from luminance data.

For colour display, one usually breaks the spectrum into the CIE tri-stimulus values X, Y and Z. The reflection distribution functions are considered as tripartite and modify the tri-stimulus values independently. These values are treated in parallel while ray-tracing.

Simplified Methods

Simplified methods can be derived from the view factor equation by dividing the room inner surface into only a few surfaces and by simplifying assumptions on the distribution of the direct illuminance.

For instance, in cavity methods only three surfaces are distinguished: ceiling, floor, and the rest (6). If the floor-ceiling view factor is known, all the other factors can be calculated using the view factor properties. The floor-ceiling view factor is often calculated with an approximate formula. The method is mostly used for artificial lighting and seldom for daylighting. Even more crude is the division of the room into two surfaces, namely: the floor with a part of the vertical walls below the mid-height of the window, and the ceiling with the rest of the walls. For the direct illuminance it is assumed that the light from the sky is distributed over the lower part and the externally reflected light is distributed over the upper part. For this reason the method is called the Split Flux Method. It is a popular way to calculate daylight factors (4).

The direct (average) illuminance of the lower (floor) part is:

$$E_{sfl} = \tau (A_g/A_{fl}) E_{sky}$$

where:
τ = direct transmittance of glazing
A_g = window area
A_{fl} = floor area
E_{sky} = exterior illuminance of the window by the sky

For vertical windows in the BRE split flux method, E_{sky} is estimated by:

$$E_{sky} = 0.4 \, E_h \, C$$

where:
C = a factor for exterior obstructions
0.4 = ratio of vertical to horizontal illuminance for a CIE overcast sky

The coefficient C takes into account the blocking of the sky by obstructions. A simple formula often used is:

$$C = 1 - (\theta/90)$$

where:
θ = average altitude (in degrees) of obstructed sky

For toplighting there are tables to find E_{sky} on planes of different inclination for the CIE overcast sky (23). A simple formula similar to the one for obstructions is (24):

$$E_{sky} = E_h \, (\theta/180)$$

where:
θ = the vertical angle subtended by the visible sky

For the upper part (ceiling), the (average) illuminance is:

$$E_{sc} = \tau (A_g/A_c) E_{gr}$$

where:
A_c = ceiling area
E_{gr} = exterior illuminance of the ground

The reflections from the ground have a smaller contribution and the extra reflections of the buildings are often neglected. For a vertical window:

$$E_{gr} = 0.5 \, \rho_{gr} \, E_h = 0.05 \, E_h$$

where:
ρ_{gr} = reflectance of the ground, usually taken at a minimum value of 0.1

For isotropic diffusing windows, the illuminances of floor and ceiling follow from:

$$A_{fl} \, E_{s\,fl} = A_c \, E_{sc} = 0.5 \, \tau \, A_g \, (E_{sky} + E_{gr})$$

In order to calculate the internally reflected component, view factors must be estimated.

Horizontally infinite room. The view factors between upper and the lower part are 1. Then the illuminance of the floor is (25):

$$E_{fl} = (E_{s\,fl} + \rho_c \, E_{sc}) / (1 - \rho_{fl} \, \rho_c)$$

where:
ρ_c = average reflectance of upper part of room
ρ_{fl} = average reflectance of lower part of room

Dividing this equation by E_h yields the average daylight factor of the floor (or work plane = 0.8 m above the floor).

This approach is assumed to be valid for large rooms with toplighting.

If the direct (sky) component is calculated (or estimated with a protractor) for a point on the working plane, the approach can be used to calculate the internally reflected component. The point illuminance is:

$$E_p = E_{sp} + [(\rho_{fl} \rho_c E_{s\,fl} + \rho_c E_{sc}) / (1 - \rho_{fl} \rho_c)]$$

where:
E_{sp} = direct illuminance

Integrated Sphere Approximation

For the inner surface of a sphere, the view factors are:

$$F_{ij} = A_j / A$$

where:
A_j = area of a small surface element
A = total surface area

The solution of the view factor equation for only two surfaces yields for the average floor illuminance:

$$E_{fl} = E_{sfl} + [(A_{fl}\rho_{fl}E_{sfl} + A_c\rho_c E_{sc}) / (A(1-\rho_{ave}))]$$

where:
ρ_{ave} = average area weighted reflectance (including windows)

Dividing by E_h yields the average daylight factor for the working plane.

Also here the direct component can be estimated with a different method for a point on the working plane:

$$E_p = E_{sp} + [(A_{fl}\rho_{fl}E_{sfl} + A_c\rho_c E_{sc}) / (A(1-\rho_{ave}))]$$

This equation leads to the best results for highly reflective surfaces and small rooms. It is the one most frequently used for daylight factor calculations.

Regression Methods

The formulae treated above all have the form:

$$E_p = F_1 E_{sky} + F_2 E_{gr}$$

By measurements in scale models one can find values of F_1 and F_2 for a number of room geometries, reflectances, glazing transmittances, etc.

The best known method of this type is the Lumen Method discussed further in Chapter 9. The original Lumen Method was designed for a sidelit room and the window had its top at the ceiling level and its sill at 0.9 m above the floor. The illuminance of the working plane was given as:

$$E_p = E_v A_g \tau K_u$$

or

$$E_p = E_{sky} A_g \tau K_{sky} + E_{gr} A_g \tau K_{gr}$$

where:
E_v = vertical illuminance on the window
K_u, K_{sky}, K_{gr} = coefficients of utilisation

Coefficients of utilisation can be found in tables. They are listed for different geometries, glazing systems (including venetian blinds), sky types (clear and overcast), etc. If empirical data are lacking, the coefficients of utilisation are often calculated with the cavity or zonal cavity method.

Another approach is correcting the integrating sphere formula in order to get a better correlation with measurements (26). Examples of other models, some integrating thermal aspects with lighting (27, 28), are described in more detail elsewhere (29, 30). An overall discussion and review of tools follows in the next chapter.

REFERENCES

(1) Tregenza, P.R., "The Daylight Factor and Actual Illuminance Ratios", *Lighting Research and Technology*, vol. 1, No. 2, 1980.

(2) Tregenza, P.R. and Waters, I.M., "Daylight Coefficients", *Lighting Research & Technology*, vol. 15, No. 2, p. 65-71, 1983.

(3) Ward, G.J. and Rubinstein, F.M., "A New Technique for Computer Simulation of Illuminated Spaces", *Journal of I.E.S.*, Winter 1988.

(4) Hopkinson, R.G., Petherbridge, P. and Longmore, J., *Daylighting*, Heinemann, London, 1966.

(5) Moon, P. and Spencer, D.E., "Illumination from a Non Uniform Sky," *Illuminating Engineering*, vol. 37, p. 707-726, 1942.

(6) IES, *Lighting Handbook*, Reference Volume, Illuminating Engineering Society of North America, New York, 1981.

(7) Driscoll, W.G., Vaughan, W., *Handbook of Optics*, McGraw-Hill, New York, 1978.

(8) Beckmann, P., Spizzichino, A., *The Scattering of Electromagnetic Waves from Rough Surfaces*, Pergamon Press, New York, 1963.

(9) Torrance, K.E., Sparrow, E.M., "Theory for Off-specular Reflection from Roughened Surfaces", *Opt. Soc. Am.*, vol. 57, No. 9, 1967.

(10) Torrance, K.E., Sparrow, E.M., "Bi-angular Reflectance of an Electric Nonconductor as a Function of Wavelength and Surface Roughness", *Journal of Heat Transfer*, Transactions of the ASME, p. 283-92, 1965.

(11) Trowbridge, T.S., Reitz, K.P., "Average Irregularity Representation of a Rough Surface for Ray Reflection", *Opt. Soc. Am.*, vol 65, No. 5, 1975.

(12) Papamichael, K.M. and Winkelmann, F.C., "Solar Optical Properties of Multi-layer Fenestration Systems", *Proc. Int. Daylighting Conference*, Long Beach, California, 1986.

(13) Matsuura, K., "General Calculation Method of Illuminance from a Vertical Louver Blind System" *Proc. Int. Daylight Conference*, Phoenix, Arizona, 1983.

(14) Papamichael, K.M. and Selkowitz, S., "Simulating the luminous and thermal performance of fenestration systems", *Lighting Design and Application*, 1987.

(15) Sparrow, E.M., Cess, R.D., *Radiation Heat Transfer*, Hemisphere Publ. Corp., McGraw Hill, New York, 1978.

(16) Wiebelt, J.A., *Engineering Radiation Heat Transfer*, Holt, Rinehart and Winston, 1966.

(17) Siegel, R. and Howell J.R., *Thermal Radiation Heat Transfer*, McGraw-Hill, New York, 1972.

(18) "Description of Superlite 1.0", *LBL Documentation Draft*, Berkeley, California, January 1985.

(19) Bensasson, S. and Burgess, K., "Computer Programs for Daylighting in Buildings", *Design office consortium evaluation report*, No. 3, Cambridge, UK, 1978.

(20) Tregenza, P.R., "The Monte Carlo Method in Lighting Calculations", *Lighting Research and Technology*, vol. 15, No. 4, p. 163-170, 1983.

(21) Fontoynont, M.R., "Simulation of Complex Window Components Using a Ray-Tracing Simulation Program", *Proc. Nat. Lighting Conference*, Cambridge, UK, 1988.

(22) "DIM, Dynamic Illumination Model", Univ. of Strathclyde.

(23) "Estimating Daylight in Buildings: Part 2", *BRE Digest 310*, BRE, Garston, UK, 1986.

(24) Lynes, J.A., "A Sequence for Daylighting Design," *Lighting Research and Technology*, vol. 11, No. 2, p. 102-106, 1979.

(25) Matsuura, K. and Tanaka, H., "Optimum Turning-off Depth for Saving Lighting Energy in Side-lit Offices", *J. Light and Vis. Env.*, vol. 3, No. 2, p. 24-31, 1979.

(26) Phillips, G.M. and Littlefair, P.J., "Average Daylight Factor Under Rooflights", *Proc. Nat. Lighting Conference*, Cambridge, UK, 1988.

(27) Selkowitz, S., Kim, J.J., Navvab, M. and Winkelmann, F., "The DOE-2 and Superlite Daylighting Programs", *Conf. Proc.*, Berkeley, California, June 1982.

(28) Gilette, G., "A Daylighting Model for Building Energy Simulation", *NBS Building Science Series 152*, Washington, DC, 1983.

(29) Gillette, G., Pierpoint, W. and Treado, S. "A General Illuminance Model for Daylight Availability," *Journal of I.E.S.*, p. 330-340, 1984.

(30) Parent, M.D. and Murdoch, J.B., "The Expansion of the Zonal Cavity Method of Interior Lighting Design to Include Skylights", *Journal of I.E.S.*, p. 141-173, 1988.

Chapter 9
EVALUATION AND DESIGN TOOLS

INTRODUCTION

Designers need tools to estimate the correctness of their choices. In the case of daylight, the concept of "correctness" involves many aspects, as we have seen in the previous chapters. Basically, these aspects can be grouped into two categories:
- environmental performance
- energy performance.

The first group includes all the questions discussed in Chapter 2, concerning the quality of the luminous environment from the point of view of human reactions to it. The second group is concerned with the level of dependence on non-renewable energy of the building design, in a given external environment, to ensure comfortable conditions for its occupants. Both categories are strongly dependent on the architectural characteristics of the project.

Given the relationship between luminous and thermal phenomena, energy needs for lighting, heating and cooling need to be integrated and evaluated with respect to each other.

Due to the complexity of the problem, a number of "tools" have been developed, each one dealing with one or more different aspects. Recently, thanks to the availability of computers, efforts have been made to provide integrated evaluation tools for designers, able to estimate both types of performance.

In this chapter, as well as in Appendices D, F and G, a review of existing design tools is presented. The tools are grouped into three categories:
- scale models (Appendix D)
- simplified design tools (Appendix F)
- computer codes (Appendix G).

SCALE MODELS

Introduction

Physical scale models are design tools that architects have used for centuries to study various aspects of building design and construction. Most architects still use this design technique to visualise the shape of the building, as well as the design of its façades and interior spaces, and also use it as a communication tool between clients and consultants.

Following slightly different construction criteria from the presentation models mentioned above, scale models provide an excellent design tool for daylighting studies (Figure 1).

Physical Properties of Light and the Influence on Scaling

Unlike thermal, acoustic and structural models, it is well known that physical models for lighting do not require any scaling corrections. The wavelengths of visible light are so short, relative to the size of models, that the behaviour of light is largely unaffected. With respect to the human visual perception, any differences are not noticeable (1). Rooms which are exactly duplicated in geometry, surface reflectances and their reflectivity, will provide the same quantity and quality of natural illumination as the actual space. If the colours used in the model match the actual colours in the real/designed room, then the visual impression of the model can be very close to the actual room.

Scale Models as Design Tools

Among a variety of techniques available to the designer (2), the scale model combines several aspects which emphasise its particular attraction for daylighting studies:
- the scale model is a simple design tool which can be understood easily
- the scale model enables precise studies of design options to be carried out with a very small budget. Even a simple scale model can provide an impression of illumination in the space
- the scale model is a design tool which can address questions concerning many aspects of the building design in addition to daylighting (eg. the spatial composition, the appropriate use of colour in the room, the design and arrangement of furniture, etc)
- the scale model can be used to simulate non-rectilinear, complex geometries and configurations. Often rooms are not simple rectilinear spaces, especially those which are designed to maximise the use of daylight
- the scale model allows quick changes to be made to the geometry, surface reflectances and reflectivity of the materials employed. Thus decisions about visual effects and levels of illumination can be made rapidly
- the scale model provides qualitative data from visual observations and photographs. The given

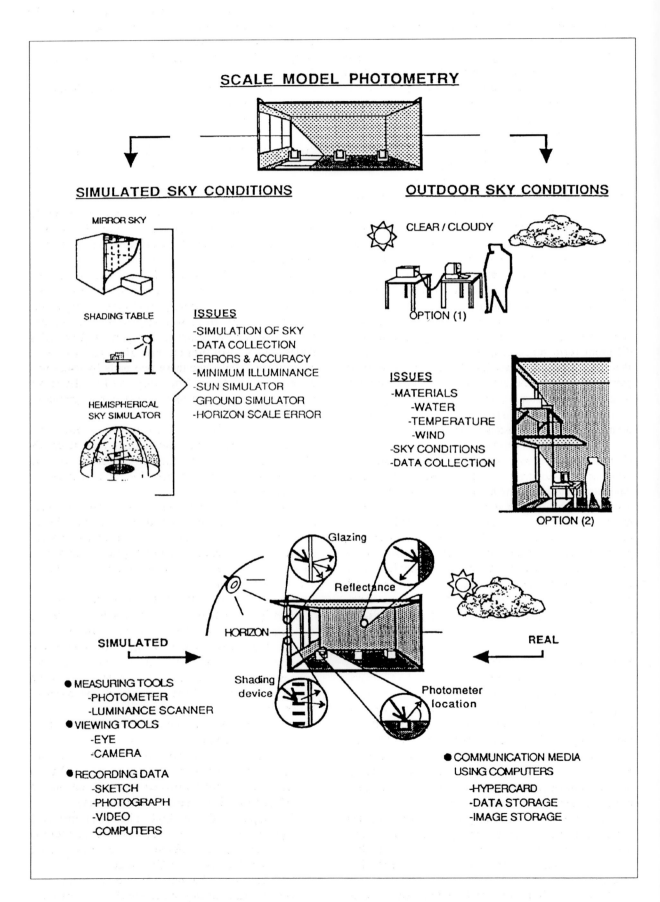

Figure 1 - Scale model photometry techniques and related issues.

information helps the designer to identify potential glare problems and other aspects of visual comfort
- the scale model also provides quantitative data of the illumination in the space. Measurements can be made to determine whether daylighting meets visual needs. From those data, projections of required electric lighting and thermal conditioning needs can be made
- the dynamic play of light within the space can be observed using the scale model and a heliodon, a sky simulator or real sky in conjunction with a video recording system. This technique provides an accurate representation of daylighting distribution changes in scale models of a space.

Other techniques, such as various computer programs, nomograms and diagrams, only allow the designer to examine or evaluate the daylight distribution for a fixed moment in time. Dynamic changes of daylight remain a neglected area.

Few computer programs can simulate non-rectilinear rooms, complex configurations and different reflectance properties (diffuse, specular). Where a program can handle these aspects it generally requires the use of a mainframe computer, which is time-consuming and not user-friendly.

Scale Models as Research Tools

In studies of the daylighting performance characteristics of new building concepts, such as atria and core-daylighting, the scale model is valuable for pre-validation. The scale model can be used to study performance characteristics of new materials, such as prismatic and holographic optical elements. In such a case the scale model acts as a pilot evaluation system to determine performance characteristics which can then be transferred to computer programs. If the performance and distribution of daylight of these devices is not fully understood, they cannot be assessed in computer programs.

Limitations of Scale Models

Scale models can be built to a very high degree of accuracy, simulating each detail within the space. However, a highly detailed scale model is expensive and in most cases does not meet the requirements for daylighting studies (such as flexibility in changing details for optimisation). It may however be used to obtain accurate data of the final design and to provide the designer and the client with the best image of the space.

A limitation of modelling arises when materials which cannot be scaled easily (eg fabrics) are involved. The use of the real material in scale models can thus cause errors in the quantitative measurements.

Electric lighting cannot be integrated in scale models for quantitative analysis. Although the intensity of artificial light can be simulated, the luminance distribution of the planned luminaires cannot. The combination of electric lighting and natural lighting can be tested only in real mock-up rooms.

More information about scale models can be found in Appendix D.

REVIEW OF EXISTING SIMPLIFIED DESIGN TOOLS

General Aspects

A brief overview of simplified methods to evaluate daylight in a building is presented here, with particular emphasis on their suitability for different design stages.

The simplified methods best suited for the calculation of the Direct Component (Sky Component and Externally Reflected Component) are not the same as those that give the most useful assessment of the Indirect Component (Internally Reflected Component). For this reason it is appropriate to calculate these two parts by using two different procedures. The simplified methods for these evaluations are either mathematical, graphical or tabular (3).

There are advantages in obtaining the total daylight illumination in a single stage procedure; this often results in reduced accuracy but saves time which may be particularly useful in the first stages of the design process.

All methods make reference to uniform, overcast or blue sky luminance distributions. None are able to take into account the luminance distribution models of various skies referred to in Chapter 3. Simplified methods that use such sky models are being studied at thr time of writing.

Methods for calculating the Daylight Factor can be provided in different forms: as equations, graphs, manual tools, nomograms, etc. Each approach is more or less useful or easy to use according to the preferences of the user, or according to the stage and accuracy of the design process. Table I shows the main characteristics of the design tools reviewed in Appendix F.

The parameters considered (shown in Table I) in order to judge the tools are:
- The position of the window plane: This allows one to take into account the method's ability to deal with windows placed on different planes (H, horizontal; V, vertical; S, slope; Any)
- The input of the method: The sky luminance distribution, which characterises different possible skies for different climatic areas (Clear; CIE, standard overcast; Unif, uniform)
- The output of the method: Direct Component (DC) (Sky Component (SC) + Externally Reflected Component (ERC)); Internally

TABLE I - Characteristics of the design tools

Category	#	Method: Flux transfert	Method: Daylight factor	Method: Lumen method	Method: Other	Flex: Max	Flex: Med	Flex: Min	Input: Clear	Input: Cie	Input: Unif.	Output: SS	Output: IRC	Output: DC ERC	Output: DC SC	Win: V	Win: H	Win: S	Win: Any
Equations	1.1	●					●				●			●	●	●			
	1.2	●					●				●			●	●	●		●	
	1.3		●				●			●	●				●	●	●			
	1.4		●				●				●		●				●			
	1.5				●	●					●			●			●			
	1.6		●							●	●				●	●				●
	1.7		●				●				●					●	●	●		
Single stage	2.1				●		●					●	●							●
	2.2				●			●			●		●				●			
	2.3				●		●				●		●				●			
	2.4				●		●				●		●				●			
	2.5				●	●					●		●							●
Lumen	3.1			●				●				●	●				●			
	3.2			●			●		●	●	●	●				●			
	3.3			●			●				●		●				●			
Tables	4.1	●					●				●	●			●	●	●	●		
	4.2		●				●				●	●			●	●	●			
	4.3				●			●			●				●	●	●			
	4.4				●		●				●				●	●	●			
	4.5	●						●			●	●				●			
Nomograms	5.1	●						●			●			●	●	●			
	5.2				●	●				●				●						●
	5.3				●	●			●							●				
	5.4				●		●			●		●				●	●			
	5.5				●					●		●				●				
	5.6				●					●			●			●				
	5.7		●					●		●	●			●	●	●	●			
	5.8		●						●	●	●			●	●	●			
	5.9		●					●			●					●				
	5.10		●				●				●			●	●	●	●			
Protractors	6.1	●					●				●			●	●	●			
	6.2		●				●				●			●	●	●		●	
	6.3				●	●			●	●				●	●	●		●	
	6.4		●				●			●	●				●	●	●			
	6.5		●				●			●	●				●	●	●			
	6.6		●				●				●	●			●	●			●	
	6.7				●	●					●	●				●		●	
	6.8		●				●				●			●				●		
Dot	7.1		●								●			●	●	●			
	7.2		●				●				●			●	●	●			
	7.3				●	●					●			●	●			●	
	7.4				●						●			●	●	●	●		
	7.5				●						●			●	●				●
Waldram	8.1		●				●			●	●			●	●				●
	8.2		●				●		●		●			●	●	●			
	8.3		●				●		●		●			●	●				●
	8.4		●					●			●			●	●	●	●		
	8.5		●					●			●			●	●	●	●		
	8.6				●						●					●			
Urban	9.1				●					●	●								
	9.2				●					●	●	●							
	9.3				●							●							
	9.4				●							●							
	9.5				●					●	●	●							

Reflected Component (IRC); Single-Stage (SS)
- The method's flexibility: This represents the method's ability to cope with complex geometrical configurations
- The principles of the method: The different physical principles on which each method is based: Lumen Method, Daylight Factor method, Flux Transfer, etc (see Chapter 8).

Other secondary, but still important procedures are often used to determine daylight reductions due to the use of fins, shelves, overhangs, venetian blinds and so on. The experimental nature of such procedures, and the fact that they are added to the other basic methods, make it inappropriate to include them in this review.

In addition to the methods to evaluate the illuminance, other methods have been developed in order to evaluate glare aspects. These aspects have been studied experimentally and related to the characteristics of the interior environment. These methods, which are referred to as Glare Control types, are useful methods of control for design choices, especially at advanced stages of the design process.

Comparison of Performance as Predicted by Different Design Tools

Daylight Factor components have been evaluated by comparing some simplified methods applied to a reference point (P) placed in a room, with two windows on two perpendicular walls (Figure 2). The geometric characteristics of the reference room are:
- height = 2.7 m
- floor surface = 42 m^2
- ceiling surface = 42 m^2
- window surfaces = 3 m^2
- wall surfaces = 67.2 m^2
- global envelope surface = 154.2 mv^2.

The reflectances are:
- windows = 20%
- ceiling = 70%
- walls = 50%
- floor = 15%.

The following average values were derived:
- average room reflectance is 45%
- average reflectance of the lower part of the room is 32%
- average reflectance of the upper part of the room is 63%

In order to compare the results provided by the methods which adopt only a uniform sky luminance distribution with those which assume a standard overcast sky luminance distribution, sky luminance ratios (Z-factors) were utilised (Table II).

A synthesis of the results obtained is reported in Table III.

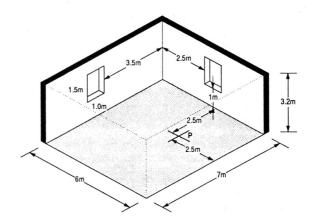

Figure 2 - Reference room.

TABLE II - Sky luminance ratios (Z-Factors) for conversion from uniform luminance sky to CIE standard overcast sky (3)

Average Angle of Elevation of Patch of Visible Sky (degrees)	Sky Luminance Ratio (Z-factor)
5	0.50
10	0.58
20	0.72
30	0.86
40	0.98
50	1.08
60	1.17
70	1.23
80	1.27
90	1.29

TABLE III - Comparison of direct component calculation methods

Method Number	Uniform Sky	Overcast Sky
8.4	1.8%	1.08%
5.7	-	1.25%
8.5	1.8%	1.08%
8.1	1.79%	1.08%
4.2	-	1.2%
7.2	-	0.9%
5.1	-	2.0%

The Direct Component values in Table III are obtained by applying some of the methods identified by the code number quoted in Appendix F. Table IV shows the values of the Internally Reflected Component (IRC), evaluated by using two of the methods which assume a CIE Standard Overcast Sky luminance distribution.

TABLE IV - Comparison of IRC calculation methods

Method Number	Overcast Sky
5.2	0.5%
1.5	0.39%

Some Single Stage methods are compared to those which directly evaluate the Daylight Factor (DF), in Table V.

TABLE V - Comparison of single stage methods

Method Number	D.F.
2.1 + 4.2	0.95%
2.1 + 5.9	0.94%
3.1 + 5.9	0.94%

All Single Stage methods give quite similar results for the reference room under consideration. This is due to the very simple (rectangular) shape of the windows. Windows of different shapes and dimensions could have been more effective in testing the flexibility of various methods, but more difficult to report.

The following observations can be made:
- As expected, the direct component values obtained with an overcast sky luminance distribution are much lower than those obtained with a uniform sky luminance distribution. This difference is due to the fact that the reference point P "sees" areas of sky near to the horizon which have a lower luminance in overcast conditions.
- The problem of an accurate and simple evaluation of the IRC within rooms of any shape still remains.
- Design tools utilising a single stage method, having been compared using a uniform luminance sky distribution as a reference, give very similar results. Nevertheless, it should be noted that the apparent regularity of the results does not correspond with precise spatial information.

User Evaluation

It is useful to know how architects control the impact of their design choices on visual comfort. To understand designers' attitudes towards daylight, questionnaires and interviews with well-known architects were carried out. One aim was to discover which design tools were used, and at what stage of the design process architects were using them in order to control daylight performance. The responses were firstly that all architects agreed on the importance of daylight in architecture but, secondly, that the emphasis is on its aesthetic consequences.

The most commonly used design tools used were:
- generative design tools, ie sketches of the internal environment, as horizontal or vertical sections, modelling shapes with "chiaroscuro" effects or shading with colours such as yellow and blue
- scale models, which are made by architects in order to control the whole building space composition as well as to understand the internal daylight distribution
- graphs, such as nomograms, and Waldram diagrams (Figure 3)
- mathematical simplified design tools, which require the designer to have a good scientific knowledge and a bias towards engineering.

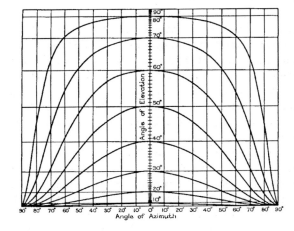

Figure 3 - Waldram diagram.

The results of the survey, carried out in southern European countries, show that only 15% of architects are concerned with daylighting issues, and are able to use adequate design tools. This is a relatively low percentage, which could be a consequence of insufficient information because daylighting issues were forgotten in the 1960s and 1970s (though they have now become important again). It is hoped that new practitioners will become more aware, due to better information provided in universities.

Daylighting issues are sometimes controlled by experts, who act as consultants for practitioners, in the same way that experts are consulted for other technical disciplines, such as structural calculations and the design of heating and cooling systems. These consultants often use computer simulation programs, but mainly for electric lighting systems design.

It must be borne in mind that it is very difficult and expensive for architects to rectify significant errors in daylighting design when the project has reached an advanced stage. Indeed, issues affecting building layout or orientation can only be addressed in future projects. It is therefore beneficial to consult experts at the early stages of design, if daylighting issues are complex or important.

Critical Comments

Although the design tools most commonly used to assess daylighting performance are drawings and scale models, it can also be very helpful to include in the design process some simplified design tools.

These simplified design tools are classified in ten categories. This allows for a more general discussion with the aim of helping the designer select those model types appropriate to the specific daylighting problem. More detailed comments and references for individual design tools are given in Appendix F.

Equations. The Daylight Factor, the SC or the IRC can be calculated by equations. These equations are based on basic physical assumptions, geometrical parameters and analytical procedures.

The evaluation of the Daylight Factor, assuming manual calculation, may be useful and relatively simple when determining a value for a single point, but it may be slow and tedious for several points. If there are a great number of points it is convenient to use the equations with an automatic data processing method. Many of the equations used in the analytical methods have been used for the development of other methods.

Single stage methods. These methods allow the Daylight Factor to be evaluated with a single stage process, without separately determining the Direct Component and the Internally Reflected Component. The main advantage of such a procedure is that it is possible to analyse rapidly the effects of a design solution for the daylighting of a room. One can evaluate whether a given size of window produces the required Daylight Factor, or conversely, the size of glazing necessary to achieve the required Daylight Factor.

The main disadvantage is that the value of the Daylight Factor, evaluated by these methods, is not a value on a single point but an average for the room. This means that these methods do not provide any information about the daylight distribution in the room.

Lumen Methods. The methods known as "Lumen Methods" are particular forms of Single Stage procedures. The original Lumen Method, from which these methods take their name, is based on a flux-transfer calculation used particularly in artificial light calculations. The first application to daylight calculations was developed by Frühling in 1928. The later development of this method allows the Daylight Factor, at three points of the room placed at different distances from the window, to be evaluated. In this way more significant information on the daylight distribution in the room is obtained. However, the advantages and disadvantages of this method are the same as those of other Single Stage procedures.

Tables. The tabular methods are based on the evaluation of the Daylight Factor, or of the Sky Component, over an extensive range of geometrical parameters that describe the window and the position of the reference point. Results are presented in tables allowing simple and rapid access. The tables also allow more complex cases to be considered by means of sums and differences of values corresponding to the geometry of a given configuration. It is possible to interpolate for values of parameters not included in the tables.

Nomograms. The nomograms are useful graphic tools for the rapid evaluation of the Daylight Factor, SC or IRC Generally, they provide the value in one or two simple stages, from basic parameters such as the dimensions of the room or window.

Nomograms are easy to use, but their accuracy may be insufficient in complex cases where the shape of the room or window is not rectangular. Nomograms are particularly useful in the initial stages of the design process.

Protractors. Manual methods, such as the protractor, may be particularly useful to evaluate daylight illumination. The use of protractors in elementary daylight situations is quickly learned, although refinements in technique are possible. Generally, protractors are versatile manual tools which allow good accuracy and provide point values.

Dot diagrams. The dot diagram method uses a graphical procedure based on the stereographic projection of the sky vault. This type of method has been adopted in many countries such as Australia, South Africa, Portugal, USA, Great Britain and Sweden.

The particularity of these methods is that the required value, normally the Sky Component, can be evaluated by counting the dots, drawn on a suitable mask, seen through a projection of the window.

These methods are accurate and can be used for cases with complex geometry. The operation of counting the dots may be tedious if the window is large.

Waldram diagrams. The basic method, developed by P.J. and J.M. Waldram in 1923, has been in continual use for at least 60 years and has advantages which are universally acknowledged (see also Chapter 1).

The Waldram Diagram is a graphical method that allows vertical and horizontal edges of windows and buildings to be plotted on the diagram. This diagram permits the area of the window to be represented in proportion to direct daylight at a reference point. The accuracy is very good and it is possible to consider very complex cases.

Methods of urban analysis. These are methods that allow the evaluation of the effects of the urban configuration on the daylight of a building, room or a single point. All methods that calculate the Direct Component may be used for this purpose.

A group of methods which specifically analyse the influence of the urban or building configuration on the daylight have been collected in Appendix F.

Glare control. Glare is a subjective phenomenon that is the result of unwanted light in the visual field, and it is usually caused by the presence of one or more artificial or natural sources of excessively bright light (see Chapter 2).

Glare, due to daylight, may be caused by a particular architectural decision, and may become critical when certain viewing conditions occur. Some simplified design tools are available to determine if and when these conditions occur and to help find a design solution to the problem.

REVIEW OF EXISTING COMPUTER CODES

A survey of some of the most interesting existing computer programs dealing with daylighting aspects is presented in Appendix G.

The programs are grouped into two categories, depending on the types of computers for which they have been developed. The thirteen programs in the first group run on micro-computers, while the six in the second group run on mini-computers or on mainframe computers. However, the rapid development of micro-computers will probably make it possible, in the near future, to run even the most complex programs on desk-top computers.

The 19 programs presented represent only a sample of the numerous products that are available. The selection has been based on the availability of information, and on the willingness of the authors to answer the questionnaires sent to them.

The information on the programs is presented both in individual forms and in summary tables. Individual forms contain information about authors, hardware and software, and modelling and reporting features, for each program. Summary tables present the same information for all programs. References are listed at the end of Appendix G.

COMPARISON AND VALIDATION

Comparison

The question most often raised regarding these tools is that of validity. When a designer uses them, it is important to know what level of confidence can be given to the results.

Light penetration in a real building depends on:
(a) type of sky (overcast, clear, without sun, etc)
(b) exterior environment (vegetation, buildings, obstructions)
(c) material properties (indoor and outdoor surface finishes)
(d) building geometry
(e) furniture.

Using a tool such as a scale model or a computer program means that some choices have to be made. in relation to these factors.

For example, with scale models:

Item (a), (skies) requires either an artificial sky (reference overcast sky) or well identified real skies. To identify a real sky, one needs to evaluate the luminance distribution which involves precise equipment and leads to a large number of parameters of little use. One can also use the half sky technique (4) which permits a quick correlation of daylight factors with parameters which define the sky "non-uniformity". Item (b), is difficult to simulate since it involves the simulation of the entire environment. Items (c) and (d) are the easiest on scale models: real building materials can be used and scale models can have the exact complex shape of the real building. For furniture simulation, (e), it is just a matter of finding or making models of furniture at the right scale.

In the case of computer tools, the analytical solving of the problem of the light penetration process requires a selection of algorithms to simulate the light source, item (a), the environment (b), materials (c), building geometry (d), and furniture (e). Each of the tools which have been presented assume a well-defined and ideal type of sky (a), usually a CIE standard overcast or clear sky. Simulation of the environment is simplified (eg to consider only obstructions) or may involve the computation of the luminances of all outdoor surfaces in a simplified manner. Material photometry is assumed to be perfectly diffusing, except in ray-tracing programs, photon generation and some radiosity algorithms, which can simulate specular surfaces. Intermediate surfaces (glossy, smooth) are simulated by the most powerful programs, but they require the knowledge of the photometric properties of the material which is being simulated. Items (d) and (e) are approached in a simplified manner. With the large majority of programs, the room shape must be rather simple and borrowed light systems cannot be simulated. With more sophisticated programs (ray-tracing, radiosity, photon generation), only the computing time as well as the time spent by the operator impose limits on the complexity of items (d) and (e).

Validation

Using a tool (computer or scale model) involves the understanding of the difference between what is simulated and reality. There are two levels of validation:
A   Checking whether illuminances are the same as in a real building where everything is similar to what has been simulated. Due to the limitations of tools, such a building is typically a test cell under an artificial sky.

B   Prediction of the daylighting performance of a real building for real climatic conditions. This requires adjustment of the results of simulation and a statistical definition of the performance. Here, the role of a daylighting expert is essential. Within the EC sponsored European Concerted Action on Daylighting, validation type A was carried out to compare design tools. The room was rectangular and two apertures were employed: one in the roof and the other in the façade. Results were very close and allowed the identification of the influence of surface finishes and construction details.

A user or a client of a design tool should remember that only type A validations have been conducted on design tools. Therefore, the results of these tools must be interpreted by designers in the light of their experience.

REFERENCES

(1) Schiler, M., "Simulating Daylight with Architectural Models", DNNA, Univ. of Southern California, LA.

(2) Navvab, M., "New Developments in Scale Model Photometry Techniques for Interior Lighting", *Proc. Conference of the Daylight and Solar Radiation Measurement*, Berlin, October 9-11, 1989.

(3) Hopkinson, R.G., Petherbridge, P. and Longmore, J., *Daylighting*, Heinemann, London, 1966.

(4) Fontoynont, M., "Daylight availability as a function of irradiation data", *Proc. Conference Lux-Europa*, Lausanne, 1985.

Chapter 10
INTEGRATED ENERGY USE ANALYSIS

INTRODUCTION

It is now recognised that artificial lighting can be a major energy consumer in some buildings. Depending on the types of lamps installed, a greater or lesser fraction of the lighting load is converted to heat and thus affects the heating or cooling load of the building. With the shift towards energy efficient design solutions, it has become important to consider the integrated management of the visual and thermal requirements, taking into account the luminous and thermal implications of daylighting strategies.

If daylighting strategies are to be accepted and implemented by the design professions, there is a need to establish those criteria, in terms of fenestration design, that have the potential to provide the requisite benefits. In general terms these benefits can be easily enumerated. Daylighting strategies can:
- provide an improved internal luminous environment with attendant benefits in visual comfort and hence productivity
- reduce energy demand for lighting
- reduce cooling loads, influencing HVAC (ventilation and air conditioning) plant size and peak electrical demand.

Not all strategies can optimise the relationship between these two areas, and possible trade-offs can only be resolved by assessing the potential cost benefits against the client's needs. To avoid design failures it is necessary to resolve some fundamental questions.
- What design strategies can be proven effective?
- What quantifiable benefits can be achieved?
- How can designers come to understand why such a strategy works?

The evolution of innovative designs now owes much to the use of advanced computer design tools but, in this case, the needs of an environmental engineer fall into the gap between the domains of energy and lighting simulations. What is needed is an integrated appraisal tool that combines the strengths of current state-of-the-art methodologies without compromising their integrity in this new context.

Two models feature in this chapter, and a third manual method is briefly mentioned, each applicable to a different stage in the design cycle. One method is based on monthly averages of daylight availability, and is useful in an indicative mode (HEATLUX), the other requires hourly values of climatic data, and functions in a deterministic manner (ESP). A simplified manual method integrates the lighting and thermal (LT) consequences of changing the glazing ratio, for a limited number of situations and European climate types.

EXAMPLE 1: ESP, ENERGY ANALYSIS WITH DAYLIGHT INTEGRATION

This example describes a methodology that enables the integration of daylight and thermal evaluation tools. The method employed is designed to function in a deterministic manner; that is, it has enough sensitivity to the parameters of any problem to enable a user to determine the cause and effect of any change in the design variables.

The scale of the problem is such that it was considered more efficient to implement a daylighting model outside the main thermal model and communicate by means of a data link, a file of predetermined coefficients, the contents of which are then utilised at run time to generate the desired variables. This methodology will allow a user to experiment with all the usual daylight design variables and, in addition, to study the implementation of integrated control strategies and their effect on both the visual and thermal environment.

Modelling Methodologies

Despite the growing acknowledgement that it is inappropriate to dissociate the subsets of the visual, thermal and psychological requirements of the overall design concept, it is apparent that these criteria can only be readily assessed using advanced computer simulation techniques. Numerous existing simulation programs have the flexibility to allow the modelling of a range of architectural design solutions and their interaction with varied climatic conditions, but few have the capability to operate in both the thermal and visual domains. The major difficulty inherent in attempting an integrated evaluation is the computational burden and its attendant time-scale. This burden must be defrayed by some method that provides a means of balancing the cost/performance ratio and yet does not compromise the integrity of the solution.

At the design stage it must be realised that the design of an optimal component specification, in global terms, is necessarily a trade-off between the requirements of optical and thermal sub-systems. An ideal working scenario would allow the integrated appraisal of conditions as, and when, any of the optical or thermal parameters were varied. Logically this would lead to the asynchronous processing of both domains with shared routines for calculating both illuminance and irradiance. Although there are many undeniable benefits implicit in this approach, the product would be a monolithic program with attendant problems in implementation and usage. Retrofitting an existing code in this manner would require fundamental changes, both in terms of data structures and the flow of logic, so such benefits that are realised would be diminished by the problems incurred. However, bearing in mind the level of overheads that a concurrent thermal and optical simulation would incur, it is more probable that each sub-system would be optimised in isolation and a cost-benefit analysis would be performed to assess the integrated performance. To this end the designer would apply the appropriate tool to a single design concept, varying the appropriate parameters to optimise the system with regard to the focus of the simulation. Suitable existing simulation programs already have the capabilities required for this method of evaluation but the appraisal is complicated by the variety of roles that a window must perform. For example, fenestration must provide both access to the external visual environment and at other times privacy for the occupants. This dynamic behaviour is continued in its thermal role with the need to both admit and exclude solar radiation, control ventilation, influence thermal comfort and, similarly, provide acoustic attenuation.

It soon becomes apparent that the control of fenestration is a function of the base design criteria of the enclosure/glazing system and the occupant reaction to its performance under dynamic boundary conditions. It is this dynamic aspect that is both the answer to the design problem and the problem in finding the answer.

Criteria for Model Choice

The dynamics of heat transfer, being fundamental to the evaluation of building energy, have long been incorporated into thermal simulation programs but in the case of lighting analysis this is rarely the case. Historically, daylighting analysis models have avoided the need to account for dynamic variations in climatic conditions by utilising sky models that do not vary with solar position and by neglecting the contribution of direct solar illumination.

However with the advent of new computer codes this is now no longer the case, although in the current machine environment it is still impossible to realise a "time series" simulation capability. The calculation of internal illuminance, especially in the case of geometries with partially occluded views, is an unwieldy problem.

Reflection algorithms, usually based on finite element techniques, are inherently CPU (central processing unit) time consuming. Therefore there is an apparent need for a methodology that is computationally efficient, yet both simple and practical.

The difficulties in re-calculating the internal illumination at every time-step to account for the rapid changes in sky luminosity, due to the change of seasons and variation in weather patterns, preclude the use of "standard" state of the art illumination models being used in a dynamic mode.

The choice of a lighting model for use in a dynamic scenario depends on the trade-offs between a number of factors such as accuracy, compatibility and computational expense.

There are perhaps three main modelling techniques that show immediate potential in this field of application. The first of these is the Lumen Method, (also known as the Libbey Owens Ford Method, see Chapters 8, 9) developed by the Illuminating Engineering Society and originally aimed at the field of artificial lighting design.

The second is the Daylight Coefficient concept (1), the coefficients being mathematical functions relating the luminance distribution of the sky to the illuminance at a given point in a room. The third is a method of using pre-processed Daylight Factors calculated for a range of climatic variables.

Note that each of these methodologies effectively relates the internal luminous conditions to some measure of those external conditions that can be readily obtained from meteorological data.

The reduction of the processes of transmission, attenuation and interreflection into a set of coefficients appears to hold the key to a computationally efficient method of producing time-series internal illuminance values.

The method of pre-calculated daylight factors can provide the link between virtually any lighting model and any thermal model and since it is a means of integrating results, rather than calculation procedures, offers a framework of great flexibility.

Daylight Factor Method

The amount of daylight that illuminates any room surface can be assumed to be a function of two factors, the geometric form and material properties of the surrounding surfaces and the luminance levels and distribution of the sky. These two factors can, given the assumptions common to this type of calculation, be assumed to be independent. For any one instance, that factor due to the built environment can be considered constant, in which case for a given luminance distribution, any increase in sky brightness will produce a proportional increase in internal illumination.

This approach gave rise to the concept of a daylight factor, this being the instantaneous ratio of the internal illumination at the measurement point, to the external illumination on the horizontal plane under the unobstructed sky hemisphere. A daylight factor, for any given sky model, can then be used to find the internal value corresponding to any level of sky brightness purely through multiplication of this factor by the current external horizontal illumination. Obviously the accuracy of the result is entirely dependent on the initial accuracy of the daylight factor calculation, but the method has the potential to recover instantly internal values that would otherwise have to be laboriously recalculated. In this manner the illumination at any point, or points, in an enclosure can be rapidly and accurately calculated.

Validation studies on the methodology (2) reveal an accuracy to within 30% for any one hour. This may not be enough to guarantee quantified levels of savings but the technique is undoubtedly viable for the identification of trends and shows enough sensitivity to the design parameters to isolate the cause and effect of any design changes.

Using a lighting model in a pre-processing mode to calculate daylight factors representing point or average surface illuminances appears to offer the greatest benefits in terms of integration and computational ease with the least run time overheads.

Time Series Internal Illumination Values

Since the calculation of internal lighting levels is such a computationally expensive task, it can be beneficial to reduce this burden by preprocessing the data before entering the thermal simulation. This can be achieved by effectively separating the geometrical processing from the energetic calculation. Considering a daylight factor as independent of the ambient external level of illumination and as being entirely representative of the geometrical constraints involved, these factors can be precalculated and stored for later use.

Within the set of physical conditions assumed valid for the majority of daylighting calculations, it is accepted that light energy emanating from a given point, or distribution on some surface, reaches a reference point with no attenuation due to capacity or storage at any intervening element. This permits the assumption that the ratio of light, between source and sink, is purely a function of the geometry of the environment and as a result the illumination at the reference point is directly proportional to the luminance of the source.

A simple algorithm can be devised that utilises a single factor combined with external illumination data calculated from solar irradiance to obtain time-related internal illumination values.

If only a single factor is used, the result can only be applicable to a source of a static distribution and cannot then account for the temporal variations in sky luminance patterns or the presence of direct sunlight. To cater for the variation in weather conditions it is necessary to provide an appropriate factor for each increment in the solar position and associated change in distribution.

In general, these daylight factors could be obtained from computer simulations, hand calculations or through experimentation on scale models.

A variation on this theme has been used in the energy simulation programs DOE-2 and NBSLD (3). In this approach weather conditions are treated as a linear combination of clear and overcast skies, and daylight factors are calculated for each sky type over a range of solar positions. At every time step the external horizontal illuminance, from sun and sky, can be determined from the irradiance values in a meteorological database, the corresponding internal illumination then being found by interpolating between the daylight factors obtained from the preprocessor. This methodology will be examined in detail.

Dynamic Sky Modelling

If it is assumed that a single daylight factor is representative of the lighting conditions produced by a unique sky luminance distribution, then different factors must be calculated for each change in this distribution. There are a number of functions that model the sky distribution for conditions varying between clear and overcast.

No one model can emulate the time-related characteristics of the dynamic variation of a partially cloud-covered sky, so this condition must be obtained by relating some perceived distribution to an analytical combination of theoretical sky types such as those discussed in Chapter 3.

As can be seen from the formulae given in that chapter, the luminance distribution of an overcast sky is presumed to be a function only of the altitude of the visible sky element, while the clear sky varies as a function of the solar altitude and azimuth. For this case the distribution will change at every time step, necessitating a new factor for each increment. This will mean that a a clear sky must be represented by a set of factors corresponding to all solar positions over the simulation period, while a single factor represents an overcast condition irrespective of time.

This set of factors relating to the clear condition can be calculated by sampling days throughout the year, say every hour in the first day in each month. This however proves to be unnecessarily complex since, over a year, the sun traces much the same path as it progresses between the winter and summer zeniths.

A more economical method is to treat time as some set of solar positions defined by the altitude and azimuth of the sun. By determining the maximum and minimum values of azimuth and altitude, that sector of the sky occupied by the sun over the simulation period can be discretised to provide the set of factors required.

To relate these factors to any given sky condition it is assumed that the sky, at any given hour, can be subdivided into two fractions, one corresponding to that area of sky with a clear distribution and the remainder being given over to an overcast distribution.

Alternatively, intermediate sky models can be provided to represent partially clear or partially cloudy conditions, but in this application any additional accuracy is offset by the increased computation time required.

Direct Sunlight

In a similar manner an analogous factor can be calculated that relates the presence of direct beam illumination to its contribution to the internal illumination.

This factor must be calculated for the same set of solar positions as were used to describe the range of clear sky luminance distributions.

The Nebulosity Index

The relative amounts of clear and overcast sky can be related to some parameter obtained from the meteorological database. This parameter, the Nebulosity Index, C_i, is rarely available as a measured value but can be approximated from solar irradiance data. There is a number of extant models for the Nebulosity Index such as the CSTB (Centre Scientifique et Technique du Bâtiment) model introduced in Chapter 3.

The index may be calculated from the current hour ratio of diffuse to global irradiance C_r, this ratio being used either as the direct equivalent of the cloud index, or the index can be expressed as some function of the cloud ratio. The CSTB methodology scales the cloud ratio by a factor derived from an empirically calculated theoretical cloud ratio:

$$C_i = (1 - C_r) / [1 - 0.12 (\sin\varphi_s)^{-0.82}]$$

where:
C_i is the Nebulosity Index
C_r is the hour ratio of diffuse to global irradiance
φ_s is the solar altitude

Basing the cloud ratio on hourly values of irradiance leads to the erroneous assumption that the sky always tends to be overcast at dawn and dusk due to refraction in the atmosphere causing diffuse illuminance to lead the direct at dawn and lag the direct at dusk.

Other methods have been proposed, ie expressing this ratio as a function of extraterrestrial irradiance, but this is a topic for future research.

Internal Illuminance

To use these factors to recover the current hour internal illuminance values, it is necessary to multiply the relevant factors by the current hour external illuminance. The values contained in a standard meteorological database may contain illuminance data or, as is more probable, values of diffuse and direct irradiance. If the latter is the case, then these values can be converted to illuminance by utilising the appropriate efficacy of the source. Given that three potential sources exist - direct beam, clear sky diffuse and overcast sky diffuse - the appropriate value of the efficacy can be found from an empirical formula. For the clear sky case, research (4) has shown that values of diffuse illuminance from a clear sky do not vary, to any great extent, as a function of the solar position and hence a single value will suffice. In the extreme overcast case, studies have shown that the efficacy does vary with solar position and thus an expression such as that given below must be used to relate the efficacy to the current hour solar altitude:

$$K_{oc} = (91.2 + 0.702\varphi_s - 0.0063\varphi_s^2) \times (1.22 - 7.96e^{-4} \Delta + 7.96e^{-7} \Delta^2)$$

where:
K_{oc} is the overcast sky efficacy
φ_s is the solar altitude
Δ is given by:

$$\Delta = I_g / \sin(\varphi_s \times \pi/180)$$

where:
I_g is global irradiance

Similarly expressions exist that define the efficacy of direct solar irradiance as a function of solar altitude:

$$K_{sun} = 17.72 + 4.4585\varphi_s - 8.7563 \times 10^{-2} \varphi_s^2 + 7.3948 \times 10^{-4} \varphi_s^3 - 2.167 \times 10^{-6} \varphi_s^4 - 8.4132 \times 10^{-10} \varphi_s^5$$

where:
K_{sun} is the efficacy of direct sunlight

The total external illuminance can be calculated as the sum of the direct beam component and that part due to the the ratio of clear to overcast diffuse sky components:

$$E_{ext} = I_d K_d + C_i (I_{dif} K_{cl}) + (1 - C_i) I_{dif} K_{oc}$$

where:
E_{ext} is the external illuminance
I_d is the direct irradiance
I_{dif} is the diffuse irradiance
K_d is the efficacy of diffuse sky
K_{cl} is the efficacy of clear sky

Recovery of internal values is simply a matter of inserting the relevant factors in the above expression:

$$E_{in} = I_d K_d DF_d + C_i (I_{dif} K_{cl} DF_{cl}) + (1 - C_i) I_{dif} K_{oc} DF_{oc}$$

where:
E_{in} is the internal illuminance
DF_d is the direct / sunlight factor
DF_{cl} is the clear sky factor
DF_{oc} is the overcast sky factor

Using this method it can be shown that it is possible to relate time series internal illuminance to the external luminous environment using data contained in the thermal model's climatic data base.

These internal values can then be used as the sensed condition required as the stimulus for control laws governing some switching strategy.

Controls

Any building control system features three functional elements; a sensor, a controller and an actuator. The versatility and efficacy of a system, whether in reality or in a simulation, is dependent on the range and scope of this implementation. Within ESP there exists a user-orientated facility that enables system states to be modified as a function of time or state variable tests (5). These elements are described below.

The sensor. A typical lighting control photosensor consists of a silicon photodiode in a housing whose geometry determines the directional sensitivity to the ambient lighting. The output of such a device, when suitably amplified, is considered proportional to the light impinging on its surface.

Within ESP the sensor is conceived as a simple photocell, able to measure illuminance at any predefined point or plane.

The sensed variable can be derived from a number of sources, thus allowing a measure of flexibility. There are three possible entry points for relevant data:
- A casual gain profile, time series values of zone casual gains, can be read from file and used to control zone flux input over the control period (this presupposes the availability of a means of capturing the relevant data).
- A run time calculation performed by an implementation of the NATLIT algorithm (6). This generates daylight factors, distributed on a user-defined plane, in relation to each window of the zone. Only diffuse values of illumination are used unless the solar position occurs within the solid angle subtended at the reference point and bounded by the dimensions of the window. This allows for the effect of room orientation and reduces the limitations of using only a standard overcast sky model.
- Finally there is the facility to use a sophisticated lighting model in a preprocessing mode to generate daylight factors relating to surface and window illuminance under a range of sky conditions, thus allowing the use of more sophisticated simulation scenarios and control regimes as previously described. This is necessitated by the need to parallel accurately the level of integrity provided by the thermal simulation as represented by ESP.

Most lighting simulation programs are restricted to certain forms of geometry description. ESP, however, is capable of accepting an entirely general description of building geometry. This has necessitated the development of a new daylighting model limited, in terms of geometry, only by the requirement that all objects are bounded by planar polygons. Each polygon can be any shape and defined by any number of vertices. This geometrical description is integrated with the computational methodology from a rigorously validated model, (7) SUPERLITE, providing a complementary degree of geometrical rigour to that embodied within ESP.

Control laws. A controller is used to govern the actuator signal based on the sensed condition. In practice a lighting control algorithm consists of a reference circuit in the control logic that supplies a voltage against which the incoming signal from the photodiode is compared.

Within ESP, controllers can govern either time or state variables. To control on the basis of time, control day types are defined during which some specified control regime is in force. Any number of day types can be specified in terms of a start and finish date. This allows many control possibilities; seasonal requiring four day types, monthly requiring 12, weekly requiring 52 and daily requiring 365. If the number of day types is given as 0 this implies the special case of weekday/Saturday/Sunday control in which only three day types are active corresponding to weekdays, Saturdays and Sundays.

Day types are then subdivided into distinct periods during which the control action is fixed. For each period a control type is specified, referencing a particular control law. This defines a control algorithm which throughout the period will represent the logic of some controller (real or imaginary).

Photosensor control. A typical control photosensor consists of a silicon photodiode in a housing whose geometry determines the directional sensitivity to the ambient lighting. The output of such a device, when suitably amplified, is considered proportional to the light impinging on its surface.

The signal from the control photosensor $S_T(t)$ can be written as follows:

$$S_T(t) = \int L_t(\Omega,t) R(\Omega) d\Omega$$

where:
$L_t(\Omega,t)$ is the spatial illuminance distribution in the enclosure at some time t.
$R(\Omega)$ is the responsivity of the sensor within the solid angle Ω.

The equation can be rewritten to reference a daylight component L_d and an electric light component L_e.

$$S_T(t) = \int L_d(\Omega,t) R(\Omega) d\Omega + \int L_e(\Omega,t) R(\Omega) d\Omega$$

The first term in this expression relates to the portion due solely to daylight and the second to that due to electric light. Replacing these terms with the nomenclature $S_d(t)$ and $S_e(t)$ referring to output signals and substituting in the above gives the following expression:

$$S_T(t) = S_d(t) + S_e(t)$$

Assuming that all the electric lights dim uniformly with respect to each other, this can be written as the product of a time-dependent function and a time-independent constant:

$$S_e(t) = \delta(t) \int L_{emax} R(\Omega) d\Omega$$

where:
L_{emax} is the illuminance of the electrical installation at 100% output
$\delta(t)$ is the fractional dimming level, assumed to vary between 0 and 1.

The integrand is time-independent and will reduce to a constant S_{emax}, the signal generated when the photosensor is exposed only to the electric light at full intensity.

Combining these equations yields an expression that relates the fractional dimming level to the total signal generated by the photosensor and to that portion due only to daylight:

$$\delta(t) = [S_T(t) - S_d(t)] / S_{emax}$$

Assuming that the response of the photosensor is linearly proportional to the amount of impinging light, this expression becomes:

$$\Omega L(t) = [L_T(t) - L_d(t)] / L_{emax}$$

If the system is calibrated at night, as is general practice, and the electrical installation provides the design illuminance, then this can be given as:

$$\Omega L(t) = [L_{emax} - L_d(t)] / L_{emax}$$

within the limits of:
$0 \leq L_d(t) \leq L_{emax}$

This equation holds only for the range of values indicated. Above L_{emax} the system switches the lights off. The fractional dimming level can then be related to the installed power, in W/m² of floor area for the electrical system.

Occupant Interaction

Any simulation of an automated control strategy provides only limited information if the interaction of the occupants is ignored. Many installations do not return the predicted benefits because the users of the space override the control system to adjust conditions to suit their own comfort requirements.

There are two readily available methods of adjusting the simulated environment in response to the predicted occupant interaction. The first relates some measure of the ambient conditions to the probability that the occupants will switch on the lights at the start of the occupied period and the second models the occupants' response as a function of visual comfort.

Probability switching function. Work at the Building Research Establishment, UK (BRE) has resulted in a method of predicting artificial lighting use based on observed patterns of behaviour over a series of field studies (8).

Data were collected from a number of schools and offices and it was found that for a manually controlled space the lights were usually either all on, or all off, and that most switching activity is confined to the extremes of a period of occupation.

The probability of someone switching on the lights within a room was found to be related to the daylight illuminance at the darkest point on the working plane.

A probability function was derived from this field data that expressed the probability of the lights being switched on as a function of the minimum daylight illuminance on the working plane.

This probability index is given by P, where:

$$P = (-0.0175 + 1.0361) / [1.0 + e^{(4.0835 \times \log E_{min}) - 1.8223}]$$

where:
E_{min} is the minimum daylight illuminance on working plane

The probability returned by the above equation can be constrained between the values of 0.0 and 1.0

and compared to a random number generated within the same range. If the lights are already off and the switching probability is greater than the random number, then the lights are deemed to be switched on.

Glare index calculation. The assessment of discomfort glare from windows is based on the contrast between the luminosity of the window and the surrounding surfaces. There is a number of existing parameterisations available, one of which, based on the work of Hopkinson, is used in DOE-2 (3). This formula returns the discomfort glare constant G as a function of the window and window surround luminances, the solid angle of the window, and a modified solid angle taking into account the principal direction of view of the occupant:

$$G = 0.48\ L_w^{1.6}\ \Omega^{0.8} / (L_b + 0.07\omega^{0.5}\ L_w)$$

where :
Ω is the modified solid angle
ω is the unmodified solid angle
L_w is the window luminance
L_b is the background luminance

Dividing the window into N by M elements gives:

$$L_w = \frac{1}{NM} \sum_{j=1}^{N} \sum_{i=1}^{M} L_w(i, j)$$

and similarly:

$$\omega = \sum_{j=1}^{N} \sum_{i=1}^{M} d\omega(i, j)$$

and the modified solid angle is:

$$\Omega = \sum_{j=1}^{N} \sum_{i=1}^{M} d\omega(i, j)\ p(i, j)$$

The values L_w, the window luminance, and L_b, the background luminance, are obtained from daylight factors in a similar manner to those related to the sensed illuminance. L_b is taken as the larger of the internally reflected components, or E_{set} the set point illuminance.

Values of the position factor p, given in Table I, are empirically determined to allow for the effect of the displacement of the principal line of sight in relation to the windows.

The net glare at the reference point can then be found from the expression:

$$GI = 10 \cdot \log_{10} \sum_{i=1}^{N_w} G$$

As the glare index, GI, is a function of the weighted ratio of the window and background illuminance, then if the predicted glare index exceeds some pre-defined value this can be used to trigger the deployment of movable window shades, thus reducing the window luminance. Alternatively, the background illuminance could be increased by increasing the provision of artificial lighting, again, until the glare index is returned to the comfort value.

Actuator

An actuator allows some system state to be changed over time. In theory an actuator could switch between banks of incandescent lamps, but commercially available systems concentrate on high-frequency ballasted systems capable of being dimmed efficiently over a large dynamic range with minimal shifts in spectral distribution.

Any artificial lighting installation generates casual gains with a characteristic radiant and convective split. ESP has the ability to define and control those casual gains that are identified as being linked to lighting and modify this flux at each time step in response to the selected control function.

Similarly, ESP has the facility to deploy blinds, shading devices or window insulation as a function of time, temperature or incident solar radiation. Each change in the perceived window properties is predefined in the Zone Blind/Shutter control file. Then for each control period, this window definition is accepted only if the sensed variable is greater than the specified actuation point.

The specified control scheme relates to each and every window in a zone but note that, at any one point in time, different window arrangements may be in place depending on the incident radiation levels. If it is necessary to impose more than one control scheme on a zone, it would first be necessary to introduce additional zoning so that each sub-zone can still possess a unique window control file. This implies that a unique set of precalculated factors would be available for each combination of shaded and unshaded windows. In the case of occupant-perceived glare, windows could be sequentially shaded until the glare index drops below the stipulated set point. In this manner, control on the basis of visual stimulae affects the thermal environment and control on thermal stimulae affects the visual environment, thus providing the integrated evaluation of both domains.

Daylighting Report Generation

While users can readily take advantage of the processing power of computers they must also constantly question the validity of the response. They must not rely only on the knowledge available in, or derived by, the computer system.

While some responsibility for accuracy always resides with the author of the software, the responsibility for the design must be with the designer. The theoretical and practical aspects of the processes involved and be must understood and the software must be operated within the limitations of the theory employed. To this end, a range of output options are provided that illustrate different levels of information on daylight availability.

This information covers both the internal and external luminous climate and also provides feedback on the usage of artificial lighting as a function of the control strategy employed.

Table II reports the hourly calculated cloud cover in tenths, the current hour solar altitude and azimuth in degrees, and the measured values of direct and diffuse solar irradiance obtained from the meteorological database.

From these data the calculated illumination due to the direct beam and diffuse sky, on an external horizontal plane, is also presented.

The percentage of daylight hours during which a minimum illuminance level, due to diffuse sky illumination, is exceeded over an annual period is tabulated in Table III. These values would appear to agree with published data showing a peak of between 35 and 40 Klux during the midsummer months.

Tables IV and V present similar data related to internal conditions. Hourly values of internal illumination due to daylight at the reference point are given in Table IV along with the percentage of artificial light required and the predicted glare index.

Table V shows the percentage of daylight hours during which a minimum value of internal illumination, at the reference point, is exceeded.

Tables VI to IX relate the percentage hourly energy usage due to the artificial lighting installation when operated in conjunction with an integrated daylighting control system. Note the distribution of usage as a function of the relative position of the sensor location. As the solar altitude increases, the penetration of daylight is reduced: this leads to a reduction of available daylight at the back of the zone and hence an increased requirement for additional artificial light.

An annual, averaged summary, of these results is given in Table X. This shows that, on the basis of energy saving, linear dimming is the most efficient. In this example, leaving the electric lighting under the control of the occupants is predicted to be more efficient than either set point or step down control regimes.

It is hoped that through the use of this methodology designers may find some of the answers to the questions posed in the Introduction to this chapter.

The ability to simulate the integrated performance of a design, quantify the net benefits and come to some understanding of why the design performs in the way in which it does, must decrease the chance of design failures.

TABLE I - Factor, p, that modifies the solid angle of a window subtended at a reference point, with respect to the vertical and horizontal line of sight of an occupant

HORIZONTAL	0	26	45	56	63	72	>72
VERTICAL 0	1.0	0.492	0.226	0.128	0.081	0.057	0.0
26	0.123	0.119	0.065	0.043	0.029	0.023	0.0
45	0.019	0.026	0.019	0.016	0.014	0.011	0.0
56	0.008	0.008	0.008	0.008	0.006	0.006	0.0
63	0.0	0.0	0.003	0.003	0.003	0.003	0.0
>63	0.0	0.0	0.0	0.0	0.0	0.0	0.0

TABLE II - Hourly values of external daylight availability

HHDDMM	CLOUD COV	ALT DEG	AZI DEG	DIR RAD	DIF RAD	SUN LUX	SKY LUX
8 4 1	*	0.	0.	0.	0.	0.	0.
9 4 1	10	5.7	139.1	0.	18.0	0.	1885.9
10 4 1	10	11.0	152.0	0	69.0	0.	7019.0
11 4 1	8	14.4	165.7	32.2	82.0	2120.7	9347.9
12 4 1	7	15.5	180.0	43.1	89.0	2947.4	10338.2
13 4 1	10	14.4	194.3	0.	91.0	0.	9417.5
14 4 1	10	11.0	208.0	0.	41.0	0.	4369.8
15 4 1	10	5.7	220.9	0.	15.0	0.	1594.4
16 4 1	*	0.	0.	0.	0.	0.	0.

TABLE III - Percentage of daylight hours that illuminance level is exceeded: climate data from ESP Test Climate

ILLUM:	0	5	10	15	20	25	30	35	40	45	50	KLUX
JAN	100	62	13	0	0	0	0	0	0	0	0	%
FEB	100	76	50	4	0	0	0	0	0	0	0	%
MAR	100	88	74	52	7	1	0	0	0	0	0	%
APR	100	86	73	56	22	7	0	0	0	0	0	%
MAY	100	87	76	64	45	15	3	0	0	0	0	%
JUN	100	92	84	70	49	30	10	5	0	0	0	%
JLY	100	90	81	67	48	29	9	3	0	0	0	%
AUG	100	81	72	58	35	16	2	0	0	0	0	%
SPT	100	65	54	40	6	9	0	0	0	0	0	%
OCT	100	55	43	24	4	0	0	0	0	0	0	%
NOV	100	37	13	0	0	0	0	0	0	0	0	%
DEC	100	27	2	0	0	0	0	0	0	0	0	%

TABLE IV - Hourly values of internal illuminance at reference point: climate data from ESP Test Climate

HH	DD	MM	INT LUX	% USE	GLARE INDEX	SET POINT
8	4	1	0.	100	0.	250
9	4	1	31.20	87	10.467	250
10	4	1	116.13	53	16.754	250
11	4	1	129.65	48	17.719	250
12	4	1	145.76	42	18.243	250
13	4	1	155.81	38	17.917	250
14	4	1	72.30	71	14.699	250
15	4	1	26.38	89	9.535	250
16	4	1	0.	100	0.	250

TABLE V - Percentage of daylight hours that a minimum illuminance level is exceeded: climate data from ESP Test Climate

ILLUM:	0	50	100	150	200	250	300	350	400	450	500	LUX
JAN	100	77	53	11	2	0	0	0	0	0	0	%
FEB	100	86	71	44	10	1	0	0	0	0	0	%
MAR	100	95	86	47	17	13	11	6	2	0	0	%
APR	100	94	88	63	34	27	20	10	2	2	0	%
MAY	100	100	98	26	17	15	14	13	7	5	0	%
JUN	100	100	92	28	15	14	12	12	6	4	0	%
JLY	100	100	97	24	15	14	13	12	7	5	0	%
AUG	100	99	93	38	24	21	17	13	8	4	0	%
SPT	100	93	89	72	39	29	17	5	2	0	0	%
OCT	100	93	85	57	31	23	12	3	0	0	0	%
NOV	100	81	59	23	3	0	0	0	0	0	0	%
DEC	100	72	40	2	0	0	0	0	0	0	0	%

TABLE VI - Percentage of hourly energy usage due to the use of daylighting, controlled on the basis of switching at a predefined set point: climate data from ESP Test Climate

HOUR:	7	8	9	10	11	12	13	14	15	16	17	18	19
JAN	0	100	100	100	100	100	100	100	100	100	100	100	0
FEB	0	100	100	100	96	96	100	96	100	100	100	100	0
MAR	0	100	87	77	74	77	77	83	83	100	100	100	0
APR	0	96	46	43	73	86	70	53	40	83	100	100	0
MAY	0	51	51	100	100	100	100	100	64	67	96	100	0
JUN	0	46	63	100	100	100	100	100	90	66	70	100	0
JLY	0	51	45	100	100	100	100	100	80	74	83	100	0
AUG	0	51	61	77	96	100	96	77	51	48	100	100	0
SPT	0	96	70	53	50	53	46	56	73	90	100	100	0
OCT	0	100	100	54	38	54	58	77	100	100	100	100	0
NOV	0	100	100	100	100	96	96	100	100	100	100	100	0
DEC	0	100	100	100	100	100	100	100	100	100	100	100	0

TABLE VII - Percentage of hourly energy usage due to the use of daylighting, controlled on the basis of a two increment step-down strategy: climate data from ESP Test Climate

HOUR:	7	8	9	10	11	12	13	14	15	16	17	18	19
JAN	0	100	100	98	74	61	67	91	98	100	100	100	0
FEB	0	100	87	57	48	48	51	51	75	100	100	100	0
MAR	0	85	43	38	37	38	38	41	58	80	98	100	0
APR	0	55	23	21	43	43	45	26	21	43	71	100	0
MAY	0	35	30	58	67	56	74	77	40	45	72	85	0
JUN	0	36	46	56	68	50	75	75	63	45	55	83	0
JLY	0	38	37	61	72	56	75	85	53	48	58	85	0
AUG	0	40	30	45	66	56	72	56	33	30	62	91	0
SPT	0	61	35	26	25	26	23	28	36	56	91	100	0
OCT	0	90	50	27	19	27	29	38	69	96	100	100	0
NOV	0	100	98	75	56	50	60	86	98	100	100	100	0
DEC	0	100	100	96	80	75	83	100	100	100	100	100	0

TABLE VIII - Percentage of hourly energy usage due to the use of daylighting, controlled on the basis of an ideal linear dimming control function: climate data from ESP Test Climate

HOUR:	7	8	9	10	11	12	13	14	15	16	17	18	19
JAN	0	100	86	62	46	39	43	58	76	98	100	100	0
FEB	0	88	59	38	26	20	24	32	47	74	98	100	0
MAR	0	50	23	28	30	28	31	37	33	46	69	99	0
APR	0	19	11	15	32	33	31	22	19	18	43	81	0
MAY	0	25	20	43	48	44	49	53	30	32	36	52	0
JUN	0	23	33	43	48	39	50	54	45	32	35	47	0
JLY	0	26	23	44	48	43	50	54	39	34	41	45	0
AUG	0	27	21	32	46	43	47	39	25	23	31	60	0
SPT	0	32	10	16	18	18	17	20	13	35	67	97	0
OCT	0	57	23	14	15	19	21	21	37	71	98	100	0
NOV	0	94	68	45	36	31	37	53	75	99	100	100	0
DEC	0	100	87	65	53	49	55	68	89	100	100	100	0

TABLE IX - Percentage of hourly energy usage due to the use of daylighting, controlled on the basis of the probability of an occupant switching lights on: climate data from ESP Test Climate

HOUR:	7	8	9	10	11	12	13	14	15	16	17	18	19
JAN	0	100	100	100	100	100	25	58	77	100	100	100	0
FEB	0	82	82	82	82	82	7	17	28	64	100	100	0
MAR	0	22	41	45	48	51	9	22	29	41	61	96	0
APR	0	13	23	26	46	50	16	33	40	46	56	80	0
MAY	0	12	22	29	45	48	29	54	54	58	64	67	0
JUN	0	6	20	33	43	46	23	53	56	60	63	66	0
JLY	0	12	19	29	41	45	19	38	41	45	48	51	0
AUG	0	16	22	32	45	48	19	35	38	41	45	51	0
SPT	0	16	26	26	30	30	3	16	16	16	63	100	0
OCT	0	38	54	58	61	64	6	19	32	70	100	100	0
NOV	0	80	80	83	83	86	23	60	86	100	100	100	0
DEC	0	100	100	100	100	100	25	70	93	100	100	100	0

TABLE X - Average artificial lighting usage under all control regimes (1. on/off at set point, 2. step down, 3. linear dimming, 4. probability switching)

CONTROL	1	2	3	4
JAN	100	89	73	87
FEB	98	74	55	66
MAR	87	59	43	42
APR	71	44	29	39
MAY	84	58	39	43
JUN	85	59	40	42
JLY	84	60	40	35
AUG	77	52	35	35
SPT	71	46	31	31
OCT	80	58	43	54
NOV	99	83	67	80
DEC	100	94	78	89
AVE	86	65	48	51

EXAMPLE 2: HEATLUX, QUICK THERMAL AND LIGHTING ENERGY ANALYSIS

Introduction

In the field of simplified programs, the code presented here has been developed. Objectives of this work are:
- the development of a mathematical model to evaluate the effect of fenestration, in residential or commercial buildings, on the overall energy balance, including illumination energy needs
- the model must be so simple as to be worked out by small desk computer on the basis of only a few synthetic climatic data
- sky luminance distributions should be in accordance with the new sky classification proposed by CSTB (see Chapter 3).

Simplified methods have been developed, based on semi-empirical correlations, found by means of complex codes, between simplified input data and general performance factors. Typical simplified methods compute building heat losses and gains for the average day of each month. A monthly "utilisability coefficient" is then evaluated as a function of the building time constant - or similar quantities - and of the ratio losses/gains. The main limit of such methods is the range of configurations they can reliably describe, thus requiring a definition of their field of applicability or reliability.

The integration of the thermal and luminous aspects is obtained by including the heat corresponding to the artificial lighting requirements among the heat gains in a simplified thermal analysis code.

SMECC is the thermal analysis code used here; it is based on analyses performed by means of the NBSLD code and permits the evaluation of energy needs for heating and/or cooling of any zone in commercial or residential buildings, on a monthly basis, taking into account the influence of the HVAC system's control laws. SMECC can be easily adapted to include the results of a luminous analysis performed on the same zone.

Thus, HEATLUX firstly computes the levels of illumination provided by natural light in a room, then estimates the necessary artificial light integration, converts it to heat, and finally performs the thermal analysis including such heat among the heat gains, together with the gains due to other sources, such as solar, occupants, appliances, etc (Figure 1).

Daylighting Algorithms

The total amount of natural light which enters an enclosure depends on the size and position of transparent openings relative to the overall area and shape of the enclosure, the luminance distribution of the sky vault and the extent and surface characteristics of external obstructions. The amount of natural light at any point in the enclosure is, in addition to the foregoing, dependent on the reflectance of internal surfaces.

To evaluate correctly the level of illumination in an hour in a day in a point inside a building zone we adopt the following model.

The inside of the zone is assumed to consist of k=1,..., N plane surfaces. These N surfaces include:
- N_W opaque walls, which reflect light diffusely
- N_C clear windows, which partially transmit light without directional scattering
- N_{CS} clear windows with sheer curtains, which transmit light partially directly, and partially diffusely (dirty windows, fly screens, etc)
- N_D diffuse windows, which scatter transmitted light to all directions (milky-texture glass, windows with shades, etc). It is assumed that these windows act as perfect diffusers.

Assigned to each window in the zone is an outside "enclosure", assumed to consist of N_E surfaces, either opaque or diffusing, or portions of the sky. It is assumed that each outside surface has a uniform luminance.

Inside the zone, luminances vary significantly across a single surface, which affects the illuminance on the work surface. Thus it is necessary to break up inside surfaces into a number of sub-surfaces, or nodes.

The illuminance at any node (j) of working surface (i), E_{ij} can be computed from:

$$E_{ij} = E_{Dij} + E_{Rij}$$

where:

E_{Dij} = direct illuminance; the light reaching the point directly from the sky, the sun, or after reflection from external surfaces;

E_{Rij} = internally reflected illuminance; the light arriving at the point after reflections from the various zone surfaces.

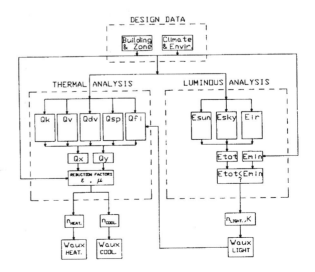

Figure 1 - Schematic flow chart of HEATLUX.

In order to evaluate light exchange factors to the sky, it is necessary to know both the direct sun illumination (E_S), and the luminance of a point in the sky ($L_{SKY,P}$) for all portions of the sky seen from the surface under consideration.

Models of luminance distribution have been developed (11) in terms of energetic parameters and of the sky type, as discussed in Chapter 3. The equation providing the luminance of point P in the sky is then:

$$L_{sky,P} = f(\eta, I_N) - g(z_P, I_N) \, h(z_0, I_N)$$

where:
I_N = Nebulosity Index, as defined in Chapter 3
z_P = zenithal angle of point P (p/2 - q in Chapter 3)
z_0 = solar zenithal angle
η = angle between P and sun:

$$\eta = \cos(z_0)\cos(z_P) + \sin(z_0)\sin(z_P)\cos(\varphi_P - \varphi_0)$$

where:
φ_0 and φ_P are the azimuthal angles of the sun and of point P, respectively

Thermal Analysis

The following algorithms allow the evaluation of energy needs for heating and/or cooling of a zone(room) in a building, the inside temperature of which is kept within a specified temperature range by an appropriate control system.

The zone's energy need is evaluated on the assumption that the HVAC system is always sufficient to maintain the temperature within these limits.

In the considered time span 't' for which the energy requirement has to be evaluated, one can list 's' quantities of energy ($Q_{(i)t}$ i = 1, 2, .., s), going into and out of the building zone, as a function of the temperature difference between the internal set temperature, assumed constant, and outside temperatures, outgoing quantities being assumed as positive.

As will be shown later, in setting the building thermal balance one can demonstrate that the heating or cooling system energy requirements cannot be taken as the algebraic sum of the above quantities, in as much as phenomena, due to the variable outside thermal conditions and to the building's thermophysical characteristics, occur preventing the internal temperature from keeping a constant value.

T is used as the time span of the average day of each month. Only mean daily values of climatic data, already available for several places, are, therefore, necessary.

Thus for a building zone :
Q_{AUXt} = average daily thermal energy need
$Q_{AUXt} = |\varepsilon \, Q_{Xt} + \mu \, Q_{Yt}|$
$Q_{Xt} = \Sigma_i \, Q_{(i)t} \cdot \delta_{Xti}$
$Q_{Yt} = \Sigma_i \, Q_{(i)t} \cdot \delta_{Yti}$

where:
$Q_{(i)t}$ are quantities of energy going into and out of the building zone
ε, μ are coefficients of reduction by semi-empirical correlations:

$\varepsilon = \varepsilon \, (t_I, K_m, \sigma)$
$\mu = \mu_1 \cdot \mu_2$
$\mu_1 = \mu_1(t_0, t_b, t_I, K_m, \sigma)$
$\mu_2 = \mu_2((Q_{Xt}/Q_{Yt}), K_m, \sigma)$

where:
K_m = mean zone transmittance
σ = mean effective thermal mass
K_m and σ are calculated on the basis of the geometrical and thermophysical characteristics of the zone (9)
t_I = plant working hours in a day;
t_0 = time of plant starting;
t_b = baricentric time of solar heat gains
δ_{Xti} and δ_{Yti} are functions that tell, in a given time span, which type of plant (heating, cooling) is working, in relation to the respective set-point temperatures:

Heating:
$\delta_{Xti} = 1$ if $Q_{(i)t} > 0$
$\delta_{Xti} = 0$ if $Q_{(i)t} \leq 0$
$\delta_{Yti} = 1$ if $Q_{(i)t} < 0$
$\delta_{Yti} = 0$ if $Q_{(i)t} \geq 0$
Cooling:
$\delta_{Xti} = 1$ if $Q_{(i)t} < 0$
$\delta_{Xti} = 0$ if $Q_{(i)t} \geq 0$
$\delta_{Yti} = 1$ if $Q_{(i)t} > 0$
$\delta_{Yti} = 0$ if $Q_{(i)t} \leq 0$
Q(i)t can be specified as:

- Q_{Kt}, heat transfer through the building envelope:

$$Q_{Kt} = f_c \cdot \left[\sum_{n=1}^{N_s} -K_n \cdot S_n \cdot (t_a - t_{sa}) + \sum_{n=1}^{N_l} k_n \cdot L_n \cdot (t_a - t_{sa}) \right] T$$

where:
N_s = number of opaque walls
N_l = number of thermal bridges
t_{sa} = sol-air temperature = $t_e + (aH)/(\alpha_e - 24)$
t_a = thermostat lower temperature, when evaluating heating needs; thermostat upper temperature, when evaluating cooling needs
t_e = mean outside temperature
a = hemispherical absorptance of the exterior surface

H = mean daily solar irradiation on the surface
α_e = surface heat transfer coefficient
K = thermal transmittance of wall
k = linear transmittance of thermal bridges
f_c = reduction coefficient to take into account the mutual long-wave radiative heat exchange between the internal surfaces of a zone: $f_c = 1 - 0.1947 K_m + 0.02077 K_{m2}$

- Q_{Vt}, heat transfer due to ventilation and/or infiltration:

$$Q_{Vt} = c_A\, G\, (t_a - t_e)\, T$$

where:
c_A = specific heat of outdoor air
G = air change mass flow rate

$$Q_{DVt} = -0.8\, f_r \sum_{n=1}^{N_v} C_n\, f_n\, H_n - S_n$$

- Q_{DVt} = heat gains due to solar radiation through glass surfaces:

where:
N_v = number of glass surfaces
0.8 = mean glass transmittance
f_r = reduction coefficient to take into account the radiant heat transfer to air and to walls (9): $f_r = 1 - 0.32\, K_m + 0.03\, K_{m2}$
C_n = shading coefficient
f_n = average fraction of the glass area exposed to the sun
H_n = mean daily solar irradiation on surface 'n'

- Q_{FIt}, heat gains due to internal energy sources:

$$Q_{FIt} = -[Q_{FCt} + f_r\, Q_{FRt}]$$

where:
Q_{FCt} = heat gain due to internal convective heat sources
Q_{FRt} = heat gain due to internal radiative heat sources

The Integrated Code
In order to achieve the above-mentioned objectives, the following limiting assumptions have been set:
- a single room of rectangular shape with horizontal and vertical rectangular surfaces
- neither external nor internal obstructions
- clear rectangular windows, horizontal or vertical, without overhangs, which partially transmit light without directional scattering
- internal reflections modelled by means of the split-flux method
- continuous controls, providing the exact service needed to keep temperatures and illuminances at the desired levels

The required input data are as follows:
- room function and occupancy characteristics
- geometry of each surface
- thermal and optical properties of each surface
- HVAC and artificial lighting system operating characteristics: set points, operating hours, efficiencies, etc
- climatic data: monthly averages of: daily total solar radiation on horizontal surface, H_h; sunshine ratio, I_S; air temperature, t_a.

The code computes, for each hour of the average day of each month, for each sky type, i, in accordance with the definition of the Nebulosity Index, as defined in (10), the ratio:

$$K_d = H_{dh} / H_h$$

where:
H_{dh} is the monthly average daily diffuse radiation

Using the Lui-Jordan procedure (11), $K_T = H_h/H_0$ is computed as a function of K_d, where H_0 is the extraterrestrial daily radiation.
Then, hourly fractions of the direct and diffuse solar radiation, respectively I_{Bh} and I_{dh}, are evaluated:

$$I_{Bh} = A\, [\exp(-B/\cos(z_0))]\cos(z_0)$$
$$I_{dh} = (\pi/24)\, K_d\, H_0\, C$$

where :
$A = (K_T - K_d)\, F$
F = coefficient dependent on the monthly average day and on the latitude (12)
B = extinction coefficient dependent on the monthly average day (15)
$C = (\cos\omega - \cos\omega_{ss}) / (\sin\omega_{ss} - \cos\omega_{ss})$
ω, ω_{ss} = respectively hour angle and sunset hour angle.

All data necessary for daylighting calculations are now available for the five sky types.
The hourly illumination at any node 'j' of working surface 'i', E_{ij}, can be computed from:

$$E_{ij} = E_{Dij} + E_{Rij}$$

where:
E_{s-ijk} = sky illuminance on surface 'i'

$$E_{s\text{-}ijk} = \int_{\varphi_{k1}}^{\varphi_{k2}} \int_{z_{k1}}^{z_{k2}} \left[L_s(\varphi,z)\, \tau_{s\text{-}ijk} \cos(\theta_{s\text{-}ij}) - \sin(z)\right] d\varphi\, dz$$

$L_s(\varphi,z)$ = sky luminance distribution = $f(I_N, \varphi_S, z_0, \varphi, z)$
$E_{S\text{-}ijk}$ = direct solar illuminance distribution

$$E_{S\text{-}ijk} = K_{Sun}\, I_{Bh}\, \cos(\theta_{S\text{-}i}) / \cos(z_0)\, \tau_{S\text{-}ijk}$$

where K_{Sun} is the luminous efficacy for direct irradiance, defined by Aydinli (13) as follows:

$$K_{Sun} = 17.72 + 4.4585\varphi_s - 8.7563 \times 10^{-2}\,\varphi_s^2 + 7.3948 \times 10^{-4}\,\varphi_s^3 - 2.167 \times 10^{-6}\,\varphi_s^4 - 8.4132 \times 10^{-10}\,\varphi_s^5$$

where
φ_s = solar altitude
$\delta_{Cijk} = 1$ if direct sunshine falls on the node 'ij' through window 'k'
$\delta_{Cijk} = 0$ if direct sunshine does not fall on the node 'ij' through window 'k'
$\tau_{s,S\text{-}ijk} = \tau(\theta_{s,S\text{-}ijk})$ is the transmittance of glass in the direction from node 'ij' to the sun through window 'k'.

It can be computed by Rivero's formula (14):

$$\tau(\theta) = 1.018\,\tau_0 \cos(\theta)\,(1-\sin\theta)$$

where:
τ_0 = transmittance at normal incidence
θ = angle of incidence

The cosine of the angle of incidence of the generic vector sky-surface 'i', can be expressed by:

$$\cos(\theta_{s\text{-}i}) = A_i \cos(\varphi_s) \cos(z_s) + B_i \sin(\varphi_s) \cos(z_s) + C_i \cos(z_s)$$

where:
A_i, B_i, C_i are the direction cosines of the normal to surface 'i'

The cosine of the angle of incidence of the vector sun-surface 'i', can be expressed by:

$$\begin{aligned}\cos(\theta_{S\text{-}i}) = &\sin(D)\sin(L)\cos(I_n) + \\ &\sin(D)\cos(L)\sin(I_n)\cos(A_z) + \\ &\cos(D)\cos(L)\cos(I_n)\cos(\omega) + \\ &\cos(D)\sin(L)\sin(I_n)\cos(A_z)\cos(\omega) + \\ &\cos(D)\sin(I_n)\sin(A_z)\sin(\omega)\end{aligned}$$

where:
D is solar declination
L is latitude
ω is hour angle
I_n is surface tilt angle
A_z is surface azimuth angle

According to the split-flux method (see Chapter 9), the daylight transmitted by the window is split into two parts, a downward-going flux, which falls on the floor and lower portions of the walls (below the "window mid-plane"), and an upward-going flux, which falls on the ceiling and upper portions of the walls. The space is assumed to behave like an integrating sphere with perfectly diffusing interior surfaces and no internal obstructions; it therefore works best for rooms which are close to cubical in shape. The resulting illuminance, E_{Rij}, due to internal reflections is, thus, uniform:

$$E_{Rij} = E_R$$

$$E_R = \frac{r_F\, A_W \sum_{k=1}^{N}(E_{s\text{-}k} + E_{S\text{-}k}) + r_C\, A_W \sum_{k=1}^{N} E_{g\text{-}k}}{A_I(1-r_M)}$$

where:

$$E_{s\text{-}k} = \int_{\varphi_{k1}}^{\varphi_{k2}} \int_{z_{k1}}^{z_{k2}} \left[L_s(\varphi,z)\,\tau_{s\text{-}ijk}\cos(\theta_{s\text{-}k}) - \sin(z)\right] d\varphi\, dz$$

$$E_{g\text{-}k} = L_g\, \tau(\theta_g)(1-\cos(I_{nk}))/2$$

$$L_g = r_g(E_{s\text{-}g} + E_{S\text{-}g})$$

$$E_{s\text{-}g} = \int_0^{2\pi}\int_0^{\pi/2} \left[L_s(\varphi,z)\cos(z)\sin(z)\right]d\varphi\, dz$$

$$E_{S\text{-}g} = K_S I_{Bh}$$

r_F = area-weighted average reflectance of the floor and those parts of the walls below the window midplane
r_C = area-weighted average reflectance of the ceiling and those parts of the walls above the window midplane
r_M = area-weighted average reflectance of the room surfaces including windows
r_g = ground reflectance
A_I = total inside surface area of the floors, walls, ceiling and windows in the room
A_W = window surface area

The integrals in the formulae above can be performed numerically by means of an "adaptive" integration scheme. The limits of integration can be defined from the geometry of the windows and working surface. For instance, in the case of sky illuminance on a point 'i' of the working surface through a vertical, rectangular window 'k', we have:

$$\varphi_{k1} \le \varphi \le \varphi_{k2}$$

where:
$\varphi_{k1} = \arctan(-A_{k1}/B_{k1})$
$\varphi_{k2} = \arctan(-A_{k2}/B_{k2})$

$$z_{k1} \le z \le z_{k2}$$

where:
$z_{k1} = \arctan[-C_{k3}/(A_{k3}\cos(\varphi) + B_{k3}\sin(\varphi))]$

z_{k2} = arctan [$-C_{k4}$ / ($A_{k4}\cos(\varphi) + B_{k4}\sin(\varphi)$)]
A_{kn}, B_{kn}, C_{kn} are the direction cosines of the normal to the planes through point 'i' on working surfaces and, respectively, the vertical edges of window (n = 1, 2), or the horizontal edges of window (n = 3, 4).

The daily energy consumption for lighting can be expressed as:

$$W_{AUX,L} = \sum_{n=1}^{N_{sky}} \left(p_n \sum_{m=1}^{N_{wh}} \left(\sum_{l=1}^{N_p} f_{nml} A_l \right) \right)$$

where:
$f_{nml} = 0$ if ($E_{SET} - E_{nml}$) ≤ 0
$f_{nml} = (E_{SET} - E_{nml})/\eta_L/K_L$ if ($E_{SET} - E_{nml}$) > 0
N_{sky} = number of sky types
p_n = probability of sky type 'n'
N_{wh} = maximum number of working hours for lighting systems
N_p = number of sub-surfaces (nodal points) on the working surface where the level of illumination has been evaluated
A_l = area of subsurface
η_L = efficiency of lighting system
K_L = visual efficacy of lighting system
E_{SET} = minimum required level of illumination in a nodal point
E_{nml} = hourly level of illumination in a nodal point for a certain sky type

The code computes thermal energy needs on a daily basis, month by month, taking into account the heat contribution from the artificial lighting system, by including $W_{AUX,L}$ among the heat gains due to internal sources, Q_{FIt}.

The total annual energy consumption for heating, cooling, and lighting, F_{tot}, is finally computed as follows:

$$F_{tot} = \sum_{t=1}^{12} N_t \left(W_{AUX,H} + W_{AUX,C} + W_{AUX,L} \right) t$$

where:
N_t = number of days in a month
$W_{AUX,H}$ = average daily energy need for heating
  = $Q_{AUX,HEAT} / \eta_H$
$W_{AUX,C}$ = average daily energy need for cooling:
  = $Q_{AUX,COOL} / \eta_C$
η_H and η_C are the average global efficiencies of the heating system and of the cooling system, respectively

Example of Application

HEATLUX has been used in the analysis of monthly energy needs of a room in a commercial building. The room has a 9 m² shape in plan and a height of 3 m, and has a rectangular window 1.5 m high, 1 m above the floor and the working plane is 1 m above the floor. Three different window widths are tested: (A) 2, (B) 5 and (C) 8 ms.

The building is located in Rome and the window room faces south. Only two vertical walls in the room, the southern and northern ones, exchange heat with the outside.

The operating characteristics of HVAC and artificial lighting systems are as follows:
- Heating: set point temperature: 20°C, activated from 8 am for 10 hours
- Cooling: set point temperature: 24°C, activated from 8 am for 10 hours
- Lighting: minimum level of illumination is 500 lux, activated from 8 am for 12 hours

As an example of graphical output, Figure 2 shows the illuminance distribution on a 9 x 9 grid in January at 9 am for a 5 m wide window (B). The room monthly energy needs are reported in Table XI. A comparison between cases (A), (B) and (C) is shown in Figure 3.

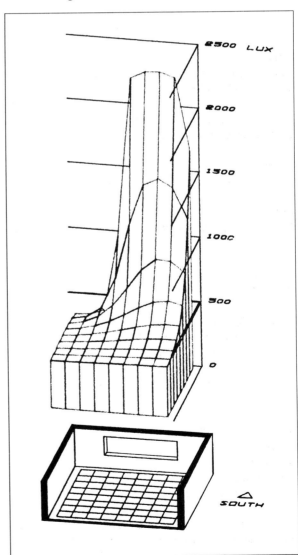

Figure 2 - Illuminance distribution in January at 9 am in a 9 x 9 m room with a 5 m wide and 1.5 m high window.

10.18 Daylighting in Architecture

TABLE XI - Monthly energy use (MJ) (case B)

MONTH	W_{HEAT}	W_{COOL}	W_{LIGHT}	W_{TOT}
Jan	119.9	0.0	232.4	352.3
Feb	99.4	0.0	139.7	239.1
Mar	77.2	0.0	91.7	168.9
Apr	34.9	0.0	43.3	78.2
May	0.0	0.0	24.0	24.0
June	0.0	0.0	17.2	17.2
Jul	0.0	1961.0	28.2	1989.2
Aug	0.0	2303.0	44.4	2347.4
Sep	0.0	0.0	80.3	80.3
Oct	4.2	0.0	118.3	122.5
Nov	66.0	0.0	226.2	292.2
Dec	106.5	0.0	265.7	372.2
TOTAL	508.1	4264.0	1311.4	6083.5

Conclusions

The code presented is a first attempt at providing designers with simple, computer-based, design tools, allowing fast and integrated evaluation of the energy implications of their choices.

The code still needs refinements and further testing, but it appears to meet the original goals.

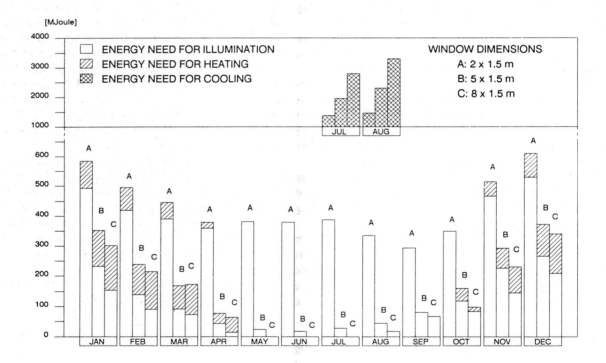

Figure 3 - Monthly energy use (MJ) in a 9 x 9 m room with a 1.5 m high window which is (A) 2, (B) 5 and (C) 8 m wide.

EXAMPLE 3: THE LT METHOD, A MANUAL LIGHTING AND THERMAL ENERGY TOOL

Introduction

The LT Method is a manual procedure to estimate energy use in non-domestic buildings. It estimates annual primary energy use for lighting, heating, ventilation and cooling (15).

Amongst the early considerations in the development of a building design, the designer is concerned with two issues: the form of the building (its plan depth, section, orientation etc) and the design of the façades (in particular the area and distribution of glazing). It is useful to know the implications for energy conservation of the designer's early decisions. The plan form, the façade design, and the arrangement of internal areas all play a crucial part.

Any calculation method employed must be quick and easy to use, in order to allow the designer to explore a number of options. It must be able to respond to the main design parameters under consideration. Energy consumption will also depend upon other parameters such as artificial lighting levels and plant efficiencies. However, these can be regarded as engineering parameters and to some extent can be considered independently.

The LT Method (LT standing for Lighting and Thermal) is an energy design tool which has been developed expressly for this purpose. A mathematical model has been used to predict annual primary energy consumption per square metre of floor area as a function of the following:
- local climatic conditions
- orientation of façade
- area and type of glazing
- obstruction due to adjacent buildings
- the inclusion of an atrium (optional)
- occupancy and vacation patterns
- lighting levels
- internal gains.

The LT Model

In the LT Method, energy use is read off from graphs which are derived by a mathematical model. The energy flows considered are indicated in Figure 4. First the model evaluates the heat conduction through the external envelope, and ventilation heat loss (or gain). Using monthly mean temperatures and an average internal temperature with a correction factor to allow for intermittent heating, a monthly gross heating load is calculated. The model then evaluates the solar gain and applies a utilisation factor to this. This factor takes account of the fact that not all the solar gains available are useful in offsetting auxiliary heating.

At the same time the monthly hours of available daylight are calculated from the average hourly sky illuminance on the façade, the daylight factor, and an internal lighting datum value. This gives a monthly electrical consumption for artificial lighting and a monthly heat gain.

The lighting heat gains and useful solar gains, together with casual gains from occupants and equipment, are then subtracted from the gross heating load to establish the net heating load. If the gains are more than the gross heating load, then the net load is zero. For air-conditioned options, it is necessary to evaluate cooling loads. If the gains are sufficient to raise the average temperature above the cooling set point, then a cooling load exists. This cooling load is added on to a fixed energy demand for fan and pump power.

Note that the model assumes that lights are only on when the daylighting value drops below a minimum datum value. In practice this would almost certainly require automatic light-sensing switching. Thus the model evaluates a technical potential performance, rather than the actual performance that would probably be found in a typical, but wasteful, building.

The monthly energy consumption is calculated for a "cell" and then reduced to the value of energy

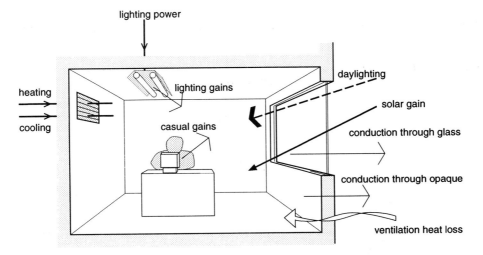

Figure 4 - Energy flows modelled for the LT Method

consumption per square metre. This is then totalled for the year and plotted as a function of glazing ratio. A cell corresponds to a room surrounded by other rooms. This implies a zero conductive heat loss through all surfaces except the external (window) wall or, in the case of rooflighting, the ceiling. Appropriate efficiency factors are applied to reduce all energy to primary energy.

The reason that primary energy is used is that it allows the different 'fuel' inputs for lighting, heating and cooling to be reduced to one common unit. Primary energy also relates better than delivered energy to CO_2 and other pollutant production.

Passive Zones

The LT Method uses the concept of passive and non-passive zones. Passive zones can be daylit and naturally ventilated and may make use of solar gains for heating in winter but may also suffer overheating by solar gains in summer. Non-passive zones have to be artificially lit and ventilated and in many cases cooled.

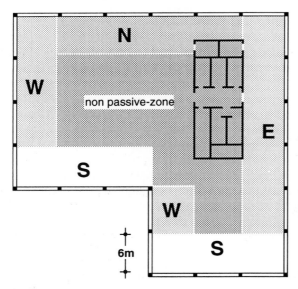

*Figure 5 - Definition of passive zones
(for 3m floor to ceiling)*

On the plan of the proposed building, the passive zones and non-passive zones are designated and their orientation defined, as in Figure 5. The depth of the passive zone from the façade is limited to twice the ceiling height. All of the top floor can be a passive zone if rooflit. The proportion of passive zone to the total building floor area is a good indicator of its potential energy performance. The energy consumption for these zones is then read off from the LT Curves.

LT Curves

An example of LT Curves is shown in Figure 6. The vertical axis represents the annual primary energy consumption in MWh/m^2, and the horizontal axis is the glazing area as a percentage of total façade area. Curves are provided for vertical glazing orientated south, east/west, and north, and for horizontal glazing (rooflights). There are sets of curves for different building types, lighting datum levels and internal gains.

Two totals are shown, heating + cooling + lighting (tot), and heating + lighting (h+l). The cooling curve includes an allowance for fan power as well as refrigeration. The total without cooling can be used to indicate the energy use of a non air-conditioned, naturally ventilated building. However, a fixed allowance for fresh air mechanical ventilation must be added for all non-passive zones.

For top floor areas daylit by roof lights, conductive heat losses through the opaque roof envelope are accounted for correctly. If the top floor is side lit, the LT curve for sidelighting assumes no losses through the ceiling and a small error results. This can be accounted for by reading off a heating load from the curve for horizontal glazing at zero glazing ratio, and adding it to the heating loads from the appropriately orientated sidelit curve.

The energy for non-passive zones can be calculated by reading from the curve at zero glazing area. The non-passive zones must always have a fan power allowance, at least for fresh air supply. For non-passive zones in buildings with high internal gains and high lighting levels, a cooling energy, obtained from the zero glazing intercept of the cooling curve, should always be added.

After the specific energy consumption for each zone has been read from the graphs, they are multiplied by the zone areas and totalled on the LT Worksheet, as shown in Figure 7.

The method includes a procedure for taking account of overshadowing from adjacent buildings, which has been shown to have a large effect on lighting energy, and smaller, but significant effects on heating and cooling energy. There is also a procedure for predicting the effect of an atrium. An atrium causes a reduction in the passive zone depth due to the obstruction to daylight, resulting in increased artificial lighting. However, there is a thermal saving due to the buffer effect and ventilation pre-heating if adopted.

The original version of the LT Method was prepared for the EC Architectural Ideas Competition *Working in the City* (16). LT Curves were provided for five European climatic zones. This was later developed to include the overshadowing procedure and published by the CEC in *Energy in Architecture: - The Passive Solar Handbook* (17) as LT Method 1.2. The most recent version 2.0, published (18) by the UK Building Research Establishment, is for mid-European and north European coastal zones only, but covers four building types and various lighting datum values and casual gains levels.

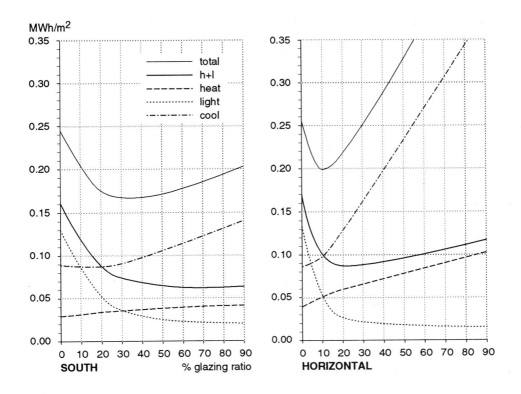

Figure 6 - LT Curves for a mid-European coastal climate, building type C (office), with a datum lighting level of 300 lux, and internal gains level of 15 W/m².

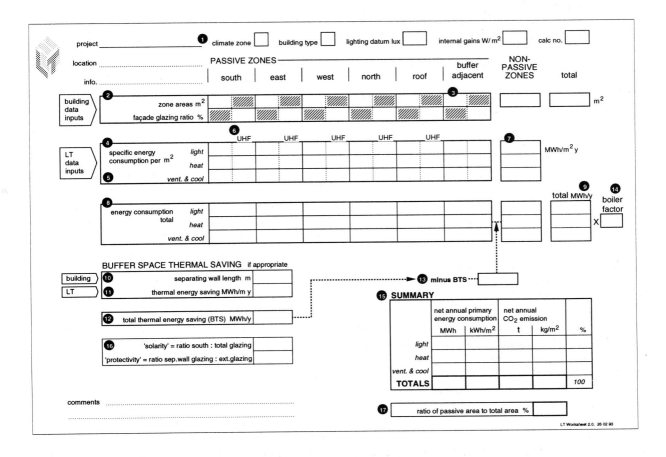

Figure 7 - The LT Worksheet.

FUTURE DIRECTIONS

These examples present methods of simulating daylighting and building energy usage in which neither the concept, nor the methodology, are either new or original. However they do represent the state of the art, both in practice and simulation, at the current time. To improve on these methods would not be difficult, but would require a greater understanding of the dynamic nature of daylight and of occupants' reaction to the time rate of change, spatial distribution and relative contrasts caused by daylight in an enclosure. Currently, accuracy is limited by the superficial treatment of dynamic skies and a less than rigorous treatment of the optical physics involved. As the state-of-the-art matures, then these problems will be overcome but it is readily acknowledged that, in practice, many integrated daylight control implementations do not produce internal conditions that satisfy the occupants' comfort criteria. The potential for energy saving and improved internal comfort is apparent, but the means through which they can be achieved is less obvious. The ability to simulate accurately a wider range of architectural design options would undoubtedly help to establish those criteria that have the potential to provide the requisite benefits.

REFERENCES

(1) Tregenza, P.R., "Daylight Coefficients", *Lighting Research and Technology*, vol. 15, No. 2, 1983.
(2) Gillette, G. and Kusada, T., "A Daylighting Computation Procedure for Use in DOE-2 and other Dynamic Building Energy Analysis Programs", *Journal of I.E.S.*, January 1983.
(3) Winkelman F. and Selkowitz S., "Daylighting Simulation in the DOE-2 Building Energy Analysis Program", *LBL-18508*, Lawrence Berkeley Laboratory, Berkeley, California, 1985.
(4) Littlefair, P., "Measurements of the Luminous Efficacy of Daylight", *Lighting Research & Technology*, vol. 20, No 4, 1988.
(5) Clarke, J., *Energy Simulation In Building Design*, Adam Hilger Ltd, Bristol and Boston, 1984.
(6) Clarke, J., "NATLIT User Documentation ABACUS", University of Strathclyde, Glasgow, Scotland, 1978.
(7) Selkowitz, S., "SUPERLITE User Documentation", Windows and Daylighting Group, Lawrence Berkeley Laboratory, Berkeley, California, 1983.
(8) Hunt, D., "Predicting Artificial Lighting Use, a Method Based on Observed Patterns of Behaviour", *Lighting Research and Technology*, vol. 11, No 2, 1980.
(9) Agnoletto, L., Brunello, P., Zecchin, R., "Simple Methods for Predicting Thermal Behaviour and Energy Consumption in Buildings", *ASHRAE-DOE Conference*, Orlando, Florida, 1979.
(10) Perraudeau, M., Chauvel, P., "One Year's Measurements of Luminous Climate in Nantes", Centre Scientifique et Technique du Bâtiment, Nantes, 1987.
(11) Winkelmann, F.C., Selkowitz, S., "Daylighting Simulation in the DOE-2 Building Energy Analysis Program", *Energy and Buildings*, No 8, p. 271-286, 1985.
(12) *"Guida al controllo energetico della progettazione"*, CNR, Rome, 1985.
(13) Aydinli, S. "On the calculation of available solar energy and daylight", Dissertation Tech.Univ.Berlin, Fortschrittber der VDI - Zeitschriften, 6 (1979), Düsseldorf, 1981
(14) Rivero, R., "Natural Lighting. The Calculation of the Direct Daylight Factor for Glazed and Unglazed Windows and for Uniform and Non-Uniform Skies", *BRS Library Communication n. 860*, London, 1958.
(15) Baker, N., Strategic Design Tools for Non-domestic Buildings. *Proc. Passive & Low Energy Architecture PLEA 91*, Seville. Kluwer, Dordrecht 1991.
(16) O'Toole, S. and Lewis, J. O. (eds.), *Working in the City*, Eblane Editions, Dublin, 1990.
(17) Goulding, J. R., Lewis, J. O. and Steemers, T. C. (eds.), *Energy in Architecture - The European Passive Solar Handbook*, Batsford, London, 1992.
(18) Baker, N., Steemers, K., "The LT Method 2.0, An Energy Design Tool for Non-Domestic Buildings", *General Information Report, Best Practice Programme,* Building Research Conservation Unit (BRECSU), Building Research Establishment, UK, 1993.

Chapter 11
CASE STUDY ANALYSIS

INTRODUCTION

There are three ways to convince designers of the advantages to be gained by improving the daylighting performance of their architectural projects:
1. show the architectural possibilities through exemplary case studies;
2. provide analyses and information describing the relevant aspects and lessons learnt from the case studies, and how this can improve the quality of current projects;
3. make available to architects the tools that can assist them in the design and analysis of daylit buildings.

The wide use of mathematical testing tools in design activity has tended to belong more to architectural engineering practice, as opposed to the formalistic approach represented by the beaux-arts tradition. It must be understood that designers operate within such streams, often led by key architectural figures and followed by the professional masses; without that understanding it is difficult to communicate research outputs on daylighting to them.

A growing body of knowledge about both design problems and design processes has been developed. Despite this, architects continue to operate in what may be described as the "traditional way". Many changes have occurred in both architectural practice and in potentially relevant and related fields, such as operational research, systems analysis, design methods, computer aided design, information technology, etc, hence one could expect some analogous changes within the architectural design process. But this is not the case.

Simplified testing tools for evaluating the daylighting performance of buildings are valuable, but architects often design with reference to architectural precedents. Therefore, what is required are up-dated precedents of building types which embody effective daylighting concepts and principles. The problem is not the degree of difficulty presupposed by the use of testing tools, but simply that there is a need for generative analysis as opposed to testing methods.

THE ARCHITECTURAL DESIGN PROCESS

Traditional Design Methods

In reading the many studies devoted to describing the design activity, it is possible to highlight the main features of the design process as followed by the majority of architects and engineers (1). These features are described below.

The initial outline of a design scheme, which is considered very important to define the daylighting performance of the resultant building, is prepared in a comparatively short time and ona rather small scale, probably by a partner or principal. This outline design is based on criteria that are predominantly intuitive, visual and aesthetic, such as form, mass, space and volume. Design ideas come from a stored repertoire of building forms visited or seen, then altered, amended and adapted to suit the task at hand.

It is assumed that there is a visual thinking process, analogous to, but different from, a literary or mathematical one, through which architectural ideas are expressed. Building plans and elevations map these notions as drawings, embodying the knowledge developed in solving the design problem. Forms are the legitimate supports of this problem-solving design activity, and typological precedents are collected and synthesised to provide this knowledge.

Architectural composition theory developed as a cumulative design tool which can be called the typological method. The stored repertoire of forms and types is contained in various treatises, and architects use them to select the appropriate building types.

Building researchers need to transfer their evaluative tools into such typological models and structure them in an architectural grammar, to make these models adaptable to specific, individual design programmes.

This chapter is devoted to developing and implementing such a typological grammar.

The current, traditional design process presents many disadvantages, mostly in relation to its relevance to one person's design skill and the lack of opportunity for consultants to contribute specialised knowledge early in the design process. Nevertheless, it is a process which few query; quantity surveyors and engineers expect to be given a set of drawings before they can commence. Even when a design team is encouraged to work together from the earliest stages of design, it has been found that design ideas are mostly generated by one person, in the form of visual concepts.

There are two types of approach to architectural design:
- the non-intuitive design: machine design, the systematic design methodology, or the engineered design process, etc,
- the traditional design: the intuitive design, or the typological design, etc.

The Proposed Design Method

The problems raised in this chapter are, in fact, those of teaching architectural design. The issue is the transfer of the competence to design satisfactory buildings. To transfer this competence requires a distinction to be made between those buildings which possess the relevant quality and those which lack it. A tool is required to draw the distinction, dividing the set of all buildings in to two separate and complementary sub-sets. There are several ways to specify sets of projects possessing or lacking specific qualities.

Firstly, one can exhibit a catalogue of all the elements in the set that should be adopted. This method is impractical for all but small sets.

Secondly, one can present the typical elements of the set and the transformation processes for generating, through various procedures, a range of elements. One can recognise this approach as the intuitive design method followed by the beaux-arts typological tradition.

The third approach is to provide some testing tools to eliminate all the elements lacking the defined quality. This approach characterises the engineering or "neo-positivist" method of non-intuitive design.

A fourth approach, one that is followed here in developing the proposal for a more efficient daylighting design tool, consists of providing the designers with a grammar for generating the elements of the set. The method presented here follows such a "constructional" cognitive system.

The Typological Approach to Heuristic Reasoning

Some examples of the use of precedents or building types within the handbooks on passive solar design, which also contain various recommendations on daylighting design tools, could become references for a study on the typological approach to heuristic reasoning (2). In his handbook (3), Mazria uses some "patterns", which come from Alexander's *et al. Pattern Language* (4) and can be interpreted as building types. But Watson and Labs also use pictures in their book *Climatic Design* (5) which could become the coupling of shapes and performances that have been called "types" in the handbook *"L'Architettura del regionalismo"* (6). In some more recent handbooks dealing directly with daylighting, Lam (7), Robbins (8), Moore (9), and Brown *et al.* (10), develop the use of pictures to bring out the daylighting performances embodied within the architectural shapes. Such handbooks describe the use of sets of design concepts covering the various types of daylit spaces as a pre-design tool. The sets include side-lighting concepts, top-lighting, angle-lighting, beam-lighting, indirect lighting, atria concepts or combinations (9), that could become part of the typological repertoire discussed here. Many of these visual or architectural concepts show the parametric transformations of a room type, and therefore open the possibility of developing some transformation rules in order to enrich the typological grammar. Recent handbooks for commercial buildings (11) recommend an "incremental change" design approach which starts from a base case building and improves it piecemeal. This process of modification can be called "parametric transformation". This is distinguished from the other approach, the non-incremental change method, that proceeds through various explorations to search for a more innovative solution, which is called a "combinatorial transformation" process. The base case building represents a design hypothesis generated to develop the analysis.

This base case building is the building type coupled with its energy performance. A set of building types already analysed could be prepared to provide the designer with a systematic reference array of base case buildings. These should be selected as samples for exemplifying the characteristics of the more recurrent types of buildings.

Studies of architectural grammar are not so diffused as many other topics in design theory. The approach adopted here derives both from the typological culture of architectural composition, developed within the treatises of architectural theories, and from the "shape grammar" developed by March and Stiny (12) dealing with design theory, artificial intelligence and computer aided design. The study of this grammar is based on two major issues:
- the repertoire of types, elaborated by adopting the "morphological box" suggested by Zwicky and Wilson (13) and applied in other studies (14);
- the rules for selecting, placing and transforming these types.

THE TYPOLOGICAL GRAMMAR OF ARCHITECTURE

Typological Grammar as a Generative Design Tool for a Multi-scale Architecture

The grammar, described later, consists of a set of composition rules and a repertoire of architectural types. It proposes a set of transformation rules to fit the types to a specific site and programme. The contribution of geometry to the definition of these rules is fundamental. A geometry of transformations has been developed after Hilbert (15), integrating the different geometries into a system of geometric transformations, within which they are characterised by the different kinds of transformation processes. The composition rules of selecting, placing and transforming the types follow these geometric transformations.

Following the studies on the building types, three levels are considered:
Level I - the room;
Level II - the building, which can be defined as a large room within which small rooms are arranged in a certain way;
Level III - the street or square as a room without a ceiling, which has the building façades as walls.

Of course, more levels could be found, but these are sufficient for the multi-scale architecture studied here.

The Room Structure

The grammar allows the selection and placing of types taken from the repertoire, and their adaptation to compose the variety of rooms required and preferred. The coupling of shapes and performances allows for a correct selection of types.

The room may be seen as a set of planes forming the floor, the walls with apertures, the roof, etc. By combining various planes we can produce many different rooms. The planes composing this room are the "parameters" of the "morphological box", the various possibilities for floors, walls, roofs, windows, doors, corners, etc, are its "variants".

To explain the concept, three classes of modules can be considered: the **plan layouts** which belong to the site group modules, the **elevations** and the **roofs** belonging to the shelter group modules.

The plan layouts considered in the case studies are those which are part of the passive solar design heritage: the **triangular**, the **linear building**, the **top-lit building**, the **courtyard building**, the **atrium building** and the **building within a building**. These can be combined with various elevation modules, or roof modules.

Several issues are relevant in a study of the wall modules:
- the ratio between the opaque portion of the wall and its apertures. Bioclimatic design recognises the climatic reasons for such differences;
- the variability of apertures related to different horizontal positions in the same façade, from ground level to the intermediate floors up to the attic floor, defining three well recognisable horizontal areas;
- the façade often acquires a depth, inside or outside its wall plane, to extend or hollow out balconies or more complex window components;
- the façade has a relationship to the street and to the building, one side connected to the urban public outdoor space, the other connected to the building's private indoor space;
- the building itself distinguishes the asymmetry between its public façade and private rear;
- the environmental dimension of construction, focused by the bioclimatic design experiences, distinguishes a south from a north elevation and an east from a west elevation, adding the environmental characters of architecture to the compositional aspects.

The Design Process

The data that describe the user requirements are organised in a design programme dealing with the definition of the design problem. To solve it, the architect may consider precedents of the possible solutions in order to generate the design hypotheses. In this work the architect is helped by typological studies (contained in the treatises and handbooks, of the past, and in the professional journals of today) showing the precedents classified and grouped together.

Books and magazines publish numerous projects without knowingly acting as treatises. No convention or system of notation is used to map these projects for a theoretical or analytical purpose.

Architectural composition includes three main operations: selecting, placing and transforming the types (or precedents). Hence the architect has to select the types correctly, bearing in mind the requirements collected from the client. After this choice the architect should place these types in the project, combining together plan layouts, elevations, sections, etc. Then, having composed the types for the design solution, the architect makes some adaptations, modifying them through parametric transformations and combinatorial transformations.

The selection and placing of types follow the base rules, by operating a pattern recognition and a place identification. Their modification follows the transformation rules, operating parametric and/or combinatorial variation of the selected and placed types.

CASE STUDIES

To achieve the task of presenting a selection of effective daylighting systems, the "typological grammar" approach introduced above is followed. The study adopts the "morphological box" proposed by Zwicky and Wilson (13) and others (16), through which, by combining "parameters" and "variations", one can represent a great number of design solutions. The "shape grammar" developed by March and Stiny (12) and Stiny (17) is employed in order to produce, through the defined rules, design hypotheses.

11.4 Daylighting in Architecture

The typological grammar collects well daylit buildings into a repertoire of building types which can be used more effectively than the simpler set of "typological precedents" theorised within the architectural culture. This grammar supplies the designer with an organised framework of types, for selecting, placing and fitting, through suitable transformations, the models that an architect needs for the specific programme and conditions of a project.

METHODOLOGY AND CRITERIA FOR CLASSIFICATION

The classification system used in this study is based on the use of the architectural envelope shape as the main tool for environmental control in the selected buildings. The system has been developed in order to provide architects with a design procedure that allows them to identify the performance of a particular solution-type.

These solution-types can be categorised in a range of possibilities between two opposite types defined as the **climate-rejecting building** and the **climate-adapted building** (11). A similar classification defines the solution types as **exclusive mode** and **selective mode** buildings to characterise different environmental control strategies (18). Although both these extremes use the form and the envelope as tools of environmental control, they are used in opposite fashions: the purpose of the form and that of the envelope is distinctly different.

As examples of the operation of these types, related to the daylighting performance, the Wainwright Building, St Louis (Arch: Sullivan) can be defined as a climate-adapted building and the Seagram Building, New York (Arch: Mies van der Rohe) as a climate-rejecting building. The Larkin Building, Buffalo (Arch: Frank Lloyd-Wright) is a combination of the two, a mix of climate-adapted and climate-rejecting buildings. In terms of investigating daylit buildings, only the climate-adapted buildings are considered.

In the compilation of the survey formats for the case studies, the above-mentioned morphological approach has been used. This morphological approach is a systematic means to pursue discovery and invention, and it includes a variety of morphological methods. The best known method is that of the "morphological box", a means of generating new combinations of existing concepts. Using a "morphological box", it is possible, when a basic knowledge of the problem area is given, to generate a very large number of morphologies, some of which would be absurd, some of which would be commonplace, but some of which could be wholly novel and of great value. This approach is selected for the mapping of well daylit buildings, as it allows for generating new concepts from the selected case study buildings. In place of simply referring to typological daylighting precedents, the designer is given the possibility of decomposing these existing daylit buildings into elements, which become the "variations" of the defined "parameters". All the selected buildings can be synthesised in terms of parameters, which can be considered as different combinations of variants. By experimenting with new combinations within the same scheme, one can achieve new concepts of daylit buildings.

THE MORPHOLOGICAL BOX

A morphological box is organised into rows, with each row corresponding to one descriptive parameter or variable. The row for each parameter contains all the possible variations of that parameter, with one position always reserved for "other". The steps in the construction of a morphological box are the listing of the parameters, the listing of the variations for each parameter, the construction of the box, and the inspection and selection of useful morphologies generated in the box.

At the beginning of this chapter the design task was defined as the process of choosing the controllable causes and adjusting them in such a way that, under the constraining circumstances defined by the uncontrollable causes, some desired effects can be obtained. Now, in order to define the controllable causes, whose dimensions are the design parameters, the morphological box is proposed. To define the controllable effects, it is proposed to couple the parameters (or envelope shapes) with related effects (or daylighting performances).

The structure of the prepared formats, following the morphological approach, has been developed in order to point out the design parameters considered as the elements which determine the lighting operation of the building shell.

The kind of morphological box that has been developed to describe the analysed buildings has been organised in a hierarchical way, placing the parameters at various levels, ranging from the level of the urban context (level III) to the level of the room (level I). Every building therefore is defined, first by characterising its relationship with the context, then by defining its internal relationships dealing with either the building as composed of rooms or the room itself as composed of elements. The street is here considered only for its influence on the daylighting of the buildings, not for the natural lighting of the urban space. The street or square is defined as the third typological level of the architectural system.

The whole building is considered in terms of how the rooms are composed with respect to daylight and envelope systems. The building is defined as the second typological level of the same architectural system.

The room represents the lighted space and is the main object of daylighting analysis. The elements which shape the room are: the floor, the walls with openings and the ceiling with openings. The floor can be considered as the lighted area which requires a specific amount of light related to the activities carried out in the room. The walls and ceiling are instead the lighting area, which can be divided into three sub-areas: the windows as light sources, the surfaces as diffusers, and the holes or dark surfaces. The room is defined as the first typological level of the architectural system.

The parameters of the first typological level, characterising the lighted room and referred to as **the room**, are:

- **A-Room plan layout**, whose variations are: A1 unilateral; A2 bilateral; A3 deep room; A4 others.
- **B-Light collector position**, whose variations are: B1 within plane central; B2 within plane skylight; B3 between plane; B4 between plane corners; B5 window walls; B6 others.
- **C-Light collecting and diffusing areas**, whose variations are: C1 side light up to 15%; C2 side light from 15% to 30%; C3 side light over 30%; C4 top light up to 15%; C5 top light from 15% to 30%; C6 top light over 30%; C7 others.
- **D-Aperture shape**, whose variations are: D1 intermediate window; D2 horizontal aperture; D3 vertical aperture; D4 total glazed wall; D5 ceiling aperture; D6 total glazed ceiling; D7 others.
- **E-Glare controls**, whose variations are: E1 light filters; E2 rigid screen; E3 others.

The parameters of the second typological level, characterising the building structure and referred to as **the building**, are:

- **F-Building plan layout**, whose variations are: F1 deep plans; F2 one level factory/warehouse; F3 unilateral/bilateral slabs; F4 courtyards; F5 atria; F6 double shell buildings; F7 others.
- **G-Wall/aperture ratio**, whose variations are G1 25% aperture; G2 50% aperture; G3 75% aperture; G4 others.
- **H-Wall apertures distribution**, whose variations are: H1 symmetric façades; H2 solar asymmetric façades: H3 urban space asymmetric façades; H4 others.
- **I-Wall shading devices**, whose variations are: I1 portico; I2 brise-soleil; I3 overhangs; I4 others.
- **J-Roof apertures**, whose variations are: J1 skylights; J2 monitor roof; J3 clerestory; J4 glazed; J5 screened glass; J6 others.

The parameters of the third typological level, characterising the building site conditions and referred to as **the civic space**, are:

- **K-Urban space layout**, whose variations are: K1 small urban blocks; K2 large urban blocks; K3 north-south slabs; K4 east-west slabs; K5 open blocks; K6 towers; K7 solar oriented blocks; K8 detached tower; M9 others.
- **L-Façades reflectances**, whose variations are: L1 high reflectance; L2 mean reflectance; L3 low reflectance; L4 others.
- **M-Street top lighting**, whose variations are: M1 profile angle 30°; M2 profile angle 60°; M3 profile angle 90°; M4 others.

Each of the above categories is illustrated over the next pages, and interrelated through the morphological box (Figure 1)

LEVEL	PARAMETERS	VARIABLES
	MORPHOLOGICAL DIAGRAM	
I	A room layout	1 2 3 4
	B collector posn.	1 2 3 4 5 6
	C collecting area	1 2 3 4 5 6 7
	D aperture shape	1 2 3 4 5 6 7
	E glare control	1 2 3
II	F plan layout	1 2 3 4 5 6 7
	G wall /aperture	1 2 3 4
	H aperture distribution	1 2 3 4
	I shading device	1 2 3 4
	J roof aperture	1 2 3 4 5 6
III	K urban layout	1 2 3 4 5 6 7 8 9
	L façades reflectances	1 2 3 4
	M street top lighting	1 2 3 4

Figure 1 - The morphological box.

SELECTION AND CLASSIFICATION OF DAYLIT BUILDINGS

A small sample of daylit buildings has been selected to represent different building typologies, climatic areas and cultural identities. They are defined following the technique outlined above and illustrated, on the following pages, using a summary format to make them comparable. A more detailed format would reveal a greater amount of information, but in this instance summary formats are used to demonstrate the principle of the morphological diagram. Further information can be obtained by referring to the sources listed.

The building examples are, in some respects, chosen to reflect the main European regional differences, covering the under-lighted and over-lighted areas, in order to show how the architects attempted to fit their buildings to these physical constraints. The solutions that the building envelopes embody deal in turn either with the light enhancement or with its shielding. Three case studies from other regions are shown to reinforce climatic differences.

11.6 Daylighting in Architecture

LEVEL I
THE ROOM

A • Room plan layout

A1 - Unilateral

A2 - Bilateral

A3 - Deep Room

A4 - Others

B • Collecting area position

B1 - Within Plane Central

B2 - Within Plane Skylight

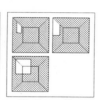
B3 - Between Planes

B4 - Between Plane Corners

B5 - Window Wall

B6 - Others

C • Light collecting and diffusing areas

C1 - Side Light up to 15%

C2 - Side Light from 15% to 30%

C3 - Side Light over 30%

C4 - Top Light up to 15%

C5 - Top Light from 15% to 30%

C6 - Top Light over 30%

D • Aperture shapes

D - Intermediate Window

D2 - Horizontal Window

D3 - Vertical Aperture

D4 - Total Glazed Wall

D5 - Ceiling Apeture

D6 - Total Glazed Ceiling

E • Glare control

E1 - Light Filters

E2 - Rigid Screen

E3 - Others

LEVEL II
THE BUILDING

F • Building plan layout

F1 - Deep Plans

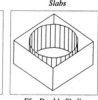
F2 - One Level Factory/Warehouse

F3 - Unilateral/Bilateral Slabs

F4 - Courtyard

F5 - Atria

F6 - Double Shell Buildings

F7 - Others

G • Wall apertures ratio

G1 - 25% Aperture

G2 - 50% Aperture

G3 - 75% Aperture

G4 - Others

H • Wall apertures distribution

H1 - Symmetric Façades

H2 - Solar Asymmetric Façades

H3 - Urban Space Asymmetric Façades

H4 - Others

I • Wall shading devices

I1 - Portico

I2 - Brise-soleil

I3 - Overhangs

I4 - Others

J • Roof apertures

J1 - Skylights

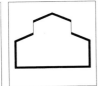
J2 - Monitor Roof

J3 - Clerestory

J4 - Glazed

J5 - Screened Glass

J6 - Others

LEVEL III
THE URBAN SPACE LAYOUT

K • Urban space layout

K1 - Small Urban Blocks

K2 - Large Urban Blocks

K3 - North-South Slab

K4 - East-West Slabs

K5 - Open Blocks

K6 - Towers

K7 - Solar Oriented Blocks

K8 - Detached Tower

K9 - Other

L • Façades reflectance

L1 - High Reflectance

L2 - Mean Reflectance

L3 - Low Reflectance

L4 - Other

M • Street top lighting

M1 - Profile Angle 30°

M2 - Profile Angle 60°

M3 - Profile Angle 90°

M4 - Others

11.8 Daylighting in Architecture

Summary Building Data

Building: Library
Location: Rovaniemi, Finland
Latitude: 66°N
Architect: Alvar Aalto
Date: 1965-68
Source: "Alvar Aalto", *Architectural Monographs 4*, Academy Editions, London, 1988.
Moore, F., *Concepts In Architectural Daylighting*, Van Nostrand Reinhold, New York, 1991.

MORPHOLOGICAL DIAGRAM		
LEVEL	PARAMETERS	VARIABLES
I	A room layout	3 deep room
	B collector posn.	2 within plane
	C collecting area	4 top light <15%
	D aperture shape	2 horizontal
	E glare control	2 rigid screen
II	F plan layout	2 one level
	G wall /aperture	2 50%
	H aperture distribution	2 solar asymmetric
	I shading device	4 other (scoop)
	J roof aperture	6 other (light scoop)
III	K urban layout	5 open blocks
	L façades reflectances	
	M street top lighting	

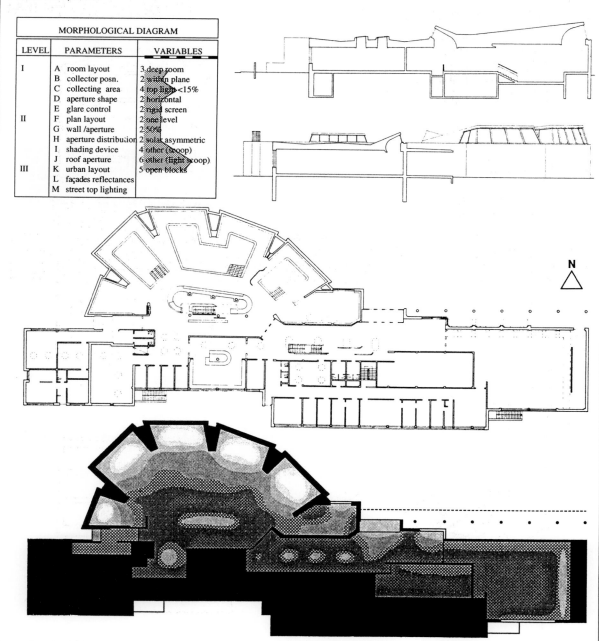

Case Study Analysis 11.9

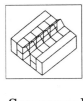

Summary Building Data

Building: BRF Headquarters
Location: Lyngby, Denmark
Latitude: 56° N
Architect: Krohn & Hartvig Rasmussen
Date: 1986
Source: *BRF hovedkontor i Lyngby,*
BRF information leaflet.
IEA, *Passive and Hybrid Solar Commercial Buildings,* International Energy Task XI, ETSU, Harwell, 1989

MORPHOLOGICAL DIAGRAM		
LEVEL	PARAMETERS	VARIABLES
I	A room layout	2 bilateral
	B collector posn.	1 within plane
	C collecting area	1 side light <15%
	D aperture shape	1 intermediate
	E glare control	1 light filters
II	F plan layout	5 atrium
	G wall /aperture	1 25%
	H aperture distribution	1 symmetric
	I shading device	4 other (louvres)
	J roof aperture	5 screened glass
III	K urban layout	1 large blocks
	L façades reflectances	
	M street top lighting	

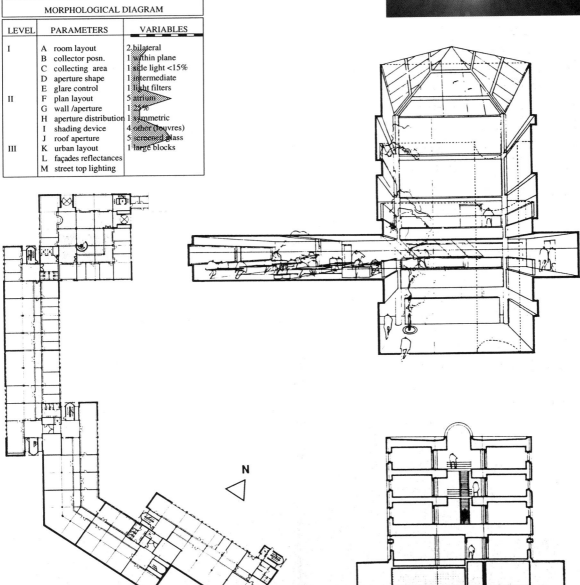

11.10 Daylighting in Architecture

Summary Building Data

Building: Stansted Airport
Location: Stansted, UK
Latitude: 52°
Architects: Foster Associates
Date: 1981-91
Source: *The Architectural Review*, No. 1131, May 1991.
The Architects Journal, vol. 193, No. 22, 29 May 1991.
Building, Issue 19, 10 May 1991.
Photo: M.Charles

MORPHOLOGICAL DIAGRAM			
LEVEL	PARAMETERS		VARIABLES
I	A	room layout	2 bilateral
	B	collector posn.	2 within plane skylight
	C	collecting area	4 top light <15%
	D	aperture shape	5 ceiling
	E	glare control	1 light filters
II	F	plan layout	2 one level
	G	wall /aperture	3 75%
	H	aperture distribution	4 asymmetric
	I	shading device	3 overhang
	J	roof aperture	5 screened glass
III	K	urban layout	9 open site
	L	façades reflectances	
	M	street top lighting	

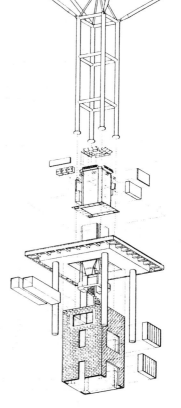

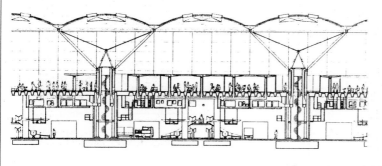

Case Study Analysis 11.11

Summary Building Data

Building: School of Architecture
Location: Lyons, France
Latitude: 46° N
Architects: Jourda and Perraudin
Date: 1985-88
Source: *The Architectural Review*, No. 1097, July 1988.
Photo: S. Couturier & D. Sucheyre.

MORPHOLOGICAL DIAGRAM			
LEVEL		PARAMETERS	VARIABLES
I	A	room layout	2 bilateral
	B	collector posn.	5 window wall
	C	collecting area	3 side light 15-15%
	D	aperture shape	4 glazed wall
	E	glare control	1 light filters
II	F	plan layout	3 bilateral slabs
	G	wall /aperture	3 15%
	H	aperture distribution	4 asymmetric
	I	shading device	3 overhang
	J	roof aperture	4 glazed
III	K	urban layout	
	L	façades reflectances	
	M	street top lighting	

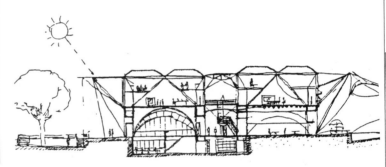

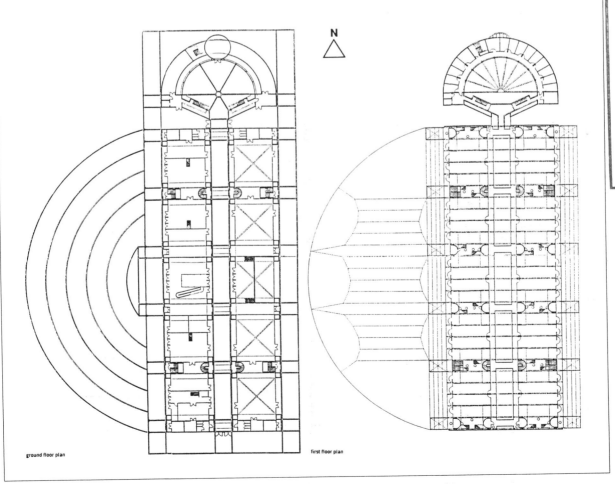

ground floor plan   first floor plan

11.12 Daylighting in Architecture

Summary Building Data

Building: Municipal Library
Location: Barcelona, Spain
Latitude: 41° N
Architects: Galí, Quintana and Solanas
Date: 1991
Source: *The Architectural Review*, No. 1133, July 1991.
Photo: H. Suzuki

MORPHOLOGICAL DIAGRAM			
LEVEL	PARAMETERS		VARIABLES
I	A	room layout	1 unilateral/ 2 bilateral
	B	collector posn.	1 within plane/ 5
	C	collecting area	2 15-30%/ 3 >30%
	D	aperture shape	1 intermediate/ 4
	E	glare control	1 light filters/ 2
II	F	plan layout	3 unilateral slabs
	G	wall /aperture	2 50%/ 3 75%
	H	aperture distribution	2 total asymmetric
	I	shading device	3 overhang
	J	roof aperture	6 other
III	K	urban layout	9 other
	L	façades reflectances	
	M	street top lighting	

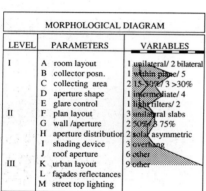

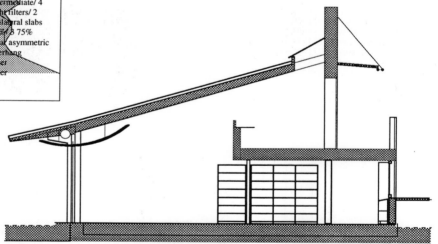

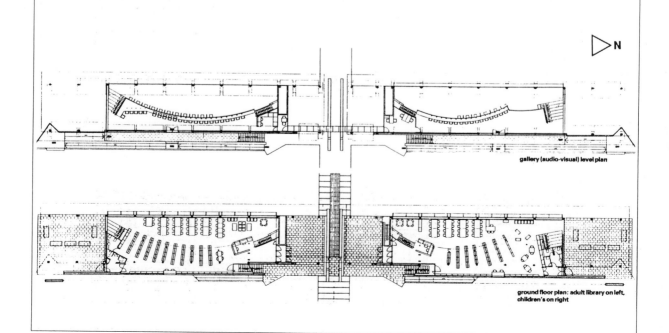

gallery (audio-visual) level plan

ground floor plan: adult library on left, children's on right

Case Study Analysis 11.13

Summary Building Data

Building: Conphoebus Laboratory
Location: Catania, Sicily, Italy
Latitude: 38° N
Architect: Sergio Los and
Natasha Pulitzer
Date: 1979-87
Source: Licata, R. *et al.*, "A test building for perimetrical enveloping components", *Proc. ISES Conf., Milan, April 1990.*

MORPHOLOGICAL DIAGRAM			
LEVEL	PARAMETERS		VARIABLES
I	A	room layout	1 unilateral/ 2 bilateral
	B	collector posn.	1 within plane/ 5
	C	collecting area	2 15-30%/ 3 >30%
	D	aperture shape	1 intermediate/ 4
	E	glare control	1 light filters/ 2
II	F	plan layout	3 unilateral slabs
	G	wall /aperture	2 50%/ 3 75%
	H	aperture distribution	2 solar asymmetric
	I	shading device	3 overhang
	J	roof aperture	6 other
III	K	urban layout	9 other
	L	façades reflectances	
	M	street top lighting	

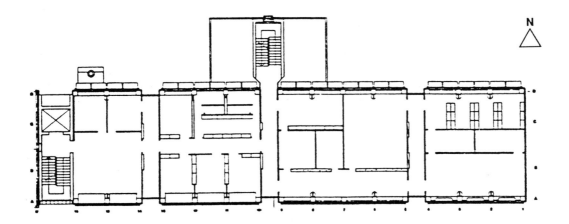

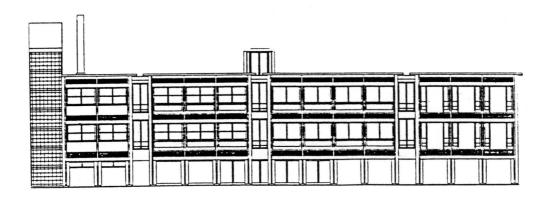

11.14 Daylighting in Architecture

Summary Building Data

Building: Tennessee Valley Authority Office Complex
Location: Chattanooga, Tennessee, USA
Latitude: 35° N
Architect: Caudill Rowlett Scott/
The Architects Collaborative/
Van der Ryn Calthorpe & Partners
Lighting Consultants:
William Lam Associates
Date: 1986
Source: Lam, W.M.C., *Sunlighting as a Formgiver for Architecture*,
Van Nostrand Reinhold Company,
New York, 1986.

	MORPHOLOGICAL DIAGRAM	
LEVEL	PARAMETERS	VARIABLES
I	A room layout	2 bilateral
	B collector posn.	1 within plane
	C collecting area	1 side light <15%
	D aperture shape	1 intermediate
	E glare control	2 rigid screens
II	F plan layout	5 atrium
	G wall /aperture	2 50%
	H aperture distribution	3 urban asymmetric
	I shading device	2 brise soleil
	J roof aperture	5 screened glass
III	K urban layout	3 north-south slabs
	L façades reflectances	
	M street top lighting	

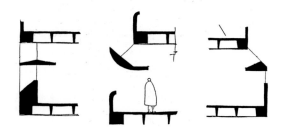

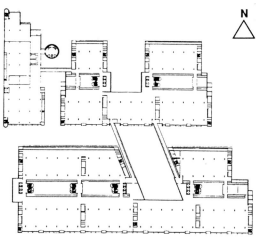

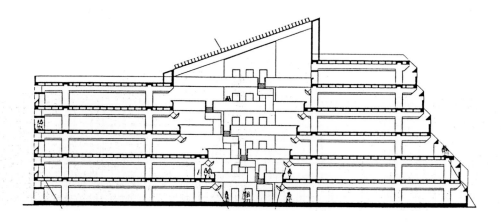

Case Study Analysis **11.15**

Summary Building Data

Building: National Assembly Hall
Location: Dacca, Bangladesh
Latitude: 24° N
Architect: Louis I. Kahn
Date: 1962-74
Source: "Louis I. Kahn.
Conception and Meaning",
Architecture and Urbanism (A+U),
November 1983.

MORPHOLOGICAL DIAGRAM		
LEVEL	PARAMETERS	VARIABLES
I	A room layout	1 unilateral/ 3 deep
	B collector posn.	2 skylight/ 4 corners
	C collecting area	2 side 15-30/ 5 top 15-30%
	D aperture shape	3 vertical/ 5 ceiling
	E glare control	3 other (sun screen)
II	F plan layout	6 double shell
	G wall /aperture	1 25%
	H aperture distribution	4 other
	I shading device	2 brise soleil
	J roof aperture	5 other
III	K urban layout	5 tower
	L façades reflectances	
	M street top lighting	

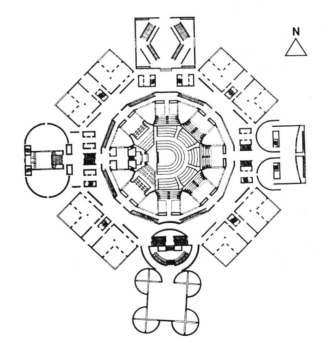

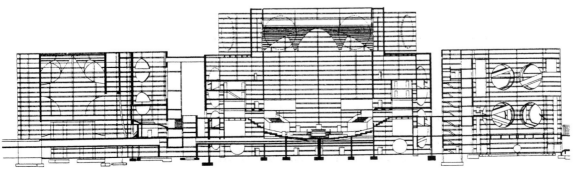

REFERENCES

(1) Lera, S., "Architects' Design Strategies: Some Justifications for Current Practices" in Powell, J.A., Cooper, I., Lera, S., *Designing for Building Utilisation*, E. & F..N Spon, London, 1984.

(2) Schon, D. A., "Designing: Rules, Types and Words", *Design Studies*, vol. 7, No. 4, July 1980.

(3) Mazria, E., *The Passive Solar Energy Book*, Rodale Press, Emmaus, Pennsylvania, 1979.

(4) Alexander, C., Ishikawa, S. and Silverstein, M., *A Pattern Language*, Oxford University Press, New York, 1977.

(5) Watson, D. and Labs, K., *Climatic Design*, McGraw-Hill, New York, 1983.

(6) Los, S. and Pulitzer, N., *L'Architettura del regionalismo, guida alla progettazione bioclimatica nel Trentino*, Provincia Autonoma di Trento, Servizio Energia, Trento, Italy, 1985.

(7) Lam, W. M. C., *Sunlighting as a Formgiver for Architecture*, Van Nostrand Reinhold Company, New York, 1985.

(8) Robbins, C. L., *Daylighting Design & Analysis*, Van Nostrand Reinhold Company, New York, 1986.

(9) Moore, F., *Concepts and Practice of Architectural Daylighting*, Van Nostrand Reinhold Company, New York, 1986.

(10) Brown, G. Z., Reynolds, J. and Ubbeholde, S., *Design Procedures for Daylighting, Passive Solar Heating and Cooling*, Teaching Passive Design in Architecture, DoE, Philadelphia, 1981.

(11) Ternoey, S., Bickle, L., Robbins, C., Busch, R. and McCord, K., *The Design of Energy-Responsible Commercial Buildings*, John Wiley and Sons, Chichester 1985.

(12) March, L. and Stiny, G., "Design Machines", *Environmental Planning B*, vol. 8, No. 3, 1981

(13) Zwicky, F. and Wilson, A. G., (eds.), *New Methods of Thought and Procedure*, Springer Verlag, New York, 1967.

(14) Los, S., "Solar Passive Architecture Design" in *Proc. 3rd National Passive Solar Conference*, ASES, San Jose, California, 1979

(15) Hilbert, D., *Grundlagen der Geometrie*, B. G. Teubner, Stuttgart, 1968.

(16) Jones, C., *Design Methods: Seeds of Human Future*, John Wiley and Sons, Chichester 1965; Grant, D., "How to construct a Morphological Box", in *Design Methods and Theories*, vol. 11, No.3, 1977.

(17) Stiny, G., "Introduction to Shape and Shape Grammar" in *Environment and Planning B*, vol. 7, No.3, 1978.

(18) Hawkes, D. and Willey, H., "User response in the environmental control system", in Steadman, P. and Owers, J. (eds.), *Transactions of the Martin Centre for Architectural and Urban Studies*, vol. 2, Cambridge, 1977.

GLOSSARY

ARCHITECTURE AND DAYLIGHTING   **Code CEC**

A. General terms   CEC 001 to 007
B. Daylighting components   CEC 008 to 033
C. Control Elements   CEC 034 to 054
D. Materials   CEC 055 to 064

LIGHTING TECHNOLOGY   **Code CIE Vocabulary**

Section 845-01 - Radiation, quantities and units   01-01 to 01-56
  CEC Extension   CEC 070

Section 845-02 - Vision, colour rendering   02-07 to 02-62

Section 845-03 - Colorimetry   03-10 to 03-49

Section 845-04 - Emission, optical properties of materials   04-04 to 04-110
  CEC extension   CEC 080 to CEC 082

Section 845-05 - Radiometric, photometric and colorimetric measurements; physical detectors   05-09 to 05-30

Section 845-09 - Lighting technology, daylighting   09-01 to 09-105
  CEC Extension   CEC 090 to CEC 100

ARCHITECTURE AND DAYLIGHT

A. GENERAL TERMS   Translation : F,G,I,S

CEC 001   **Directional lighting**
A form of lighting where the major part of light is received from a single direction

Eclairage dirigé
Gerichtete Beleuchtung
Illuminazione direzionale
Iluminación dirigida

CEC 002   **Diffuse lighting**
A form of lighting where approximately the same intensity of light comes from different directions

Eclairage diffus
Diffuse Beleuchtung
Iluminazione diffusa
Iluminación difusa

CEC 003   **Indirect lighting**
Lighting achieved by reflection, usually from wall and/or ceiling surfaces

Eclairage indirect
Indirekte Beleuchtung
Iluminazione indiretta
Iluminación indirecta

CEC 004   **Direct sunlight**
Portion of daylighting coming directly from the sun at a specific location which is not diffused on arrival

Lumière solaire directe
Direktes Sonnenlicht
Luce solare diretta
Iluminación solar directa

CEC 005   **Natural lighting system**
Component or a series of components joined in a specific building for natural daylighting

Système d'éclairage naturel
Natürliches Beleuchtungs-Syste
Listema di illuminazione natur
Sistema de iluminación natural

CEC 006   **Lateral light**
Light which enters laterally to interior spaces through windows, curtain walls, etc...

Lumière latérale
Seitliches Licht
Luce laterale
Luz lateral

CEC 007   **Toplighting**
Light which enters through the top part of interior space such as clerestories, light ducts or skylights, etc...

Eclairage zénithal
Oberlicht
Illuminazione zenitale
Iluminación cenital

B. DAYLIGHTING COMPONENTS

Translation : F,G,I,S

CEC 008   **Conduction component**
Space designed to guide and/or distribute daylight towards the interior of a building, from one pass-through component to another

Composant de conduction
Konduktions-Komponente
Componente diffusore di luce
Componente de conducción

CEC 009   **Intermediate light space**
Conduction component which is part of the perimetral zone of a building guiding and distributing daylight into attached interior spaces

Espace lumineux intermédiaire
Zwischenlichtraum
Spazio luminoso intermedio
Espacio de luz intermedio

CEC 010   **Interior light space**
Conduction component which is part of the interior zone of a building guiding and distributing daylight into specific zones of a building separated from the outside

Espace lumineux intérieur
Innenlichtraum
Spazio luminoso interno
Espacio de luz interior

CEC 011   **Pass-through component**
Architectonic part of a building which connects two light environments, permetting light to pass from one to another

Composant de transmission
Transmissions-Komponente
Componente collettore di luce
Componente de paso

CEC 012   **Lateral pass-through component**
Pass-through component which is situated in the interior or exterior vertical enclosures of a building. It separates two light environments, permetting lateral penetration of light

Composant latéral de passage
Seitliche Transmissions-Komponente
Componente collettore di luce laterale
Componente de paso lateral

CEC 013   **Zenithal pass-through component**
Pass-through component which is situated in both interior and exterior horizontal separations of a building. It separates two light environments allowing zenithal entry of daylight to the lower space

Composant zénithal de passage
Zenithale Transmissions-Komponente
Componente collettore di luce zenitale
Componente de paso cenitale

CEC 014   **Global pass-through component**
Pass-through component which is part of the enclosure of a constructed volume composed of a surface with transparent or translucent material. It surrounds totally or partially a luminic environment, permitting a global entry of daylight

Composant global de passage
Globale Transmissions-Komponente
Componente collettore di luce generale
Componente de paso global

CEC 015   **Gallery**
An intermediate light space attached to a building. Its may be open to the exterior (open gallery) or closed off by glass (closed gallery), in order to let daylight into the inside parts of the building

Galerie
Galerie
Galleria
Galeria

CEC 016   **Porch**
An intermediate light space attached to a building at ground level, open to the exterior environment which permits the entry of daylight to the parts of the building directly connected to the porch by pass-through components

Veranda
Veranda
Portico
Porche

CEC 017 **Greenhouse** Serre
An intermediate light space attached to a building by one of its Glashaus
faces, the others being separated from the exterior by a frame Serra
supporting transparent of translucent surfaces. It can be opened Invernadero
in certain places to allow ventilation. It permits the entry of light
and direct solar radiation towards the interior space through its
surfaces

CEC 018 **Courtyard** Patio
An exterior light space enclosed laterally by the walls of one or Hof
several buildings and open to the exterior at the top and sometimes Corte
laterally, which allows natural ventilation and the entry of light Patio
to the spaces related to it

CEC 019 **Atrium** Atrium
An interior light space enclosed laterally by the walls of a building Atrium
and covered with transparent or translucent material which Atrium
permits the entry of light to the other interior spaces linked to Atrio
it by pass-through components

CEC 020 **Light-duct** Conduit de lumière
An interior light space which conducts natural light to interior Lichtschacht
zones of a building. Its surfaces are finished with high reflection Condotto luminoso
materials Conducto de luz

CEC 021 **Sun-duct** Conduit de soleil
A non-habitable interior light space with a prominent dimension, Sonnenlichtschacht
designed to conduct solar beams to interior zones which surfaces Condotto solare
are covered with high reflection finishings Conducto de sol

CEC 022 **Window** Fenêtre
A vertical opening in a wall with an inferior limit above floor Fenster
level. Its permit lateral penetration of light or direct solar Finestra
radiation, interchange of view, and natural ventilation Ventana

CEC 023 **Balcony** Balcon
A vertical opening in a wall with an inferior limit at floor level Balkon
which allows people to go from the interior to the exterior. It Balcone
permits the lateral penetration of light or direct solar radiation, Balconera
interchange of view, passage and natural ventilation

CEC 024 **Translucent wall** Mur translucide
Walls constructed with translucent materials making up part of Transluzente Wand
a vertical enclosure in a building. It separates two light envi- Parete traslucida
ronments permitting the lateral penetration of natural light and Muro translucido
diffusing it through the translucent material

CEC 025 **Curtain wall** Mur rideau
A continuous translucent or transparent vertical or almost vertical Vorhangfassade
surface, with no structural function, that separates the interior Facciata continua
from the exterior of a building. It permits lateral penetration of Muro cortina
natural light, interchange of view and usually does not allow
ventilation

CEC 026	**Clerestory** An upper elevation with vertical or tilted openings in one or several laterals over the roof plane. It permits zenithal penetration of daylight, protecting against direct radiation and/or redirecting it towards lower spaces. It can allow natural ventilation without an exterior view	Lucarne Oberlicht Lucernario o lanternino Lucernario
CEC 027	**Monitor roof** A raised section of a roof, including the ridge, with lateral openings on it. It permits the zenithal entry of daylight towards the lower zone increasing luminic level and allowing ventilation through the apertures which can be opened or closed	Lanterneau vertical First-Lichtband Lucernario verticale Tejado monitor
CEC 028	**North-light roof** Series of successive parallel north orientated slopes with vertical or tilted lineal openings in each slope. It permits zenithal entry of daylight providing diffuse light and high luminic level without contrast t lower space. The space can be ventilated by the apertures which can be opened or closed	Toiture en sheds ou dents de scie Sheddach nach Norden Copertura a denti di sega Cubierta en diente de sierra
CEC 029	**Translucent ceiling** A horizontal closing partially constructed with translucent materials, which separates the inside from the outside space or two inside superposed spaces. It permits the zenithal entry of daylight diffused through the translucent material to the lower space providing a homogeneous luminic level	Plafond translucide Transluzente Decke Soffitto traslucido Forjado translúcido
CEC 030	**Skylight** An opening situated in a horizontal or tilted roof. It permits the zenithal entry of daylight increasing the luminic level of the lower space under the skylight. It can be opened to permit ventilation	Lanterneau horizontal Dachflächenfenster Lucernario orizzontale Claraboya
CEC 031	**Dome** A hemispheric roof which may have perforations or can be constructed in its totallity with translucent materials. It permits the zenithal illumination of the space under it covering all or most of the inferior area	Dôme Kuppel Cupola Cúpula
CEC 032	**Lantern** A covered elevation on the highest point of a roof with lateral openings, through which light enters. It permits zenithal entry of daylight to the central zone of the lower space	lanterne Laterne Lanterna Linterna
CEC 033	**Membrane** A translucent or transparent surface totally or partially involving a luminic environment. It permits a global entry of light to the luminic space and provides a high non-contrastung inside luminic level	Membrane Membrane Membrane Membrana

C. CONTROL ELEMENTS

Translation : F,G,I,S

CEC 034 **Control element**
Particular device specially designed to admit and/or control the entry of light through a pass-through component

Elément de contrôle
Steuer/Regel-Element
Elemento di controllo
Elemento de control

CEC 035 **Separator surface**
Control element of a transparent or translucent material which separates two light environments permitting light to pass through while detaining air and sometimes view

Surface séparatrice
Trennfläche
Divisorio trasparente
Superficie separadora

CEC 036 **Flexible screen**
Control element which partially or totally detains sunlight and also diffuses natural light, allowing natural ventilation and impedes any view from the exterior. It can be opened or closed to eliminate the above mentioned

Ecran flexible
Bewegliches Gitter
Schermatura flessibile
Pantalla flexible

CEC 037 **Rigid screen**
Rigid and opaque control element which redirects and/or detains direct solar radiation falling upon a pass-through component, which is normally a fixed structure and cannot be regulated

Ecran rigide
Starres Gitter
Schermatura rigida
Pantalla rígida

CEC 038 **Solar filter**
Control element which covers the entire surface of an opening, protecting the interior zones against direct solar radiation and allowing ventilation. It can be fixed, open/close or adjustable

Filtre solaire
Sonnenfilter
Filtro solare
Filtro solar

CEC 039 **Solar obstructor**
Control element composed of opaque and adjustable surfaces which covers the whole of the opening

Pare-soleil
Sonnenverschattung
Frangisole
Obstructor solar

CEC 040 **Conventional division**
Control element placed in a pass-through component, which shares two environments while allowing view and light to through

Partition conventionnelle
Konventionelle Lichtverteilung
Divisorio convenzionale
Separador convencional

CEC 041 **Treated division**
Control element placed in a pass-through component, which shares two environments modifying the radiation characteristics passing through it diffusing, redirecting or controlling its intensity depending on the specific treatment of the division

Partition sélective
Modifizierte Lichtverteilung
Divisorio selettivo
Separador corrector

CEC 042 **Prismatic division**
Control element placed in a pass-through component which share two environments redirecting light by its optical-geometrical characteristics

Partition prismatique
Prismatische Lichtverteilung
Divisorio prismatico
Separador prismatizado

CEC 043 **Active division**
Control element placed in a pass-through component which share two environments changing optical absorption properties of certain materials by an externally applied electric field

Partition active
Aktive Lichtverteilung
Divisorio attivo
Separador activo

CEC 044	**Awning** A control element made of opaque or diffusing flexible material placed on the exterior of a pass-through component to protect it agiants the sun	Store Sonnensegel / Markise Tenda Toldo
CEC 045	**Curtain** A control element made of opaque or diffusing flexible material placed inside a pass-through component to protect against sun, light or view	Rideau Vorhang Cortina Cortina
CEC 046	**Overhang** A control element which is part of the building itself protruding horizontally from the facade above a vertical pass-through component. It protects the zones close to the openings of the building, obstructing direct solar radiation which falls upon the overhang	Auvent Überhang Aggetto Alero
CEC 047	**Light-shelf** A control element usually placed horizontally above eye level in a vertical pass-through component, dividing it into a superior an inferior area. It protects the interior zones to the openings against direct solar radiation, obstructing and redirecting light to the interior celing	Console lumineuse Light-shelf/Lichtblende Schermo riflettore Repisa de luz
CEC 048	**Sill** A control element placed horizontally on the bottom of a window opening. It reflects and redirects the natural light which falls upon the sill increasing the luminic level in the interior zone	Allège Fensterbank Davanzale Alfeizar
CEC 049	**Fin** A control element placed on the exterior facade of a building and fixed vertically on the sides of the opening. It reflects and redirects natural light which falls laterally upon the fin to the inside	Alette ou aileron Vertikale Lamelle Cornice Aleta
CEC 050	**Baffle** A fixed single opaque or translucent element, which reflects or protects a pass-through component against direct solar radiation at certain angles, and may reflect daylight to the interior	Ecran Vertikaler Sonnenschutz Diaframma Pantalla solar
CEC 051	**Blind** An exterior or interior element composed of slatted screens placed over the whole of a window. The slats may be fixed or movable. When movable it may be adjusted according to sun angle and shading requirements. This device can be moved along the opening, drawn to the side, or rolled up to the top	Persienne Jalousie Persiana Persiana
CEC 052	**Louver** A series of exterior and fixed slats which may be adjustable. It covers the whole of the opening outside the window. Depending on the orientation of the slats, solar radiation which falls upon them may be obstructed and/or reflected and/or redirected to the interior	Paralumes Lamelle Frangisole Lamas

CEC 053   **Jalousie or brise-soleil**   Jalousie
           An exterior fixed element composed of a perforated frame   Jalousie
           covering the whole pass-through component that permits natural   Gelosie ; brise-soleil
           light and ventilation through the apertures. Depending on the   Celosía
           geometric design of the frame it may detain direct solar radiation
           at certain sun angles

CEC 054   **Shutter**   Volet
           A control element which is a continuous exterior or interior   Laden
           opaque surface which totally detains solar radiation. It can be   Scuri
           folded or drawn towards the side of the opening   Porticón

D. MATERIALS

Translation : F,G,I,S

CEC 055 **Diffuser**
A device object or surface used to alter the spacial distribution of light by diffusing it by reflection or transmission

Diffuseur
Diffusor
Diffusore
Difusor

CEC 056 **Reflector**
A device which returns incident visible radiation used to alter the spacial distribution of light

Reflecteur
Reflektor
Riflettore
Reflector

CEC 057 **Refractor**
A device which redirects light changing its direction when passing from one medium to another

Refracteur
Refraktor
Rifrattore
Refractor

CEC 058 **Translucent glass**
A glass with the property of transmiting light diffusely, and through which vision varies from almost clear to almost obscure

Verre translucide
Transluzentes Glas
Vetro traslucido
Vidrio translúcido

CEC 059 **Frosted glass**
A non transparent glass treated in order to allow the entry of daylight and to protect against the view from the exterior

Verre dépoli
Genörpeltes Glas
Vetro smerigliato
Vidrio esmerilado

CEC 060 **Glass-block**
Construction element of translucent material placed in a closing with non load-bearing function

Pavé de verre
Glasbaustein
Vetro-cemento
Bloques de vidrio

CEC 061 **Glass slate or glass tile**
A piece of glass, similar in size to a tile or slate, placed on a roof which permits daylight into lower space

Tuile de verre
Glasziegel
Vetro-tegola
Teja de vidrio

CEC 062 **Matte surface**
A surface with the property of diffusing the reflected light

Surface mate
Matte Oberfläche
Superficie opaca
Superficie mate

CEC 063 **Specular surface**
A surface with reflective properties where the angle of visible incident radiation is equal to the angle of reflection.

Surface spéculaire
Spiegelnde Oberfläche
Superficie speculare
Superficie especular

CEC 064 **Glossy material**
Material which spreads light towards a predominant direction.

Matériau brillant
Glänzendes Material
Materiale lucido
Material brillante

GL.10 Daylighting in Architecture

SECTION 845-01 - RADIATION, QUANTITIES AND UNITS

A. GENERAL TERMS   Translation : F, G, I, S

01-01   **(Electromagnetic) radiation**   Rayonnement (ou radiation)
        1. Emission or transfer of energy in the form of electromagnetic Strahlung
        waves with the associated photons   Radiazione
        2. These electromagnetic waves or these photons   Radiación
        NOTE : The French term "radiation" applies preferably to a single element of
        any radiation, characterized by one wavelength or one frequency (see
        01-07)

01-02   **Optical radiation**   Rayonnement optique
        Electromagnetic radiation at wavelengths between the region of   Optische Strahlung
        transition to X-rays ($\lambda \approx 1\,nm$) and the region of transition to   Radiazione ottica
        radio waves ($\lambda \approx 1\,mm$)   Radiación óptica

01-03   **Visible radiation**   Rayonnement visible
        Any optical radiation capable of causing a visual sensation directly   Sichtbare Strahlung
                                                                          Radiazione visibile
        NOTE : There are no precise limits for the spectral range of visible radiation since   Radiación visible
        they depend upon the amount of radiant power reaching the retina and
        the responsivity of the observer. The lower limit is generally taken
        between 360 nm and 400 nm and the upper limit between 760 nm and
        830 nm

01-06   **Light**   Lumière
        1. Perceived light (see 02-17)   Licht
        2. Visible radiation (see 01-03)   Luce
        NOTES : 1 - The world light is sometimes used in sense 2 for optical radiation   Luz
                      extending outside the visible range, but this usage is not
                      recommended
                 2 - The terms "light" in English and "Licht" in German are also used,
                      especially in visual signalling, for certain lighting devices and for
                      light signals

01-07   **Monochromatic radiation**   Radiation monochromatique
        Radiation characterized by a single frequency. In practice,   Monochromatische Strahlung
        radiation of a very small range of frequencies which can be   Radiazione monocromatica
        described by stating a single frequency   Radiación monocromática
        NOTE : The wavelength in air or in vacuo is also used to characterize a
        monochromatic radiation

01-08   **Spectrum** *(of a radiation)*   Spectre (d'un rayonnement)
        Display or specification of the monchromatic components of the   Spektrum (einer Strahlung)
        radiation considered   Spettro
        NOTES : 1 - There are line spectra, continous spectra, and spectra exhibiting   Espectro
                      both these characteristics
                 2 - This term is also used for spectral efficiencies (excitation
                      spectrum, action spectrum)

01-11 **Coherent radiation**
Monochromatic radiation whose electromagnetic oscillations maintain constant phase differences from one point to another

Rayonnement cohérent
Kohärente Strahlung
Radiazione coerente
Radiación coherente

01-20 **Steradian** (*sr*)
SI unit of solid angle : Solid angle that, having its vertex at the centre of a sphere, cuts off an area of the surface of the sphere equal to that of a square with sides of length equal to the radius of the sphere (ISO, 31/1-2.1, 1978)

Stéradian
Steradiant
Steradiante
Estereorradian

B. RADIANT, AND LUMINOUS QUANTITIES AND THEIR UNITS

Translation : F, G, I, S

PRELIMINARY NOTES

1. *Photopic and scotopic quantities* - Luminous (photometric) quantities are two kinds, those used for photopic vision and those used for scotopic vision. The wording of the definitions in the two cases being almost identical, a single definition is generally sufficient with the appropriate adjective, *photopic* or *scotopic* added where necessary. The symbols for scotopic quantities are prime
(Φ', $V'(\lambda)$ etc), but the units are the same in both case.
For mesopic vision, the CIE has not yet defined the relevant quantities
2. *Radiant, and luminous (photometric) quantities* - These two kinds of quantities have the same basic symbols, identified respectively, where necessary, by the subscript e (energy), or v (visual), e.g. Φ_e, Φ_v
3. The adjective *luminous* used here is also used in section 02 (Vision), but with a different meaning

01-22 **Spectral luminous efficiency** (*of a monochromatic radiation of wavelength* λ) (*$V(\lambda)$ for photopic vision ; $V'(\lambda)$ for scotopic vision*)
Ratio of the radiant flux at wavelength λ_m to that wavelength λ such that both radiations produce equally intense luminous sensations under specified photometric conditions and λ_m is chosen so that the maximum value of this ratio is equal to 1

NOTE : Unless otherwise indicated, the value used for the spectral luminous efficiency in photopic vision are the values agreed internationally in 1924 by the CIE (Compte Rendu 6ᵉ session, p. 67), completed by interlopation and extrapolation (Publications CIE N° 18 (1970), p. 43 and N° 15 (1971), p. 93), and recommended by the International Committee of Weights and Measures (CIPM) in 1972.
For scotopic vision, the CIE in 1951 adopted, for young observers, the values published in Compte Rendu 12ᵉ session, Vol. 3, p. 37, and ratified by the CIPM in 1976.
These values define respectively the $V(\lambda)$ or $V'(\lambda)$ functions represented by the $V(\lambda)$ or $V'(\lambda)$ curves
(see the curve, following page)

Efficacité lumineuse relative spectrale
Spektraler Hellempfindichkeitsgrad
Fattore spettrale di visibilità
Eficiencia luminosa espectral

GL.12 Daylighting in Architecture

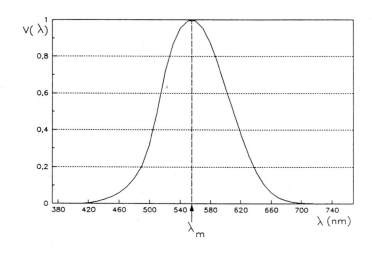

01-23   **CIE standard photometric observer**
Ideal observer having a relative spectral responsivity curve that conforms to the $V(\lambda)$ function for photopic vision or to the $V'(\lambda)$ function for scotopic vision, and that complies with the summation law implied in the definition of luminous flux

Observateur de référence phototrique CIE
Photometrischer NormalbeobachCIE
Osservatore fotometrico CIE
Observador fotométrico patrón C

01-24   **Radiant flux ; radiant power** $(\Phi_e ; \Phi ; P)$
Power emitted, transmitted or received in the form of radiation
Unit : W

Flux énergétique ; puissarayonnante
Strahlungsleistung ; Strahlunfluss
Flusso energetico
Flujo radiante

01-25   **Luminous flux** $(\Phi_V ; \Phi)$
Quantity derived from radiant flux Φ_e by evaluating the radiation according to its action upon the CIE standard photometric observer. For photopic vision $\Phi_v = K_m \int_0^\infty \frac{d\Phi_e(\lambda)}{d\lambda} \cdot V(\lambda) d\lambda$ where $\frac{d\Phi_e(\lambda)}{d\lambda}$ is the spectral distribution of the radiant flux and $V(\lambda)$ is the spectral luminous efficiency
Unit : lm
NOTE : For the values of K_m (photopic vision) and K_m' (scotopic vision), see 01-56

Flux lumineux
Lichstrom
Flusso luminoso
Flujo luminoso

01-31   **Luminous intensity** *(of a source, in a given direction)* $(I_v ; I)$
Quotient of the luminous flux $d\Phi_v$ leaving the source and propagated in the element of solid angle $d\Omega$ containing the given direction, by the element of solid angle

Intensité lumineuse
Lichtstärke
Intensità luminosa
Intensidad luminosa

01-34 **Radiance** *(in a given direction, at a given point of a real or imaginary surface)* $(L_e\ ;\ L)$

Quantity defined by the formula $L_e = \dfrac{d\Phi_e}{dA \cdot \cos\theta \cdot d\Omega}$ where $d\Phi_e$ is the radiant flux transmitted by an elementary beam passing through the given point and propagating in the solid angle $d\Omega$ containing the given direction ; dA is the area of a section of that beam containing the given point ; θ is the angle between the normal to that section and the direction of the beam.
Unit : $W \cdot m^{-2} \cdot sr^{-1}$

Luminance énergétique ; radiance
Strahldichte
Radianza
Radiancia

NOTES 1 to 2. In the two following notes the symbols for the quantities are without subscripts because the formulas are also valide for the term 01-35

1. For an area dA of the surface of a source, since the intensity dI of dA in the given direction is $dI = d\Phi/d\Omega$, then an equivalent formula is $L = dI/dA \cdot \cos\theta$, a form mostly used in illuminating engineering
2. For an area dA of a surface receiving the beam, since the irradiance or illuminance dE produced by the beam on dA is $dE = d\Phi/dA$, then an equivalent formula is $L = dE/d\Omega \cdot \cos\theta$, a form useful when the source has no surface (e.g. the sky, the plasma of a discharge)

01-35 **Luminance** *(in a given direction, at a given point of a real or imaginary surface)* $(L_v\ ;\ L)$

Quantity defined by the formula $L_v = \dfrac{d\Phi_v}{dA \cdot \cos\theta \cdot d\Omega}$, where $d\Phi_v$ is the luminous flux transmitted by an elementary beam passing through the given point and propagating in the solid angle $d\Omega$ containing the given direction ; dA is the area of a section of that beam containing the given point ; θ is the angle between the normal to that section and the direction of the beam.
Unit : $cd \cdot m^{-2} = lm \cdot m^{-2} \cdot sr^{-1}$
NOTE : See notes 1 to 2 to 01-34

Luminance (lumineuse) ; luminance visuelle
Leuchtdichte
Luminanza
Luminancia

01-37 **Irradiance** *(at a point of a surface)* $(E_e\ ;E)$
Quotient of the radiant flux $d\Phi_e$ incident on an element of the surface containing the point, by the area dA of that element
Equivalent definition. Integral, taken over the hemisphere visible from the given point, of the expression $L_e \cdot \cos\theta \cdot d\Omega$ where L_e is the radiance at the given point in the various directions of the incident elementary beams of solid angle $d\Omega$, where θ is the angle between any of these beams and the normal to the surface at the given point.

$$E_e = \frac{d\Phi_e}{dA} = \int_{2\pi sr} L_e \cdot \cos\theta \cdot d\Omega$$

Unit : $W \cdot m^{-2}$

Eclairement énergétique
Bestrahlungsstärke
Irradiamento
Irradiancia

CEC 070 **Irradiation (radiant exposure - CIE 01-42)**
Surface density of the radiant energy received
Unit : $W.S.m^{-2}$

Irradiation
Bestrahlung
Irradiazione
Irradiación

GL.14 Daylighting in Architecture

01-38 **Illuminance** *(at a point of a surface)* $(E_v ; E)$
Quotient of the luminous flux $d\Phi_v$ incident on an element of the surface containing the point, by the area dA of that element
Equivalent definition. Integral, taken over the hemisphere visible from the given point, of the expression $L_v \cdot \cos\theta \cdot d\Omega$ where L_v is the luminance at the given point in the various directions of the incident elementary beams of solid angle $d\Omega$, and θ is the angle between any of these beams and the normal to the surface at the given point.

$$E_v = \frac{d\Phi_v}{dA} = \int_{2\pi sr} L_v \cdot \cos\theta \cdot d\Omega$$

Unit : $lx = lm \cdot m^{-2}$

Eclairement (lumineux)
Beleuchtungsstärke
Illuminamento
Iluminancia

01-40 **Spherical irradiance ; radiant fluence rate** *(at a point)* $(E_{e,o} ; E_o)$
Quantity defined by the formula $E_{e,o} = \int_{4\pi sr} L_e d\Omega$ where $d\Omega$ is the solid angle of each elementary beam passing through the given point and L_e its radiance at that point
Unit : $W \cdot m^{-2}$

NOTES :
1 - This quantity is the quotient of the radiant flux of all the radiation incident on the outer surface of an infinitely small sphere centered at the given point, by the area of the diametrical cross-section of that sphere
2 - The analogous quantitiues <u>spherical illuminance</u> $E_{e,o}$ and <u>photon spherical irradiance</u> $E_{p,o}$ are defined in a similar way, replacing radiance L_e by luminance L_v or photon radiance L_p
3 - The term "spherical irradiance", or <u>scalar irradiance</u>, or similar terms may be found in the literature, in the definition of which the area of the cross-section is sometimes replaced by the surface area of the spherical element which is four times larger

Eclairement sphérique énergétic
débit de fluence énergétique
Raumbestrahlungsstärke ; Energ
fluβdichte
Irradiamento sferico
Irradiancia esférica

01-41 **Cylindrical irradiance** *(at a point, for a direction)* $(E_{e,z} ; E_z)$
Quantity defined by the formula

$$E_{e,z} = \frac{1}{\pi} \int_{4\pi sr} L_e \sin\epsilon \cdot d\Omega$$

where $d\Omega$ is the solid angle of each elementary beam passing through the given point, L_e its radiance at that point and ϵ the angle between it and the given direction ; unless otherwise stated, that direction is vertical
Unit : $W \cdot m^{-2}$

NOTES :
1 - This quantity is the quotient of the radiant flux of all radiation incident on the outer curved surface of an infinitely small cylinder containing the given point and whose axis is in the given direction, by π times the area of the cross-section of that cylinder measured in a plane containing its axis
2 - The analogous quantities <u>cylindrical illuminance</u> $E_{v,z}$ and <u>photon cylindrical irradiance</u> $E_{p,z}$ are defined in a similar way, replacing radiance L_e by luminance L_v or photon radiance L_p

Eclairement cylindrique énergé
que
Zylindrische Betrahlungsstärke
Irradiamento cilindrico
Irradiancia cilíndrica

01-50   **Candela** (*cd*)   Candela
         SI unit of luminous intensity : The candela is the luminous   Candela
         intensity, in a given direction, of a source that emits mono-   Candela
         chromatic radiation of frequency 540 x 10^{12} hertz and that has   Candela
         a radiant intensity in that direction of 1/683 watt per steradian.
         (16th General Conference of Weights and Measures, 1979)
         $$1\ cd = 1\ lm \cdot sr^{-1}$$

01-51   **Lumen** (*lm*)   Lumen
         SI unit of luminous flux : Luminous flux emitted in unit solid Lumen
         angle (steradian) by a uniform point source having a luminous   Lumen
         intensity of 1 candela. (9th General Conference of Weights and Lumen
         Measures, 1948)
         Equivalent definition. Luminous flux of a beam monochormatic
         radiation whose frequency is 540 x 10^{12} hertz and whose radiant
         flux is 1/683 watt

01-52   **Lux** (*lx*)   Lux
         SI unit of illuminance : Illuminance produced on a surface of   Lux
         area 1 square metre by a luminous flux of 1 lumen uniformly   Lux
         distributed over that surface   Lux
         $$1\ lx = 1\ lm \cdot m^{-2}$$
         NOTE : Non-metric unit : lumen per square foot ($lm.ft^{-2}$) or footcandle (fc)
                 (USA) = 10.764 lx

01-53   **Candela per square metre** ($cd.m^{-2}$)   Candela par mètre carré
         SI unit of luminance   Candela pro Quadratmeter
         NOTE : This unit was sometimes called the nit (nt) (name discouraged).   Candela al metro quadrato
                 Other units of luminance :   Candela por metro cuadrado

                 metric, non-SI : Lambert $(L) = \frac{10^4}{\pi} cd \cdot m^{-2}$

                 non-metric : footlambert $(fL) = 3.426\,cd \cdot m^{-2}$

01-56   **Luminous efficacy of radiation** (*K*)   Efficacité lumineuse d'un rayon-
         Quotient of the luminous flux Φ_v by the corresponding radiant   nement
         flux Φ_e
                                                                         Photometrisches Strahlungsäquiva-
         $$K = \frac{\Phi_v}{\Phi_e}$$ lent

         Unit : $lm \cdot W^{-1}$ Coefficiente di visibilità di
                                                                         una radiazone
         NOTE : When applied to monochromatic radiations, the maximum value of   Eficacia luminosa de una radiación
                 $K(\lambda)$ is denoted by the symbol K_m

                 $K_m = 683\,lm \cdot W^{-1}$ for $v_m = 540 \times 10^{12} Hz (\lambda_m \approx 555\,nm)$
                 for photopic vision
                 $K_m' = 1700\,lm \cdot W^{-1}$ for $\lambda_m' = 507\,nm$ for scotopic vision
                 For other wavelengths :
                 $K(\lambda) = K_m V(\lambda)$ and $K'(\lambda) = K_m' V'(\lambda)$

SECTION 845-02 - VISION, COLOUR RENDERING

B. LIGHT AND COLOUR

Translation : F, G, I, S

02-07 **Adaptation**
The process by which the state of the visual system is modified by previous and present exposure to stimuli that may have various luminances spectral distributions and angular subtenses
  NOTES : 1 - The terms light adaptation and dark adaptation are also used, the former when the luminances of the stimuli are at least several candelas per square metre, and the latter when the luminances are of less than some hundredths of a candela per square metre
  2 - Adaptation to specific spatial frequencies, orientations, sizes, etc. are recognized as being included in this definition

Adaptation
Adaptation
Adattamento
Adaptación

02-17 **(Perceived) light**
Universal and essential attribute of all perceptions and sensations that are peculiar to the visual system
  NOTES : 1 - Light is normally, but not always, perceived as a result of the action of a light simulation on the visual system
  2 - See 01-06

Lumière (perçue)
(Wahrgenommenes) Licht
Luce (percepita)
Luz (percibida)

02-28 **Brightness ; luminosity** *(obsolete)*
Attribute of a visual sensation according to which an area appears to emit more or less light

Luminosité
Helligkeit
Brillanza
Luminosidad

02-29 **Bright**
Adjective used to describe high levels of brightness

Lumineux
Hell
Brillante
Luminoso

02-31 **Lightness** *(of a related colour)*
The brightness of an area judged relative to the brightness of a similarly illuminated area that appears to be white or highly transmitting
  NOTE : Only related colours exhibit lightness

Clarté
Helligkeit
Chiarore
Claridad

02-32 **Light**
Adjective used to describe high levels of lightness

Clair
Hell
Chiaro
Claro

C. VISUAL PHENOMENA

Translation : F, G, I, S

02-47 **Contrast**
1. In the perceptual sense : Assessment of the difference in appearance of two or more parts of a field seen simultaneously or successively (hence : *brightness contrast, lightness contrast, colour contrast, simultaneous contrast, successive contrast*, etc)
2. In the physical sense : Quantity intended to correlate with the perceived brightness contrast, usually defined by one of a number of formulae which involve the luminances of the stimuli considered, for example : $\Delta L/L$ near the luminance threshold, or L_1/L_2 for much higher luminances

Contraste
Kontrast
Contrasto
Contraste

02-52 **Glare**
Condition of vision in which there is discomfort or a reduction in the ability to see details or objects, caused by an unsuitable distribution or range of luminance, or to extreme contrasts
   NOTE : In Russian, the terms 02-52 to 57 relate to the properties of the light sources and other luminous surfaces which disturb the condition of vision, and not to the changed condition of vision caused by an unsuitable distribution of luminance in the visual field

Eblouissement
Blendung
Abbagliamento
Deslumbramiento

02-53 **Direct glare**
Glare caused by self-luminous objects situated in the visual field, especially near the line of sight

Eblouissement direct
Infeldblendung ; direkte Blendung
Abbagliamento diretto
Deslumbramiento directo

02-54 **Glare by reflection**
Glare produced by reflections, particularly when the reflected images appear in the same or nearly the same direction as the object viewed
   NOTE : Formely reflected glare

Eblouissement par réflexion
Reflexblendung
Abbagliamento da luce riflessa
Deslumbramiento reflejado

02-55 **Veiling reflections**
Specular reflections that appear on the object viewed and that partially or wholly obscure the details by reducing contrast

Réflexions-voile
Schleierreflexionen
Riflessioni di velo
Reflejos velantes

02-56 **Discomfort glare**
Glare that causes discomfort without necessarily impairing the vision of objects

Eblouissement inconfortable
Psychologische Blendung
Abbagliamento psicologico
Deslumbramiento molesto

02-57 **Disability glare**
Glare that impairs the vision of objects without necessarily causing discomfort

Eblouissement perturbateur
Physiologische Blendung
Abbagliamento fisiologico
Deslumbramiento perturbador

D. COLOURS RENDERING
(See also CIE Publication N° 13.2 (1974))

Translation : F, G, I, S

02-59 **Colour rendering**
Effect of an illuminant on the colour appearance of objects by conscious or subconscious comparison with their colour appareance under a reference illuminant
NOTE : In German, the term "Farbwiedergabe" is also applied to colour reproduction

Rendu des couleurs
Farbwiedergabe
Resa dei colori
Rendimiento en color

02-60 **Reference illuminant**
An illuminant with other illuminants are compared
NOTE : A more particular meaning may be needed in the case of illuminants for colour reproduction

Illuminant de référence
Bezugslichtart
Illuminante di riferimento
Iluminante de referencia

02-61 **Colour rendering index [R]**
Measure of the degree to which the psychophysical colour of an object illuminated by the test illuminant conforms to that of the same object illuminated by the reference illuminant, suitable allowance having been made for the state of chromatic adaptation
NOTE : In German, the term "Farbwiedergabe-Index" is also applied to colour reproduction

Indice de rendu des couleurs
Farbwiedergabe-Index
Indice di resa dei colori
Indice de rendimiento en color

02-62 **CIE 1974 special colour rendering index [R_i]**
Measure of the degree to which the psychophysical colour of a CIE test colour sample illuminated by the test illuminant conforms to that of the same sample illuminated by the reference illuminant, suitable allowance having been made for the state of chromatic adaptation

Indice de rendu des couleurs 1974
Spezieller Farbwiedergabe-In CIE 1974
Indice speciale di resa dei col CIE 1974
Indice especial de rendimiento color CIE 1974

SECTION 845-03 - COLORIMETRY

B. ILLUMINANTS

Translation : F, G, I, S

03-10 **Illuminant**
Radiation with a relative spectral power distribution defined over the wavelength range that influences object colour perception
NOTE : In everyday English this term is not restricted to this sense, but is also used for any kind of light falling on a body or scene

Illuminant
Lichtart
Illuminante
Iluminante

03-13 **CIE standard sources**
Artificial sources specified by the CIE whose radiations approximate CIE standard illuminants A, B and C (see CIE Publication N° 15)

Sources normalisées CIE
CIE-Normlichtquellen
Sorgenti CIE
Fuentes patrones CIE

C. TRICHROMATIC SYSTEMS

Translation : F, G, I, S

03-28 **CIE 1931 standard colorimetric system** (XYZ)
A system for determining the tristimulus values of any spectral power distribution using the set of reference colour stimuli $[X]$, $[Y]$, $[Z]$ and the three CIE colour-matching functions $\bar{x}(\lambda), \bar{y}(\lambda), \bar{z}(\lambda)$ adopted by the CIE in 1931 (see CIE Publication N° 15)
NOTES : 1 - $\bar{y}(\lambda)$ is identical to $V(\lambda)$ and hence the tristimulus values Y are proportional to luminances
2 - This standard colorimetric system is applicable to centrally-viewed fields of angular subtense between about 1° and about 4° (0.017 and 0.07 rad)

Système de référence colorimétrique CIE 1931
CIE-Normvalenzsystem 1931
Sistema colorimetrico CIE 1931
Sistema colorimétrico patrón CIE 1931

03-29 **CIE 1964 supplementary standard colorimetric system** ($X_{10} Y_{10} Z_{10}$)
A system for determing the tristimulus values of any spectral power distribution using the set of reference colour stimuli $[X_{10}]$, $[Y_{10}], [Z_{10}]$ and the three CIE colour-matching functions $\bar{x}_{10}(\lambda), \bar{y}_{10}(\lambda), \bar{z}_{10}(\lambda)$ adopted by the CIE in 1964 (see CIE Publication N° 15)
NOTES : 1 - This standard colorimetric system is applicable to centrally-viewed fields of angular subtences greater than about 4° (0.07 rad)
2 - When this system is used, all symbols that represent colorimetric measures are distinguished by use of the subscript 10
3 - Values of Y_{10} are not proportional to luminances

Système de référence colorimétrique supplémentaire CIE 1964
10°-CIE-Normvalenzsystem 1964 ;
CIE-Großfeld-Normvalenzsystem 1964
Sistema colorimetrico supplementare CIE 1964
Sistema colorimétrico patrón CIE 1964

D. CHROMATICITY

Translation : F, G, I, S

03-41 **Planckian locus**
The locus of points in a chromaticity digram that represents chromaticities of the radiation of Planckian radiators at different temperatures

Lieu des corps noirs
Planckscher Kurvenzug
Luogo del corpo nero
Lugar de los estímulos (de colo planckianos

03-42 **Daylight locus**
The locus of points in a chromaticity diagram that represents chromaticities of phases of daylight with different correlated colour temperatures

Lieu des lumières du jour
Tageslichtkurvenzug
Luogo della luce diurna
Lugar de los estímulos (de colo luz de día

03-49 **Colour temperature** *(T_c)*
The temperature of a Planckian raidator whose radiation has the same chromatity as that of a given stimulus
Unit : K
NOTE : The reciprocal colour temperature is also used, unit K^{-1}

Température de couleur
Farbtemperatur
Temperatura di colore
Temperatura de color

SECTION 845-04 - EMISSION, OPTICAL PROPERTIES OF MATERIALS

A. EMISSION   **Translation : F, G, I, S**

04-04 **Planckian radiator ; blackbody**
Ideal thermal radiator that absorbs completely all incident radiation, whatever the wavelength, the direction of incidence or the polarization. This radiator has, for any wavelength and any direction, the maximum spectral concentration of radiance for a thermal radiator in thermal equilibrium at a given temperature

Radiateur de Planck ; corps noir
Planckscher Strahler ; Schwarzer Körper
Radiatore di Planck ; corpo nero
Radiador planckiano

B. OPTICAL PROPERTIES OF MATERIALS   **Translation : F, G, I, S**

04-42 **Reflection**
Process by which radiation is returned by a surface or a medium, whithout change of frequency of its monochromatic components

Réflexion
Reflexion
Riflessione
Reflexión

NOTES :
1 - Part of the radiation falling on a medium is reflected at the surface of the medium (surface reflection) ; another part may be scattered back from the interior of the medium (volume reflection)
2 - The frequency is unchanged only if there is no Doppler effect due to the motion of the materials from which the radiation is returned

04-43 **Transmission**
Passage of radiation through a medium without change of frequency of its monochromatic components

Transmission
Transmission
Trasmissione
Transmisión

04-44 **Diffusion ; scattering**
Process by which the spatial distribution of a beam of radiation is changed when it is deviated in many directions by a surface of by a medium, without change of frequency of its monochromatic components

Diffusion
Streuung
Diffusione
Difusión

NOTES :
1 - A distinction is made between selective diffusion and non-selective diffusion according to whether or not the diffusion properties vary with the wavelength of the incident radiation
2 - See Note 2 to 04-42

04-45 **Regular reflection ; specular reflection**
Reflection in accordance with the laws of geometrical optics, without diffusion

Réflexion régulière ; réflexion spéculaire
Gerichtete Reflexion
Riflessione regolare ; riflessione speculare
Reflexión regular

04-46 **Regular transmission ; direct transmission**
Transmission in accordance with the laws of geometrical optics, without diffusion

Transmission régulière
Gerichtete Transmission
Trasmissione regolare
Transmisión regular

04-47	**Diffuse reflection** Diffusion by reflection in which, on the macroscopic scale, there is no regular reflection	Réflexion diffuse Gestreute Reflexion ; diffu| Reflexion Riflessione diffusa Reflexión difusa
04-48	**Diffuse transmission** Diffusion by transmission in which, on the macroscopic scale, there is no regular transmission	Transmission diffuse Gestreute Transmission ; diffu| Transmission Trasmissione diffusa Transmisión difusa
04-49	**Mixed reflection** Partly regular and partly diffuse reflection	Réflexion mixte ; réflexion sem| diffuse ; réflexion semi-réguliè| Gemischte Reflexion Riflessione mista ; riflessio| semidiffusa ; riflessione semir| golare Reflexión mixta
04-50	**Mixed transmission** Partly regular and partly diffuse transmission	Transmission mixte ; transmissi| semi-diffuse ; transmission sem| régulière Gemischte Transmission Trasmissione mista ; trasmissio| semidiffusa ; trasmissione sem| regolare Transmisión mixta
04-56	**Lambert's (cosine) law** For a surface element whose radiance or luminance is the same in all directions of the hemisphere above the surface $$I(\theta) = I_n \cos\theta$$ where $I(\theta)$ and I_n are the radiant or luminous intensities of the surface element in a direction at an angle θ from the normal of the surface and in the direction of that normal, respectively	Loi (du cosinus) de Lambert Lambertsches (Cosinus-)Gesetz Legge (del coseno) di Lambert Ley de Lambert
04-57 & CEC 080	**Lambertian surface** **Perfectly diffusing surface** Ideal surface for which the radiation coming from that surface is distributed angularly according to Lambert's cosine law NOTE : For a Lambertian surface, $M = \pi L$ where M is the radiant or luminous exitance, and L the radiance or luminance	Surface lambertienne Lambertfläche ; vollkommen mat| Fläche Superficie di Lambert Superficie lambertiana
04-58	**Reflectance** *(for incident radiation of given spectral composition, polarization and geometrical distribution)* (ρ) Ratio of the reflected radiant or luminous flux to the incident flux in the given conditions Unit : 1 NOTE : See Note 1 to 04-62	Facteur de réflexion Reflexionsgrad Fattore de riflessione Reflectancia

04-59	**Transmittance** *(for incident radiation of given spectral composition, polarization and geometrical distribution)* (τ) Ratio of the transmitted radiant or luminous flux to the incident flux in the given conditions *Unit : 1* NOTE : See Note 1 to 04-63	Facteur de transmission Transmissionsgrad Fattore di trasmissione Transmitancia
04-60	**Regular reflectance** (ρ_r) Ratio of the regularly reflected part of the (whole) reflected flux, to the incident flux *Unit : 1* NOTE : See Notes 1 and 2 to 04-62	Facteur de réflexion régulière Grad der gerichteten Reflexion Fattore di riflessione regolare Reflectancia regular
04-61	**Regular transmittance** (τ_r) Ratio of the regularly transmitted part of the (whole) transmitted flux, to the incident flux *Unit : 1* NOTE : See Notes 1 and 2 04-63	Facteur de transmission régulière Grad der gericheten Transmission Fattore di trasmissione regolare Transmitancia regular
04-62	**Diffuse reflectance** (ρ_d) Ratio of the diffusely reflected part of the (whole) reflected flux, to the incident flux *Unit : 1* NOTES : 1 - $\rho = \rho_r + \rho_d$ 2 - The results of the measurements of ρ_r and ρ_d depend on the instruments and the measuring techniques used	Facteur de réflexion diffuse Grad der gestreuten Reflexion Fattore di riflessione diffusa Reflectancia difusa
04-63	**Diffuse transmittance** (τ_d) Ratio of the diffusely transmitted part of the (whole) transmitted flux, to the incident flux *Unit : 1* NOTES : 1 - $\tau = \tau_r + \tau_d$ 2 - The results of the measurements of τ_r and τ_d depend on the instruments and the measuring techniques used	Facteur de transmission diffuse Grad der gestreuten Transmission Fattore di trasmissione diffusa Transmitancia difusa
04-68	**Radiance factor** *(at a surface of a non-self-radiating medium, in a given direction, under specified conditions of irradiation)* ($ß_e$; $ß$) Ratio of the radiance of the surface element in the given direction to that of a perfect reflecting or transmitting diffuser identically irradiated *Unit : 1* NOTE : For photoluminescent media, the radiance factor is the sum of two portions, the <u>reflected</u> <u>radiance</u> <u>factor</u> $ß_S$ and the <u>luminescent</u> <u>radiance</u> <u>factor</u> $ß_L$: $ß_e = ß_S + ß_L$	Facteur de luminance énergétique Strahldichtefaktor Fattore di radianza Factor de radiancia

04-69   **Luminance factor** *(at a surface element of a non-self-radiating medium, in a given direction, under specified conditions of illumination)* (β_v ; β)
Ratio of the luminance of the surface element in the given direction to that of a perfect reflecting or transmitting diffuser identically illuminated
Unit : 1

NOTES : 1 -   For photoluminescent media, the radiance factor is the sum of two portions, the reflected luminance factor β_S and the luminescent luminance factor β_L : $\beta_v = \beta_S + \beta_L$

2 -   In Germany, instead of the luminance factor β (or transmittance τ) it is usual in colorimetry to use the quantity "Hellbezugswert" $A = 100\,\beta$ (or $100\,\tau$)

Facteur de luminance (lumineuse)
Leuchtdichtefaktor
Fattore di luminanza
Factor de luminancia

04-70   **Radiance coefficient** *(at a surface element of a medium, in a given direction, under specified conditions of irradiation)* [q_e ; q]
Quotient of the radiance of the surface element in the given direction by the irradiance on the medium
unit : sr^{-1}

NOTE :   In the USA the concept bi-directional reflectance distribution function (BRDF) is similar to the above coefficient except that it is defined for directional incident radiation

Coefficient de luminance énergétique
Strahldichtekoeffizient
Coefficiente di radianza
Coeficiente de radiancia

04-71   **Luminance coefficient** *(at a surface element, in a given direction, under specified conditions of illumination)* [q_v ; q]
Quotient of the luminance of the surface element in the given direction by the illuminance on the medium
unit : sr^{-1}

NOTE :   See Note to 04-70

Coefficient de luminance (lumineuse)
Leuchtdichtekoeffizient
Coefficiente di luminanza
Coeficiente de luminancia

04-74   **Absorption**
Process by which radiant energy is converted to a different form of energy by interaction with matter

Absorption
Absorption
Assorbimento
Absorción

04-75   **Absorptance** (α)
Ratio of the absorbed radiant or luminous flux to the incident flux under specified conditions
Unit : 1

Facteur d'absorption
Absorptionsgrad
Fattore di assorbimento
Absortancia

04-89   **Diffusion factor** *(of a diffusing surface, by reflection or by transmission)*
Ratio of the mean of the values of luminance measured at 20° and 70° (0.35 and 1.22 rad) to the luminance measured at 5° (0.09 rad) from the normal, when the surface considered is illuminated normally

$$\frac{L(20)+L(70)}{2L(5)}$$

NOTES : 1-   The diffusion factor is intended to give an indication of the spatial distribution of the diffused flux. It is equal to 1 for every isotropic diffuser, whatever the value of the diffuse reflectance or transmittance

2 -   This way of defining the diffusion factor can be applied only to materials for which the indicatrix of diffusion does not differ appreciably from that of ordinary opal glasses

Facteur de diffusion
Streuvermögen
Indice di diffusione
Factor de difusión

04-91	**Indicatrix of diffusion ; scattering indicatrix** *(for a specified incident beam)* Representation in space, in the form of a surface expressed in polar coordinates, of the angular distribution of (relative) radiant or luminous intensity or of (relative) radiance or luminance of an element of surface of a medium that diffuses by reflection or transmission NOTES : 1 - For a narrow incident beam of radiation, it is convenient to represent the indicatrix of diffusion in Cartesian coordinates. If the angular distribution has rotational symmetry, a meridian section of the surface is sufficient 2 - The term indicatrix is often used to denote, instead of the surface, the curve obtained in a similar manner in a plane normal to the element concerned	Indicatrice de diffusion Streuindikatrix Indicatrice di diffusione Indicatriz de difusión
CEC 081	**Dispersion** Spatial dispersion of light occuring on the occasion of a transmission (or reflexion) process, light is scattered in various directions	Dispersion Dispersion Dispersione Dispersión
CEC 082	**Dispersion angle** Angle for which the intensity has half the value of the maximum intensity, when the indicatrix of diffusion is symmetrical about the direction of this maximum intensity	Angle de dispersion Dispersionwinkel Angolo di dispersione Angulo de dispersión
04-108	**Transparent medium** Medium in which the transmission is mainly regular and which usually has a high regular transmittance in the spectral range of interest NOTE : Objects may be seen distincly trhough a medium which is transparent in the visible region, if the geometric form of the medium is suitable	Milieu transparent Durchsichtiges Medium Mezzo trasparente Medio transparente
04-109	**Translucent medium** Medium which transmits visible radiation largely by diffuse trabsmission, so that objects are not seen distinctly through it	Milieu translucide Durchscheinendes Medium Mezzo traslucido Medio translúcido
04-110	**Opaque medium** Medium which transmits no radiation in the spectral range of interest	Milieu opaque Lichtundurchlässiges Medium Mezzo opaco Medio opaco

SECTION 845-05 - RADIOMETRIC, PHOTOMETRIC AND COLORIMETRIC MEASUREMENTS ; PHYSICAL DETECTORS

A. GENERAL TERMS AND INSTRUMENTS   Translation : F, G, I, S

05-09 **Photometry**
Measurement of quantities referring to radiation as evaluated according to a given spectral luminous efficiency function, e.g. $V(\lambda)$ or $V'(\lambda)$
NOTE : In Russian publications, photometry refers only to the evaluating according to $V(\lambda)$. The term photometry is also often used in publications in Russian and in other languages in a broader sense covering the science of optical radiation measurement (radiometry)

Photométrie
Photometrie
Fotometria
Fotometría

05-13 **Physical photometry**
Photometry in which physical detectors are used to make the measurements

Photométrie physique
Physikalische Photometrie
Fotometria fisica
Fotometria física

05-16 **Illuminance meter**
Instrument for measuring illuminance

Luxmètre
Beleuchtungsstärkemesser
Luxmetro
Iluminancímetro

05-17 **Luminance meter**
Instrument for measuring luminance

Luminancemètre
Leuchtdichtemesser
Luminanzometro
Luminancímetro

05-22 **Goniophotometer**
Photometer for measuring the directional light distribution characteristics of sources, luminaires, media or surfaces

Goniophotomètre
Goniophotometer
Goniofotometro
Goniofotómetro

05-24 **Integrating sphere ; Ubricht sphere**
Hollow sphere whose internal surface is a diffuse reflector, as non-selective as possible
NOTE : An integrating sphere is used frequently with a radiamoter or photometer

Sphère d'Ulbricht ; sphère intégrante
Ulbrichtsche Kugel
Sfera di Ulbricht ; sfera integratrice
Esfera integrante

05-25 **Integrating photometer**
Photometer for measuring luminous flux, generally incorporating an integrating sphere

Lumenmètre
Lichtstrommeßgerät
Fotometro integratore
Lumenómetro

05-26 **Reflectometer**
Instrument for measuring quantities pertaining to reflection

Réflectomètre
Reflektometer
Riflettometro
Reflectómetro

05-27 **Densitometer** Densitomètre
Photometer for measuring reflectance or transmittance optical Densitometer
density Densitometro
 Densitómetro

05-30 **Glossmeter** Luisancemètre
Instrument for measuring the various photometric properties of Glanzmesser
a surface giving rise to gloss Lucentimetro
 Brillómetro

SECTION 845-09 - LIGHTING TECHNOLOGY, DAYLIGHTING

A. GENERAL TERMS   Translation : F, G, I, S

09-01   **Lighting ; illumination**
Application of light to a scene or their surroundings so that they may be seen
NOTE : This term is also used colloquially with the meaning "lighting system" or "lighting installation"

Eclairage
Beleuchtung
Illuminazione
Iluminación

09-02   **Lighting technology ; illuminating engineering**
Applications of lighting considered under their various aspects

Eclairagisme
Lichttechnik ; Beleuchtungstechn
Illuminotecnica
Luminotecnia

09-03   **Luminous environment**
Lighting considered in relation to its physiological and psychological effects

Ambiance lumineuse
Beleuchtetes Umfeld
Ambiente luminoso
Ambiente luminoso

09-04   **Visual performance**
Performance of the visual system as measured for instance by the speed and accuracy with which a visual task is performed

Performance visuelle
Sehleistung
Prestazione visiva
Rendimiento visual

B. TYPES OF LIGHTING   Translation : F, G, I, S

09-09   **Permanent supplementary artificial lighting** *(in interiors)*
Permanent artificial lighting intended to supplement the natural lighting of premises, when the natural lighting is insufficient or objectionable if used alone

NOTE : This type of lighting is generally denoted in brief by the initial letters PSALI of the words of the English term

Eclairage artificiel compléme
taire permanent
Tageslichtergänzungsbeleuchtung
Illuminazione artificiale compl
mentare permanente
Alumbrado artificial suplementar
permanente

09-14   **Direct lighting**
Lighting by means of luminaires having a distribution of luminous intensity such that the fraction of the emitted luminous flux directly reaching the working plane, assumed to be unbounded, is 90 % to 100 %

Eclairage direct
Direkte Beleuchtung
Illuminazione diretta
Alumbrado directo

09-18   **Indirect lighting**
Lighting by means of luminaires having a distribution of luminous intensity such that the fraction of the emitted luminous flux directly reaching the working plane, assumed to be unbounded, is 0 to 10 %

Eclairage indirect
Indirekte Beleuchtung
Illuminazione indiretta
Alumbrado indirecto

09-19	**Directional lighting** Lighting in which the light on the working plane or on an object is incident predominantly from a particular direction	Eclairage dirigé Gerichtete Beleuchtung Illuminazione direzionale Alumbrado dirigido
09-20	**Diffused lighting** Lighting in which the light on the working plane or on an object is not incident predominantly from a particular direction	Eclairage diffusé Diffuse Beleuchtung ; gestreute Beleuchtung Illuminazione diffusa Alumbrado difuso

C. TERMS USED IN LIGHTING CALCULATIONS

Translation : F, G, I, S

09-23	**Illuminance vector** *(at a point)* Vector quantity equal to the maximum difference between the illuminances on opposite sides of an element of surface through the point considered, that vector being normal to and away from the side with the greater illuminance	Vecteur d'éclairement Lichtvektor Vettore d'illuminamento Vector de iluminación
09-49	**Reference surface** Surface on which illuminance is measured or specified	Surface de référence Bezugsfläche : Meßfläche Superficie di riferimento Superficie de referencia
09-50	**Work plane ; working plane** Reference surface defined as the plane at which work is usually done NOTE : In interior lighting and unless otherwise indicated, this plane is assumed to be a horizontal plane 0.85 m above the floor and limited by the walls of the room. In the USA the work plane is usually assumed to be 0.76 m above the floor, in the USSR 0,8 m above the floor	Plan utile ; plan de travail Nutzebene Piano utile Plano de trabajo
09-55	**Room index ; installation index** $[K]$ Number representative of the geometry of the part of the room between the working plane and the plane of the luminaires, used in calculation of utilization factor or utilance NOTES : 1 - Unless otherwise indicated, the room index is given by the formula $$K = \frac{a \cdot b}{h(a+b)}$$ in wich a and b are the dimensions of the sides of the room and h the <u>mounting</u> <u>height</u>, that is the distance between the working plane and the plane of luminaires 2 - In British practice, the <u>ceiling</u> <u>cavity</u> <u>index</u> is calculated from the same formula except that h is the distance from ceiling to luminaires 3 - In the USA, the term <u>room</u> <u>cavity</u> <u>ratio</u> is currently used. This is equal to five times the reciprocal of the room index defined by the formula in Note 1. Two supplementary terms are used : <u>ceiling</u> <u>cavity</u> <u>ratio</u> and <u>floor</u> <u>cavity</u> <u>ratio</u> which are derived in the same way as the room cavity ratio except that h is respectively the distance from ceiling to luminaires and from floor to working plane	Indice du local ; indice d'installation Raum-Index Indice di locale Indice del local

09-58 **Uniformity ratio of illuminance** *(on a given plane)*
Ratio of the minimum illuminance to the average illuminance on the plane

> NOTE : Use is made also of a) the ratio of the minimum to the maximum illuminance and b) the inverse of either of these two ratios

Facteur d'uniformité de l'éclairement
Gleichmäßigkeitsgrad Beleuchtungsstärke
Fattore di uniformità dell'illuminamento
Grado de uniformidad de iluminancia

09-59 **Light loss factor ; maintenance factor** *(obsolete)*
Ratio of the average illuminance on the working plane after a certain period of use of a lighting installation to the avergae illuminance obtained under the same conditions for the installation considered conventionally as new

> NOTES : 1 - The term <u>depreciation</u> factor has been formely used to designate the reciprocal of the above ratio
> 2 - The light losses take into account dirt accumulation on luminaire and room surfaces and lamp depreciation

Facteur de dépréciation ; facteur de maintenance
Verminderungsfaktor
Fattore di manutenzione
Factor de depreciación

09-60 **Service illuminance** *(of an area)*
Mean illuminance during one maintenance cycle on an installation averaged over the relevant area

> NOTE : The area may be either the whole area of the working plane in an interior or the working areas

Eclairement en service
Betriebswert der Beleuchtungsstärke
Illuminamento in esercizio
Iluminancia en servicio

E. TERMS RELATING TO INTERREFLECTION

Translation : F, G, I, S

09-70 **Interreflection ; interflection** *(USA)*
General effect of the reflections of radiation between several reflecting surfaces

Réflexions mutuelles ; interflexions
Mehrfachreflexion ; Interflexion
Riflessione mutua
Interreflexión

F. DAYLGHTING

Translation : F, G, I, S

09-76 **Solar radiation**
Electromagnetic radiation from the Sun

Rayonnement solaire
Sonnenstrahlung
Radiazone solare
Radiación solar

09-79 **Direct solar radiation**
That part of extraterrestrial solar radiation which as a collimated beam reaches the Earth's surface after selective attenuation by the atmosphere

Rayonnement solaire direct
Direkte Sonnenstrahlung
Radiazone solare diretta
Radiación solar directa

09-80 **Diffuse sky radiation**
That part of solar radiation which reaches the Earth as a result of being scattered by the air molecules, aerosol particles, cloud particles or other particles

Rayonnement diffus du ciel
Diffuse Himmelsstrahlung
Radiazone diffusa dal cielo
Radiación del cielo

09-81	**Global solar radiation** Combined direct solare radiation and diffuse sky radiation	Rayonnement solaire global Globalstrahlung Radiazone solare globale Radiación solar global
09-82	**Sunlight*** Visible part of direct solar radiation *NOTE : When dealing with actinic effects of optical radiations, this term is commonly used for radiations extending beyond the visible region of the spectrum	Lumière solaire Sonnenlicht Luce solare Luz solar
09-83	**Skylight*** Visible part of diffuse sky radiation *NOTE : When dealing with actinic effects of optical radiations, this term is commonly used for radiations extending beyond the visible region of the spectrum	Lumière du ciel Himmelslicht Luce dal cielo Luz del cielo
09-84	**Daylight*** Visible part of global solar radiation *NOTE : When dealing with actinic effects of optical radiations, this term is commonly used for radiations extending beyond the visible region of the spectrum	Lumière du jour Tageslicht Luce del giorno ; luce diurna Luz de dia
09-85	**Reflected (global) solar radiation** Radiation that results from reflection of the global solar radiation by the surface of the Earth and by any surface intercepting that radiation	Rayonnement solaire (global) réfléchi Reflektierte Globalstrahlung Radiazione solare (globale) riflessa Radiación solar (global) reflejada
09-87	**Total turbidity factor** *(according to Linke)* $[T]$ Ratio of the vertical optical thickness of a turbid atmosphere to the vertical optical thickness of the pure and dry atmosphere (Rayleigh atmosphere), related to the whole solar spectrum $$T = \frac{\delta_R + \delta_A + \delta_Z + \delta_W}{\delta_R}$$ where δ_R is the optical thickness with respect to Rayleigh scattering at the air molecules, $\delta_A, \delta_Z, \delta_W$ are the optical thicknesses with respect to Mie scattering and absorption at the aerosol particles, to ozone absorption, and to water vapour absorption respectively	Facteur total de trouble Trübungsfaktor Fattore totale di torbidità Factor total de turbidez
09-89	**Global illuminance** (E_g) Illuminance produced by daylight on a horizontal surface on the earth	Eclairement (lumineux) global Globalbeleuchtungsstärke Illuminamento globale Iluminancia global

09-90	**CIE standard overcast sky** Completely overcast sky for which the ratio of its luminance L_γ in the direction at an angle γ above the horizon to its luminance L_Z at the zenith is given by the relation $$L_\gamma = \frac{L_Z(1+2\sin\gamma)}{3}$$	Ciel couvert normalisé CIE Bedeckter Himmel nach CIE Cielo coperto secondo CIE Cielo cubierto patrón CIE
09-91	**CIE standard clear sky** Cloudless sky for which the relative luminance distribution is described in CIE Publication N° 22(1973)	Ciel serein normalisé CIE Klarer Himmel nach CIE (genorm* Cielo sereno secondo CIE Cielo despejado patrón CIE
09-92	**Total cloud amount** Ratio of the sum of the solid angles subtended by clouds to the solide angle 2π steradians of the whole sky NOTE : The total cloud amount is frequently called <u>fractional cloud cover</u> in the USA	Nébulosité Gesamtbewölkungsgrad Nuvolosità Factor de nebulosidad
CEC 090	**Cloud ratio [CR]** Ratio of diffuse irradiance to global irradiance	Rapport de couverture nuageuse Bewölkungsgrad Rapporto di copertura nuvolosa Relación de tiempo cubierto
CEC 091	**Nebulosity index $[I_N]$** $$I_N = \frac{1-CR_M}{1-CR_T}$$ with CR_M: measured value of Cloud ratio and CR_T theoretical value of Cloud ratio	Indice de nébulosité Bewölkungsindex Indice di nuvolosita Indice de nebulosidad
09-93	**Sunshine duration [S]** Sum of time intervals within a given time period (hour, day, month, year) during which the irradiance from direct solar radiation on a plane normal to the sun direction is equal to or greater than 200 watts per square metre	Durée d'ensoleillement Sonnenscheindauer Soleggiamento Tiempo de insolación
09-96	**Relative sunshine duration** Ratio of sunshine duration to possible sunshine duration within the same time period	Durée relative d'ensoleillemer Relative Sonnenscheindauer Soleggiamento relativo Tiempo de insolación relativo
CEC 092	**Relative dayly sunshine duration [σ]** Ratio of measured (or calculated) sunshine duration during *one day* to possible sunshine duration within the same day	Fraction journalière d'insolat Relative Tägliche Sonnenschei ner Indice di soleggiamento Fracción diaria de insolación

CEC 093　**Glazing index** (for a room)
Ratio of glazing areas to room area

Indice de vitrage
Verglasungsanteil
Indice di vetratura
Indice de acristalamiento

09-97　**Daylight factor** [D]
Ratio of the illuminance at a point on a given plane due to the light received directly or indirectly from a sky of assumed or known luminance distribution, to the illuminance on a horizontal plane due to an unobstructed hemisphere of this sky. The contribution of direct sunlight to both illuminances is excluded
　NOTES : 1 - Glazing, dirt effects, etc. are included
　　　　　2 - When calculating the lighting of interiors, the contribution of direct sunlight must be considered separately

Facteur de lumière du jour
Tageslichtquotient
Fattore di luce diurna
Factor de luz de dia

CEC 097　**Sunlight factor** [SB]
Ratio of the illuminance at a point on a given plane due to the light received directly or indirectly from the sun, to the illuminance on a horizontal plane due to the sun only
　NOTE : Glazing, dirt effects, etc. are included

Facteur de lumière du soleil
Sonnenlichtquotient
Fattore di luce solare
Factor de luz del sol

09-98　**Sky component of daylight facteur** [D_s]
Ratio of that part of the illuminance at a point on a given plane which is received directly (or through clear glass) from a sky of assumed or known luminance distribution, to the illuminance on a horizontal plane due to an unosbtructed hemisphere of this sky. The contribution of direct sunlight to both illuminances is excluded
　NOTE : See Note 2 to 09-97

Composante de ciel du facteur de lumière du jour
Himmelslichtanteil des Tageslichtquotienten
Componente cielo del fattore di luce diurna
Componente celeste del factor de luz de dia

CEC 098　**Sun component of sunlight factor** [SB_s]
Ratio of that part of the illuminance at a point on a given plane which is received directly (or through clear glass) from the sun, to the illuminance on a horizontal plane due to the sun too

Composante directe du facteur de lumière du soleil
Sonnenlichtanteil des Sonnenlichtquotienten
Componente diretta del fattore di luce solare
Componente directa del factor de luz del sol

09-99　**Externally reflected component of daylight factor** [D_e]
Ratio of that part of the illuminance at a point on a given plane in an interior which is received directly from external reflecting surfaces illuminated directly or indirectly by a sky of assumed or known luminance distribution, to the illuminance on a horizontal plane due to an unosbtructed hemisphere of this sky. The contribution of direct sunlight to both illuminances is excluded
　NOTE : See Note 2 to 09-97

Composante réfléchie externe du facteur de lumière du jour
Außenreflexionsanteil des Tageslichtquotienten
Componente riflessa esterna del fattore di luce diurna
Componente reflejada externa del factor de luz de dia

CEC 099	**Externally reflected component of sunlight factor** [SB_e] Ratio of that part of the illuminance at a point on a given plane in an interior which is received directly from external reflecting surfaces illuminated directly or indirectly by the sun, to the illuminance on a horizontal plane due the sun too	Composante réfléchie externe facteur de lumière du soleil Außenreflexionsanteil des Sonnenlichtquotienten Componente riflessa esterna de fattore di luce solare Componente reflejada externa de factor de luz del sol
09-100	**Internally reflected component of daylight factor** [D_i] Ratio of that part of the illuminance at a point on a given plane in an interior which is received directly from internal reflecting surfaces illuminated directly or indirectly by a sky of assumed or known luminance distribution, to the illuminance on a horizontal plane due to an unosbtructed hemisphere of this sky. The contribution of direct sunlight to both illuminances is excluded NOTE : See Note 2 to 09-97	Composante réfléchie interne facteur de lumière du jour Innenreflexionsanteil des Tageslichtquotienten Componente riflessa interna d fattore di luce diurna Componente reflejada interna d factor de luz de dia
CEC 100	**Internally reflected component of sunlight factor** [SB_i] Ratio of that part of the illuminance at a point on a given plane in an interior which is received directly from internal reflecting surfaces illuminated directly or indirectly by the sun, to the illuminance on a horizontal plane due the sun too	Composante réfléchie interne facteur de lumière du soleil Innenreflexionsanteil des Sonnenlichtquotienten Componente riflessa interna d fattore di luce solare Componente reflejada interna d factor de luz del sol
09-101	**Obstruction** Anything outside a building which prevents the direct view of part of the sky	Obstruction Verbauung Ostruzione Obstrucción
09-102	**Daylight opening** Area, glazed or unglazed, that is capable of admitting daylight to an interior	Prise de jour Tageslichtöffnung Presa di luce diurna Abertura a la luz de dia
09-105	**Shading** Device designed to obstruct, reduce or diffuse solar radiation	Brise-soleil ; écran solaire Sonnenschutzeinrichtung Schermo solare Brise-soleil

Appendix A
SKY TYPE PROBABILITY

Daily probability of each sky type as function of daily relative sunshine duration

Results for the European Stations of the Test Reference Years (TRY) and four German Stations

The meaning of the symbols used in the graphs is as follows:

1. Overcast sky
2. Intermediate overcast sky
3. Intermediate mean sky
4. Intermediate blue sky
5. Blue sky

F   Daily relative sunshine duration
P   Daily probability of each sky type

A.2 Daylighting in Architecture

Sky type probability: Belgium, Denmark, France.

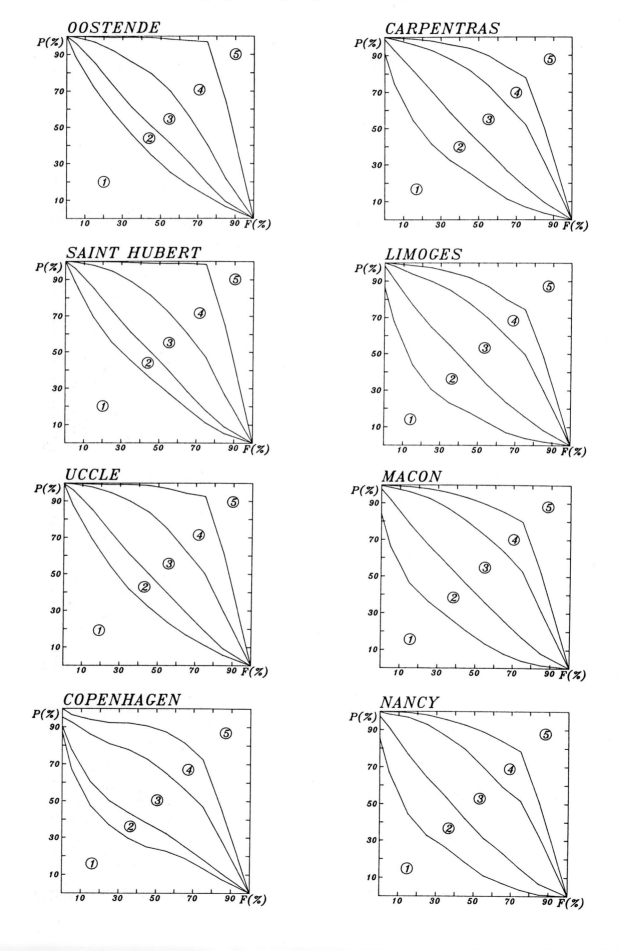

Sky type probability: France, Germany, Ireland.

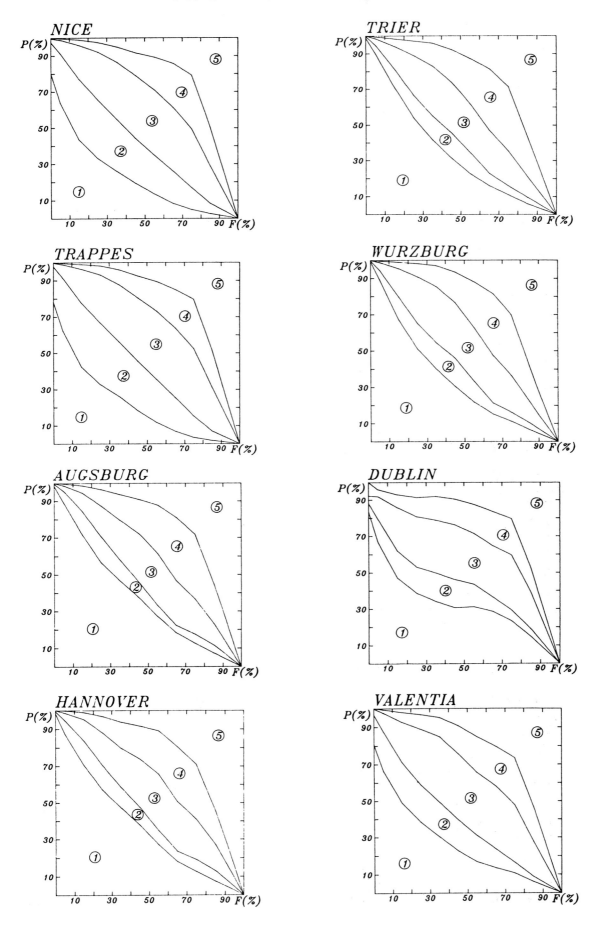

A.4 Daylighting in Architecture

Sky type probability: Italy.

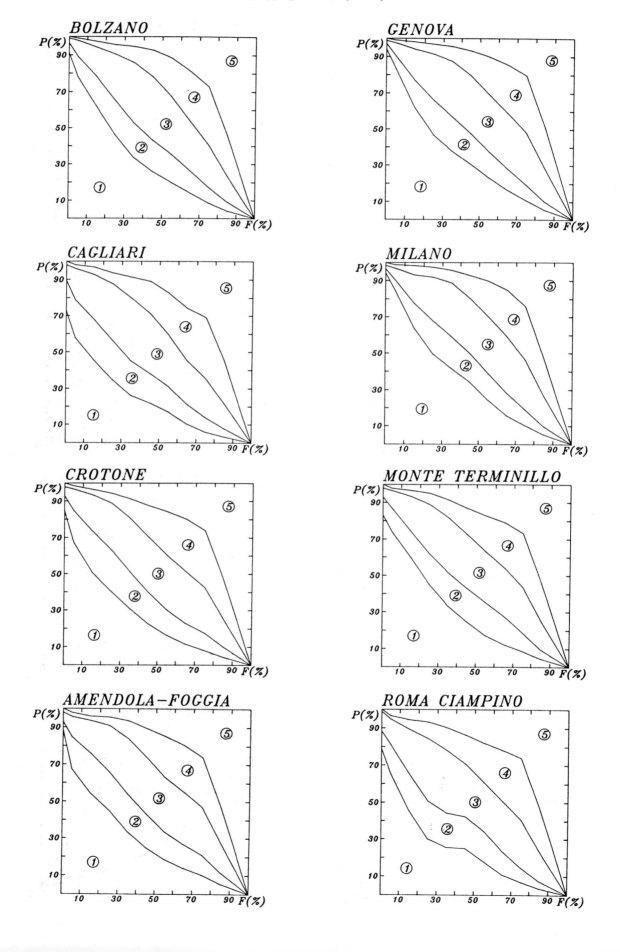

Sky type probability: Italy, Netherlands, United Kingdom.

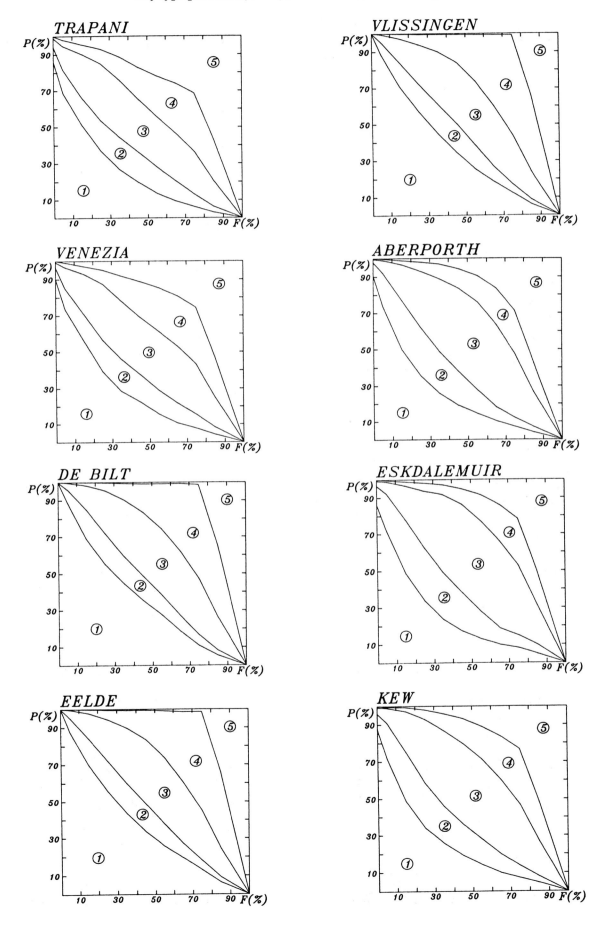

A.6 Daylighting in Architecture

Sky type probability: United Kingdom.

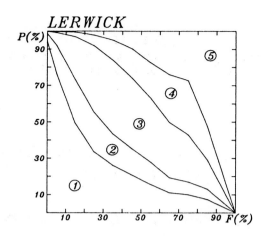

Appendix B
DAYLIGHT AVAILABILITY

Probability of illuminance lower than datum level on horizontal and vertical planes

Results for the European Stations of the Test Reference Years (TRY) and four German Stations

Belgium:	Ostend, Saint Hubert, Uccle.
Denmark:	Copenhagen.
France:	Carpentras, Limoges, Mâcon, Nancy, Nice, Trappes.
Germany:	Augsburg, Hanover, Trier, Wurzburg.
Eire:	Dublin, Valentia.
Italy:	Amendola-Foggia, Bolzano, Cagliari, Crotone, Genoa, Milan, Monte Terminillo, Roma Ciampino, Trapani, Venice.
Netherlands:	De Bilt, Eelde, Vlissingen.
United Kingdom:	Aberporth, Eskdalemuir, Kew, Lerwick.

Results are given for:

  diffuse illuminance
  global illuminance

five planes:
  horizontal
  vertical north, east, south and west

four periods:
  Winter = November + December + January + February
  Mid-season = March + April + September + October
  Summer = May + June + July + August
  Year

and
  am
  pm
  am + pm

Key to graphs:

Winter	———
Mid season	- - - - - - -
Summer	················
Year	—·—·—

B.2 Daylighting in Architecture

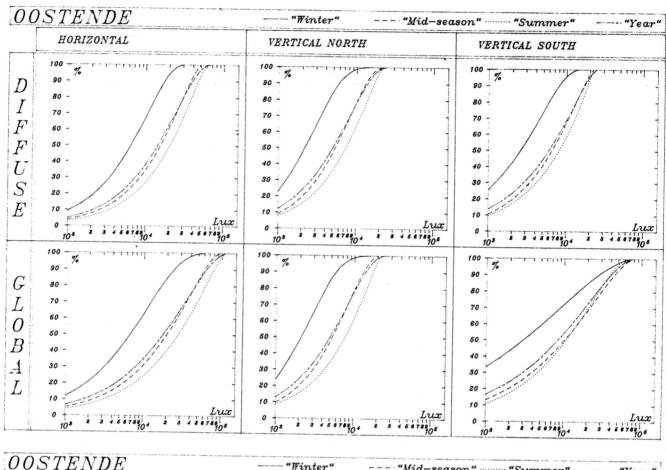

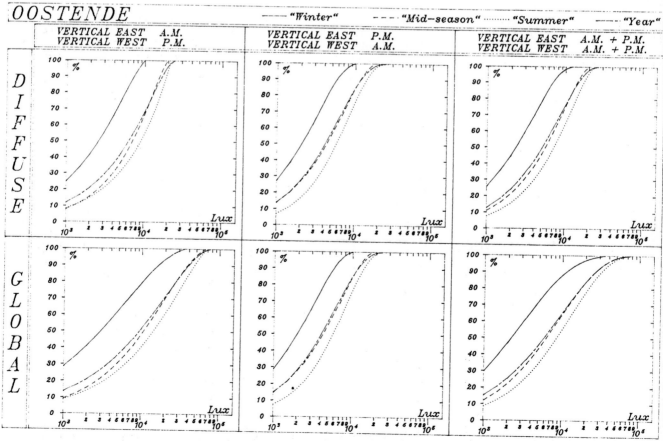

Daylight Availability B.3

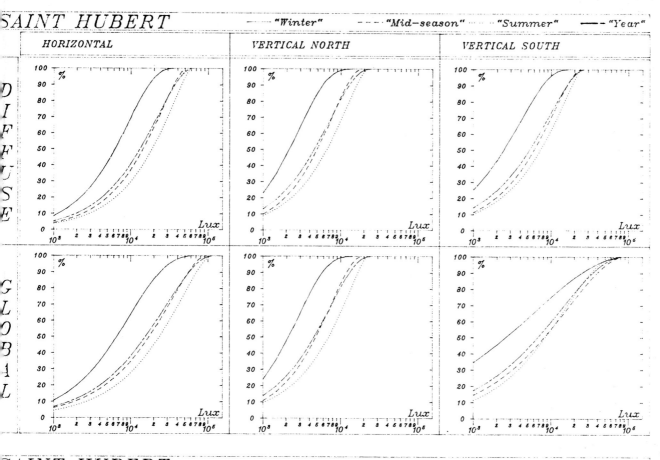

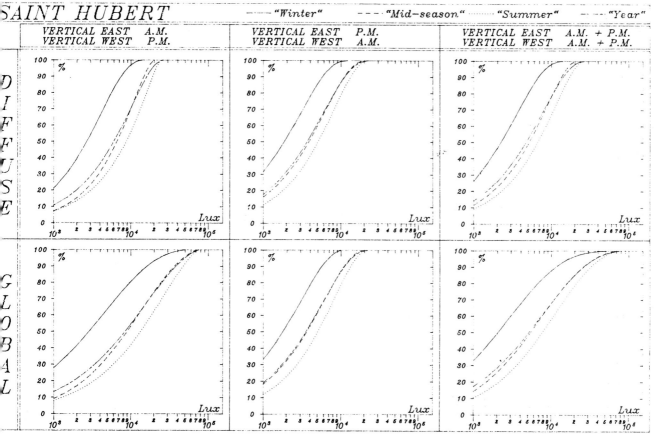

B.4 Daylighting in Architecture

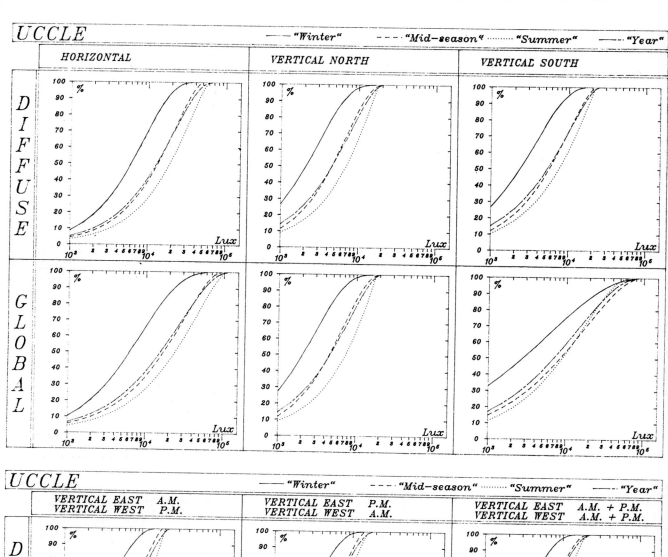
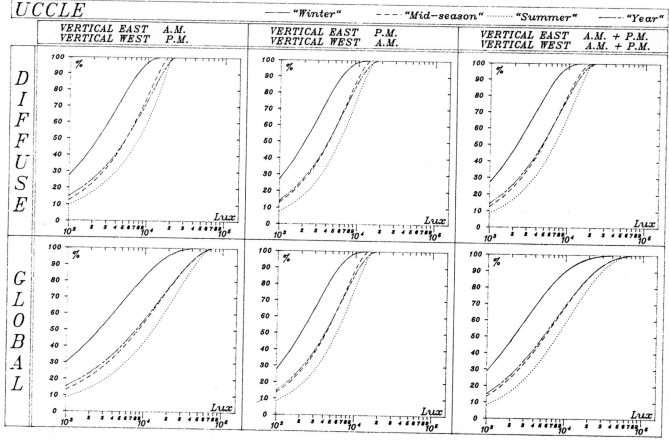

Daylight Availability B.5

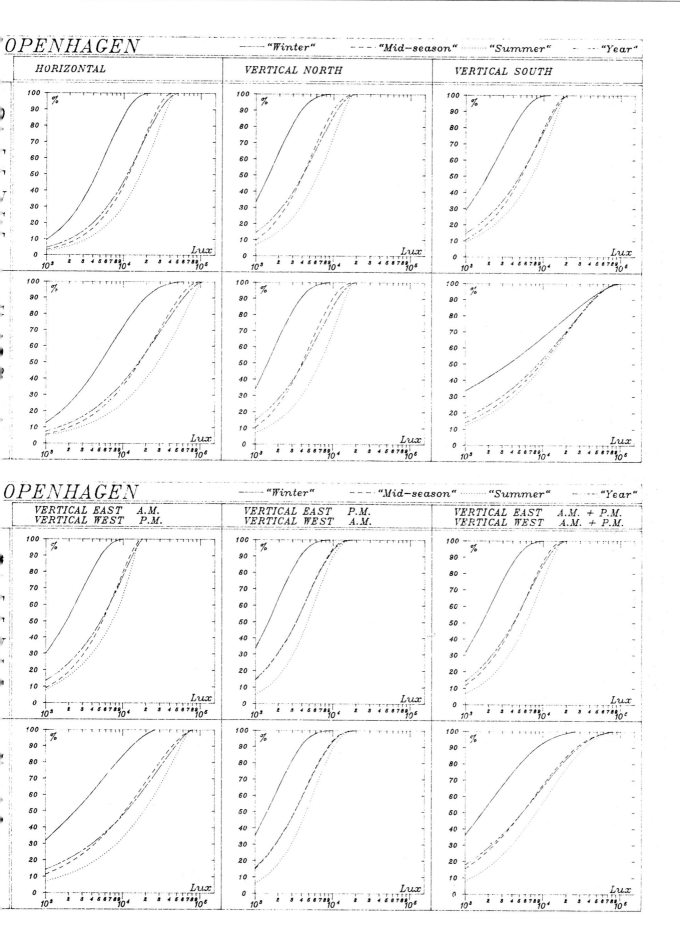

B.6 Daylighting in Architecture

Sky type probability: United Kingdom.

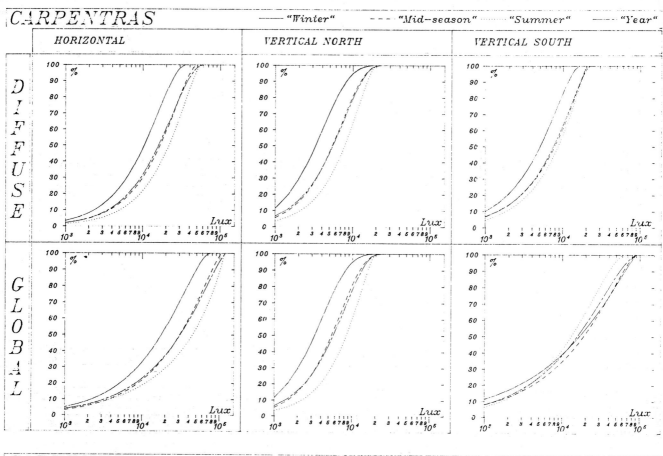

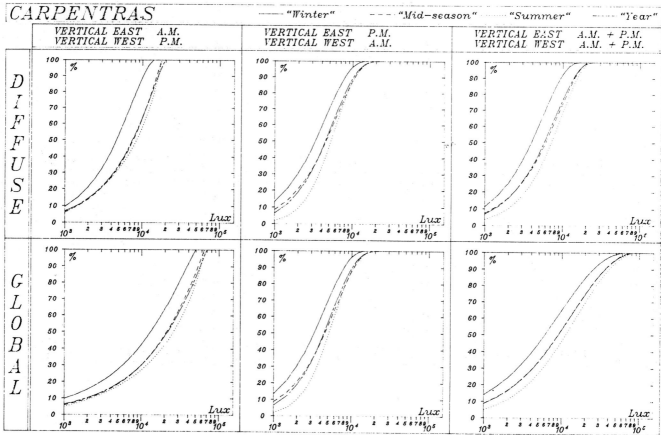

Daylight Availability B.7

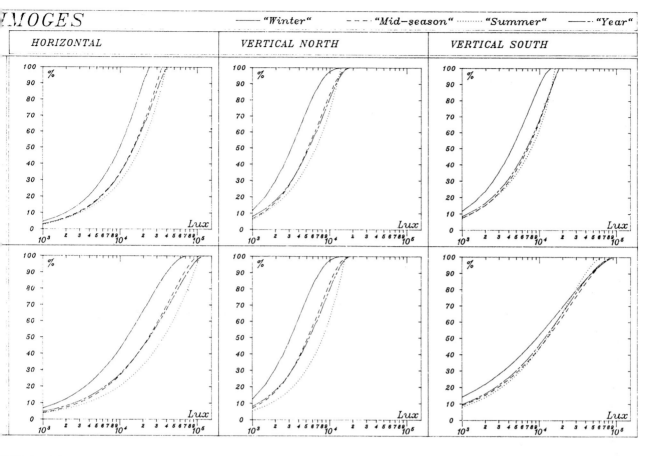

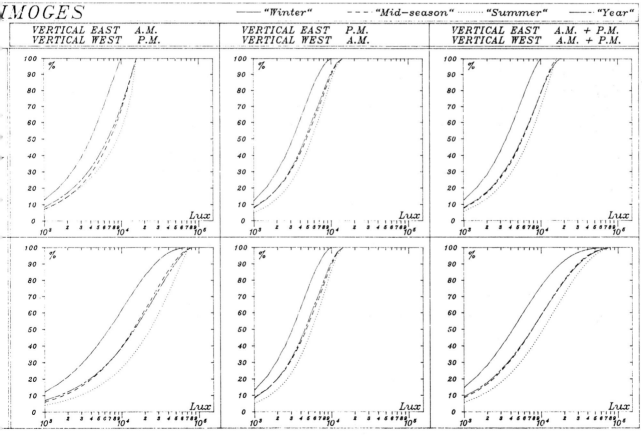

B.8 Daylighting in Architecture

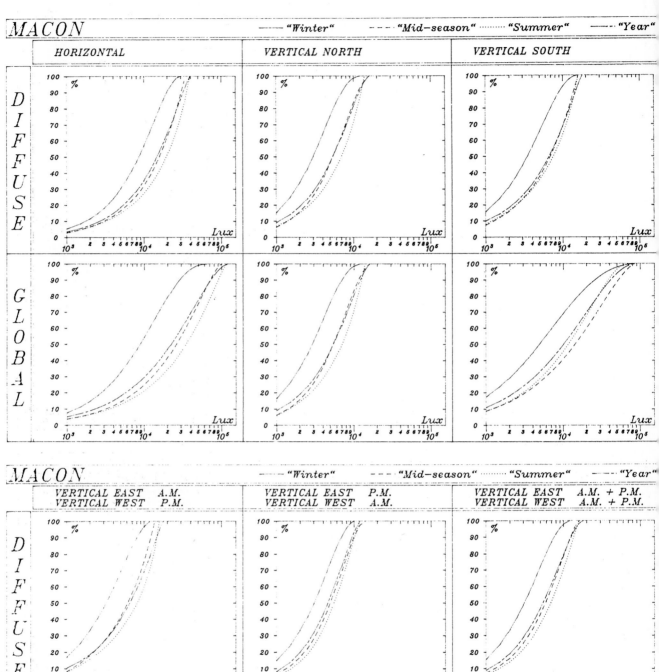

Daylight Availability B.9

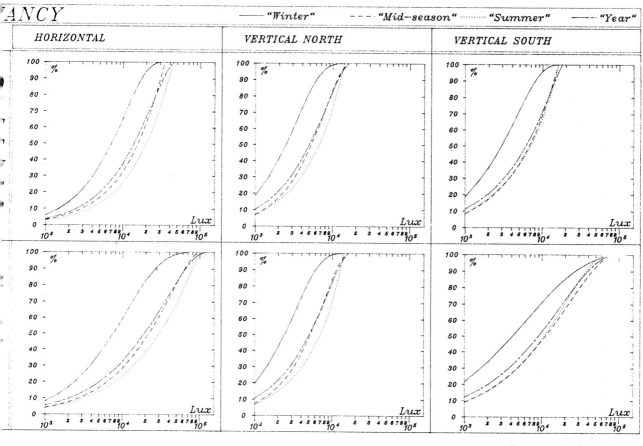
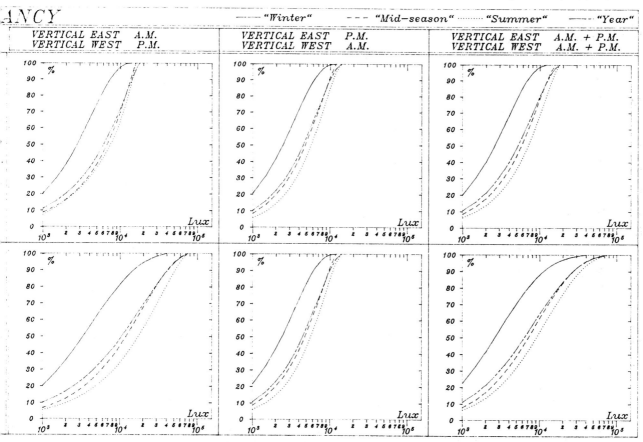

B.10 Daylighting in Architecture

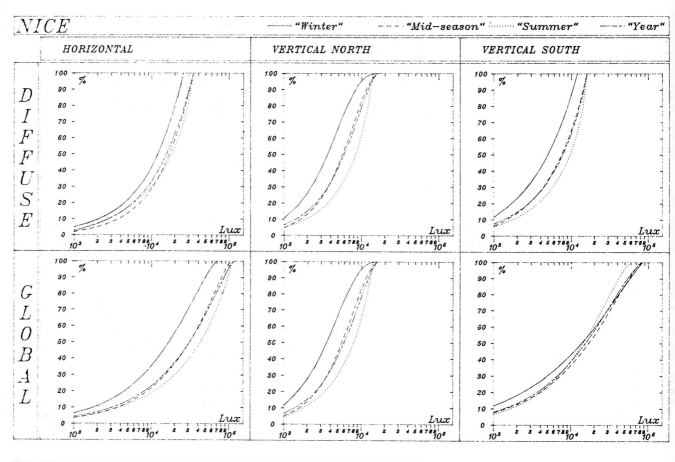

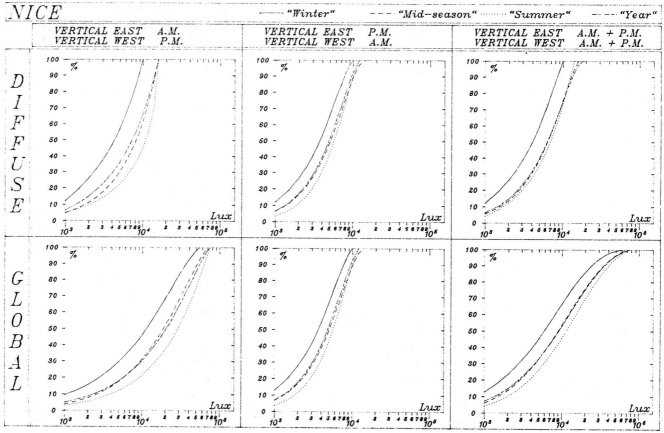

Daylight Availability **B.11**

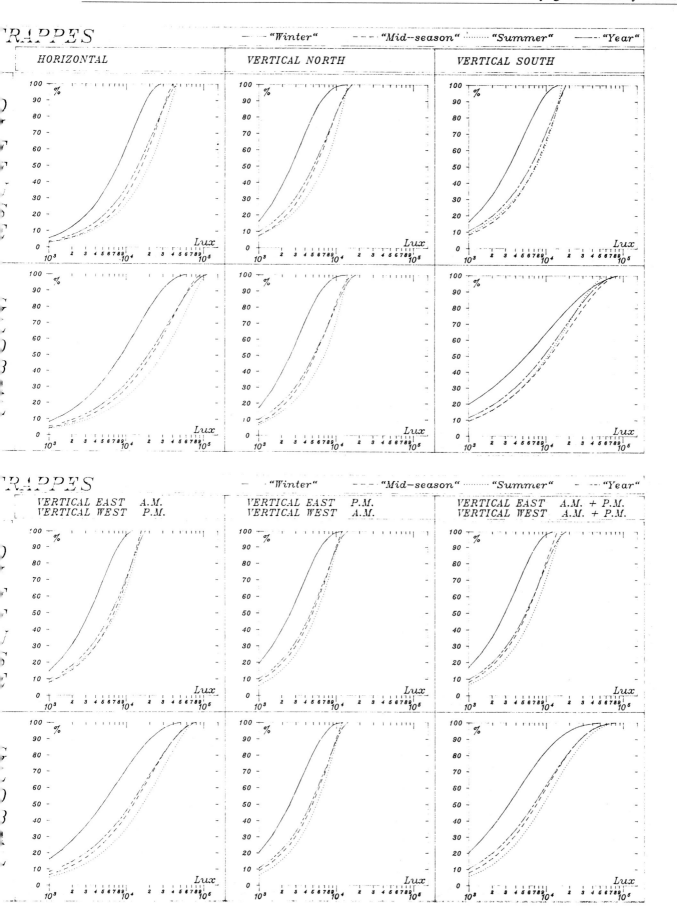

B.12 Daylighting in Architecture

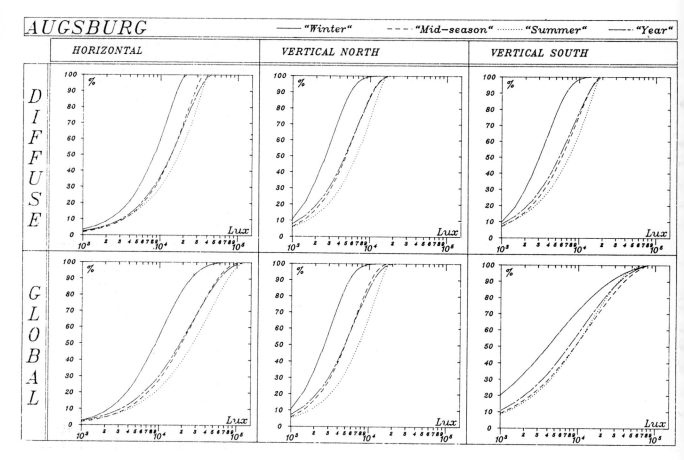

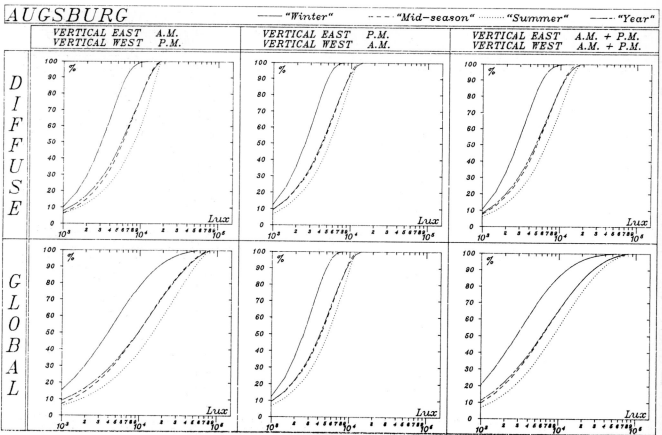

Daylight Availability B.13

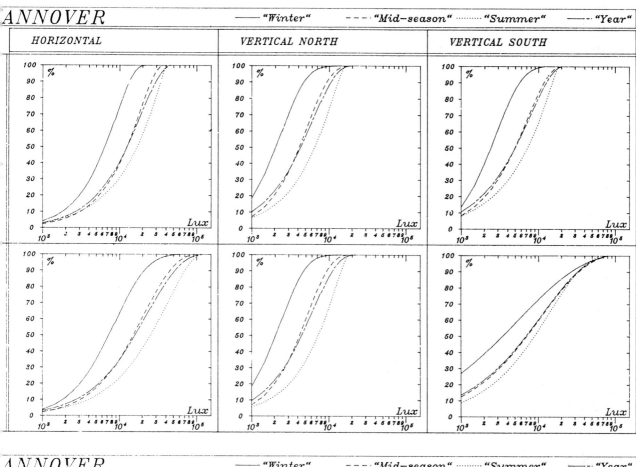

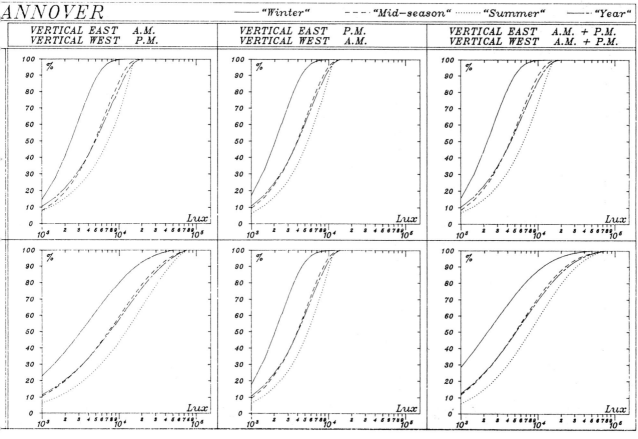

B.14 Daylighting in Architecture

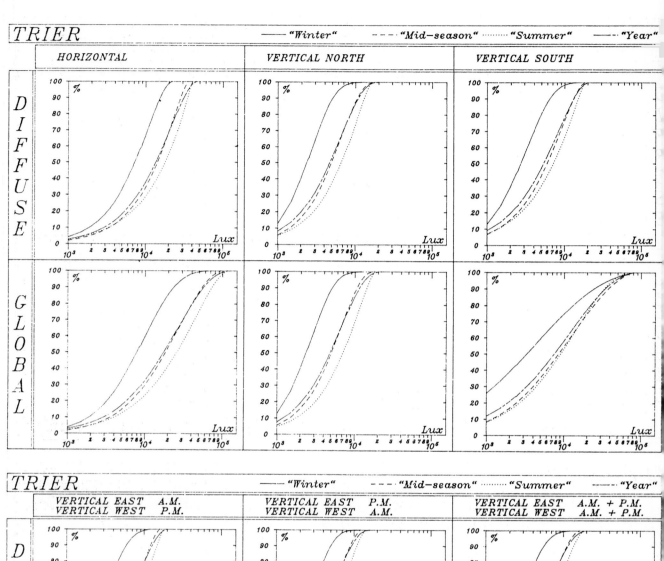
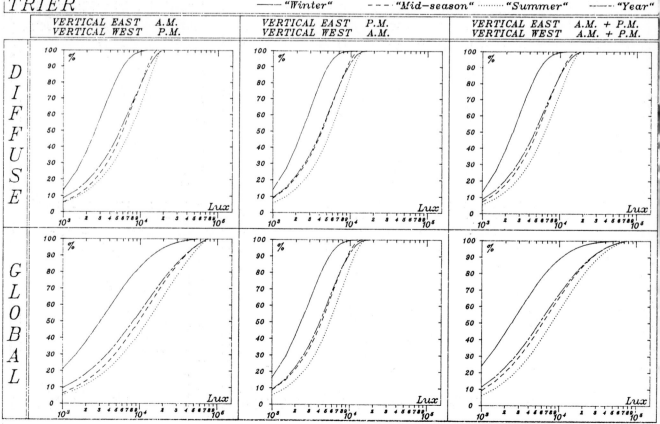

Daylight Availability B.15

WÜRZBURG

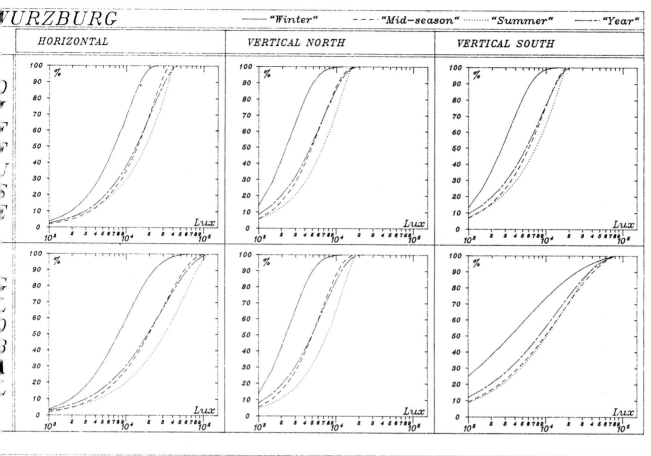

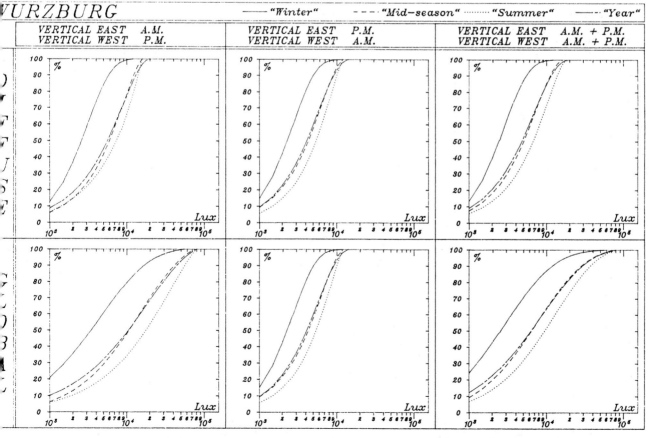

B.16 Daylighting in Architecture

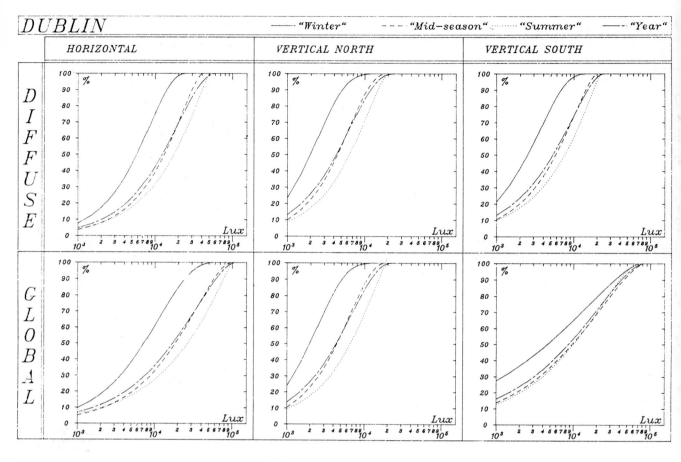

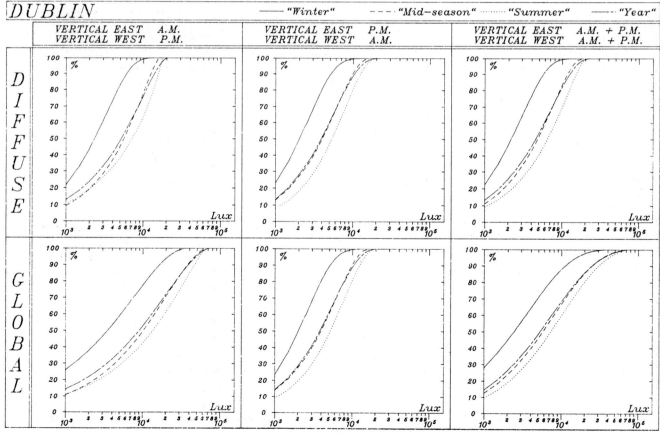

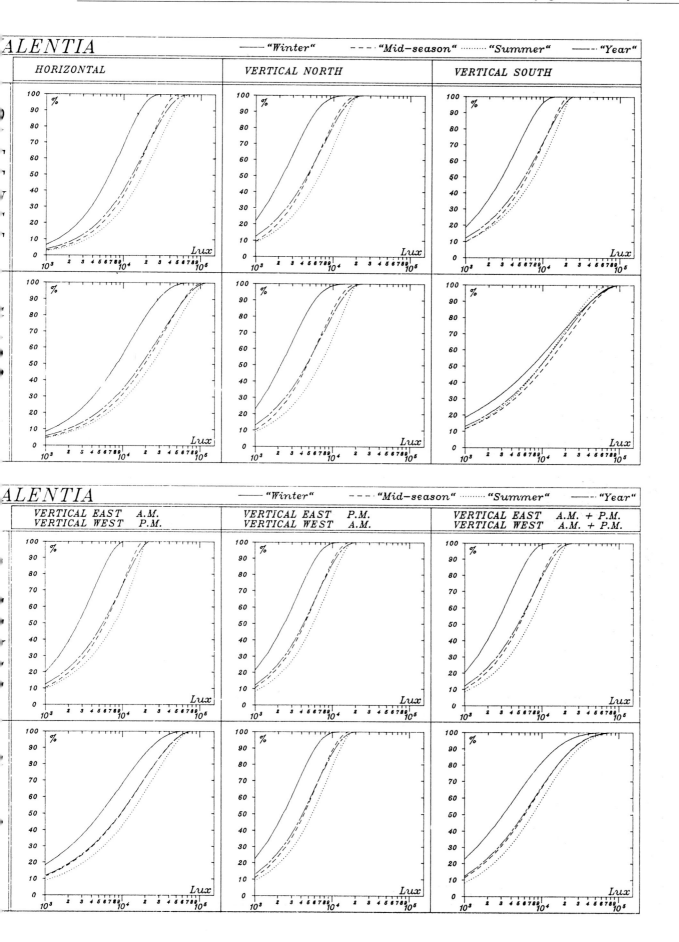

B.18 Daylighting in Architecture

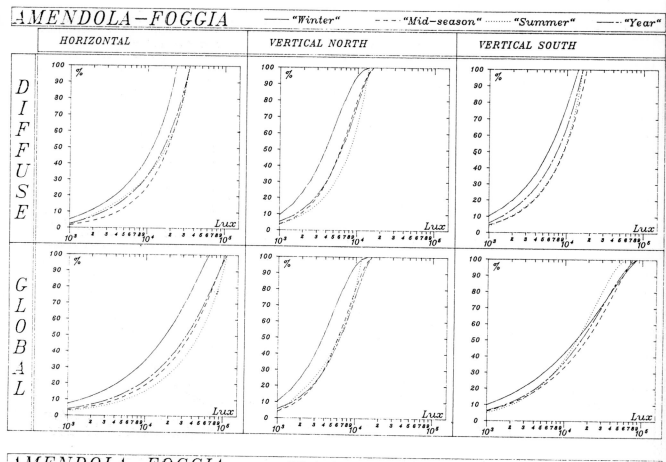

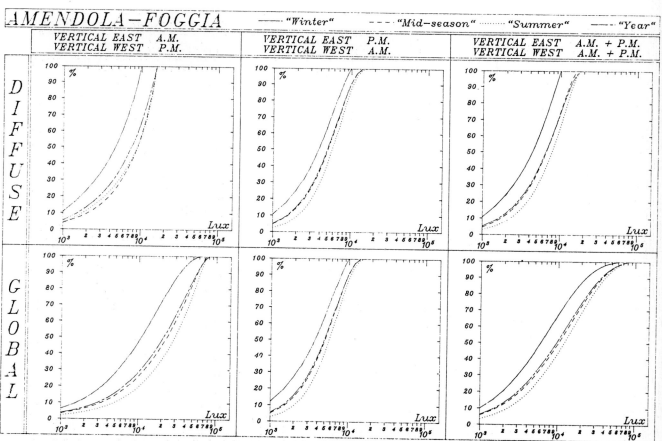

Daylight Availability B.19

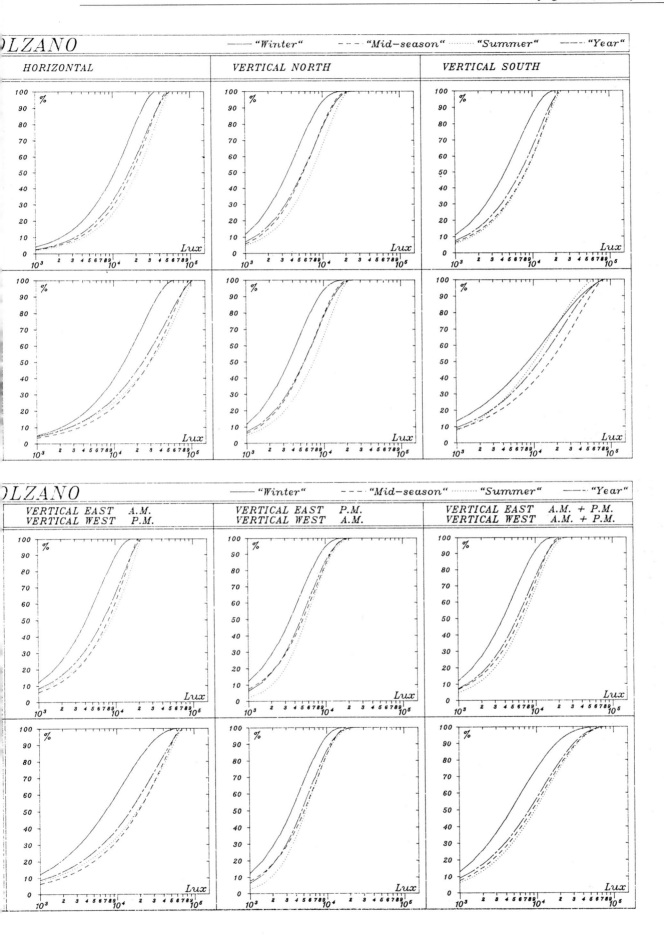

B.20 Daylighting in Architecture

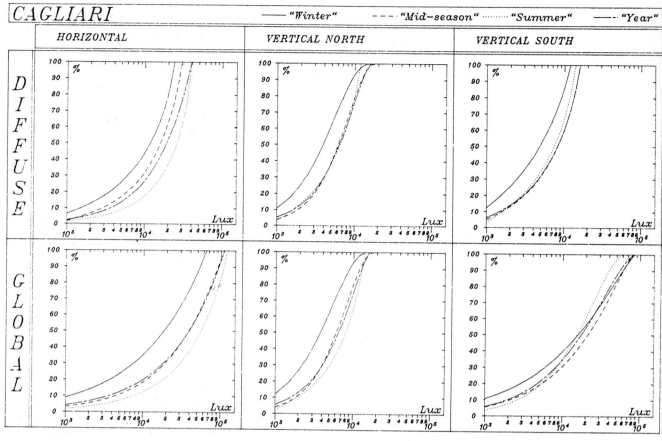

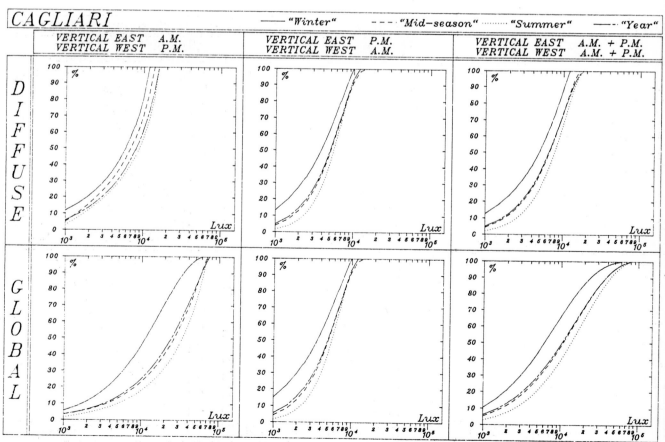

Daylight Availability B.21

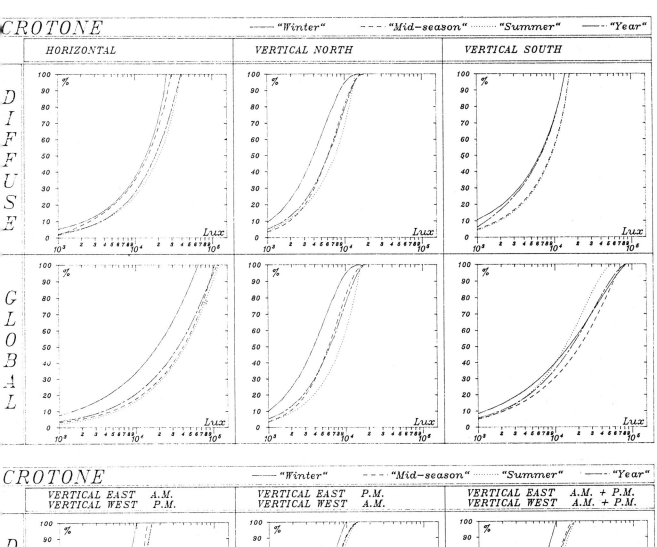

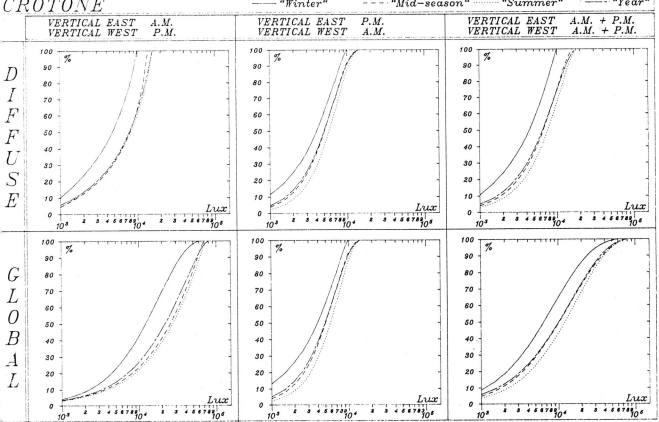

B.22 Daylighting in Architecture

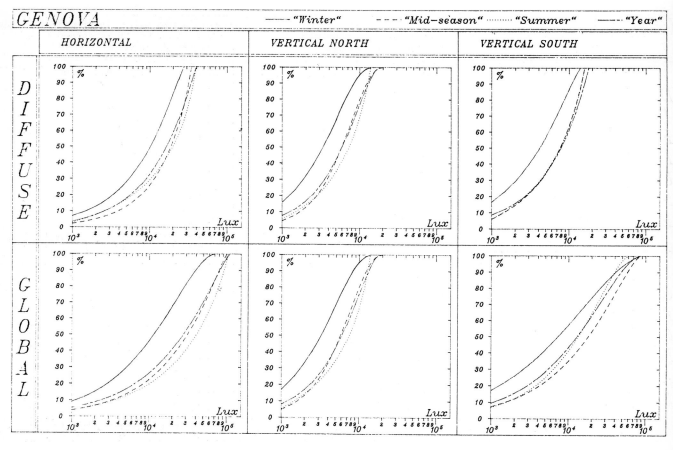

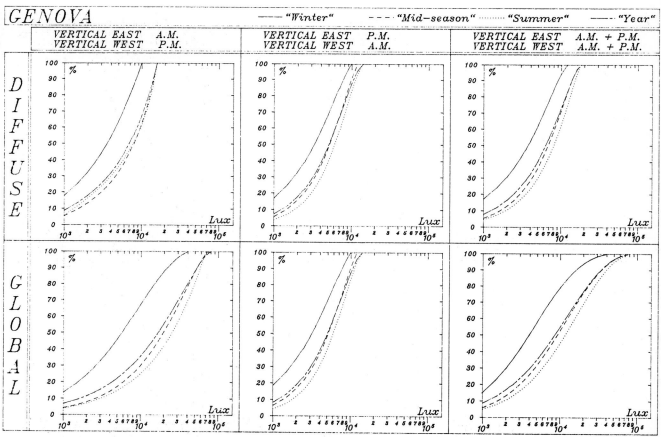

Daylight Availability B.23

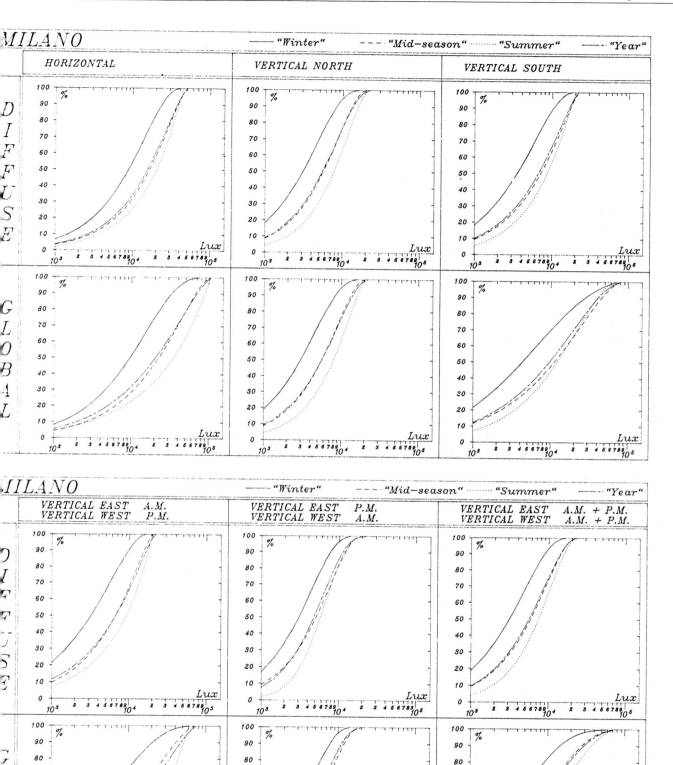

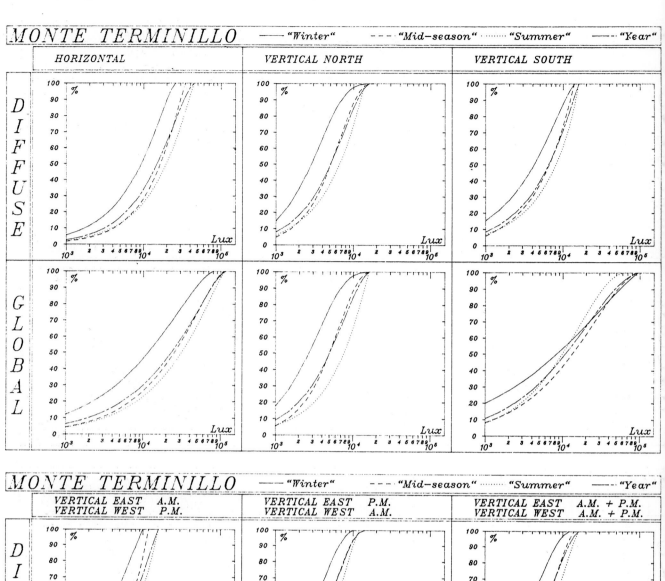
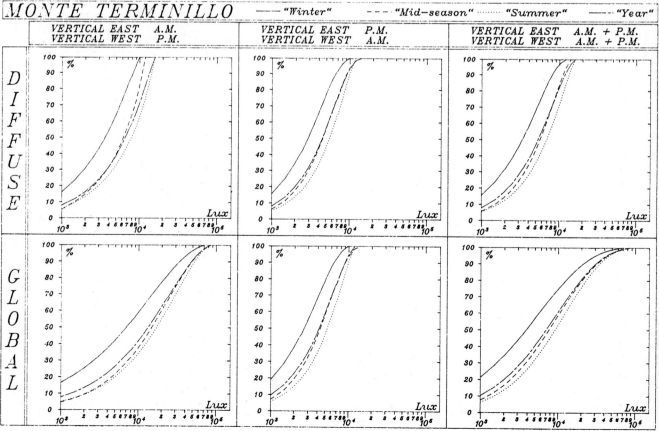

Daylight Availability B.25

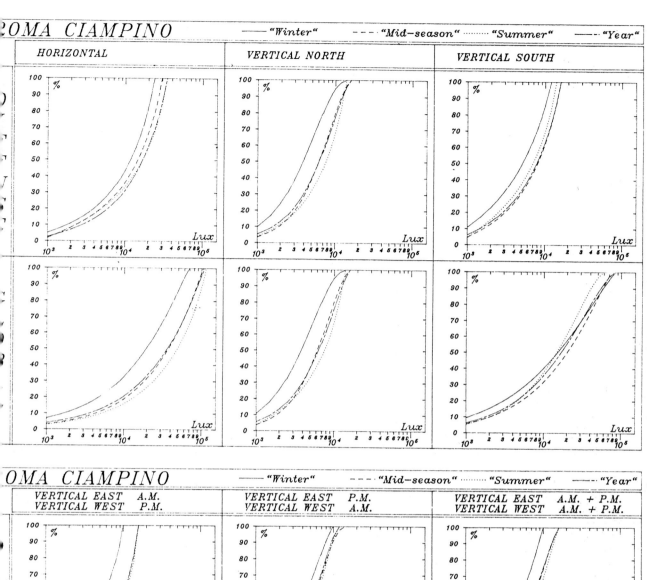

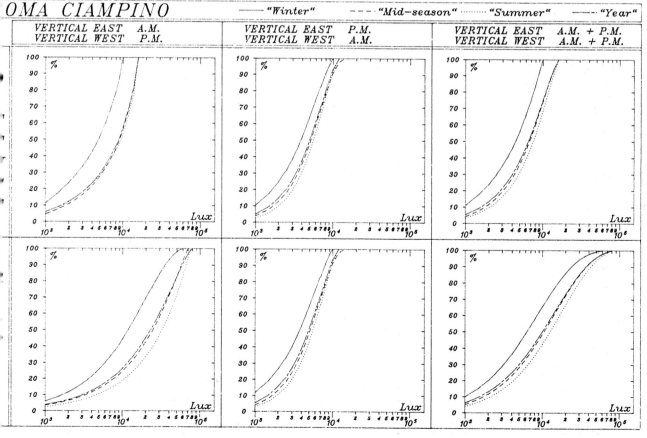

B.26 Daylighting in Architecture

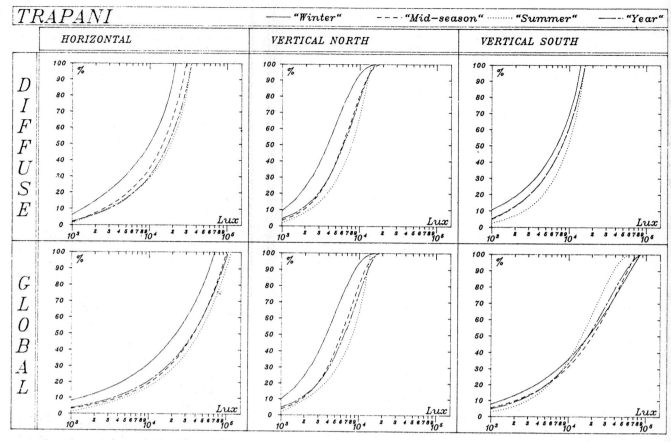
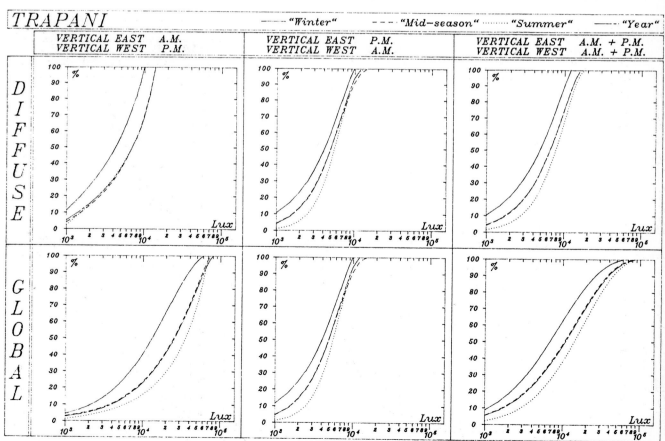

Daylight Availability **B.27**

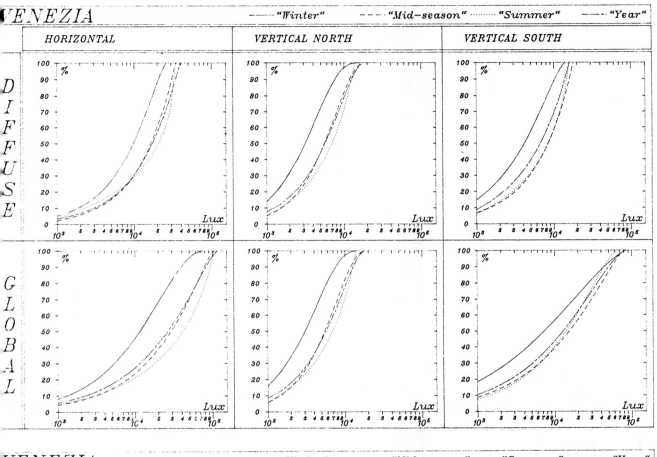

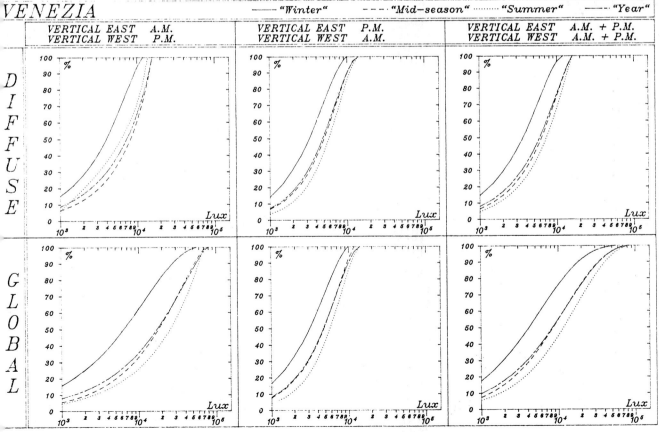

B.28 Daylighting in Architecture

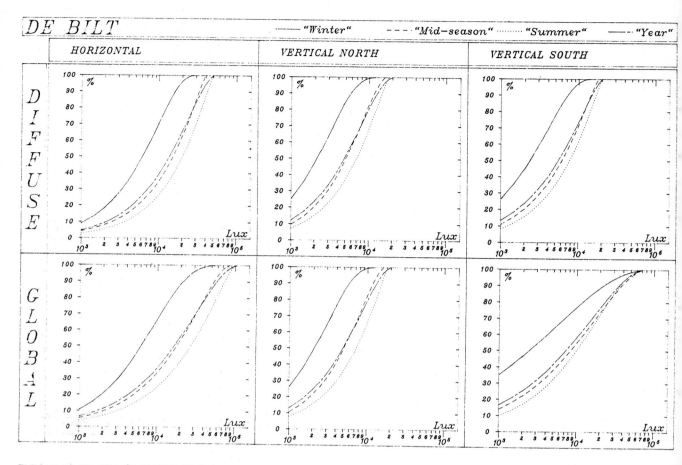
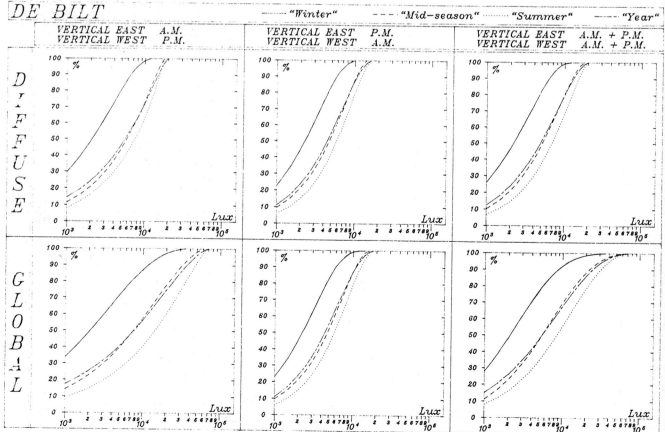

Daylight Availability **B.29**

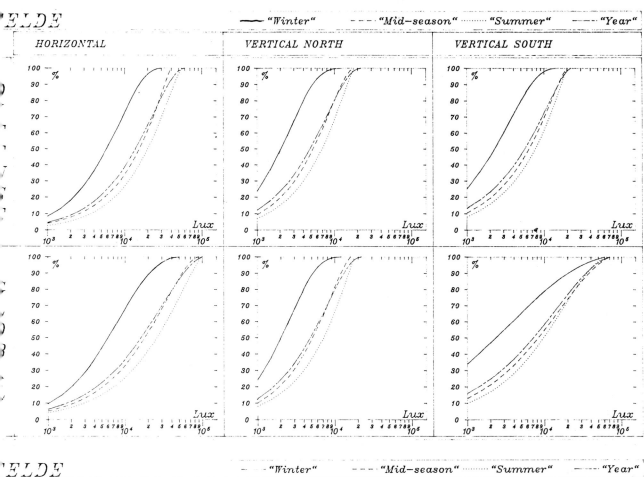

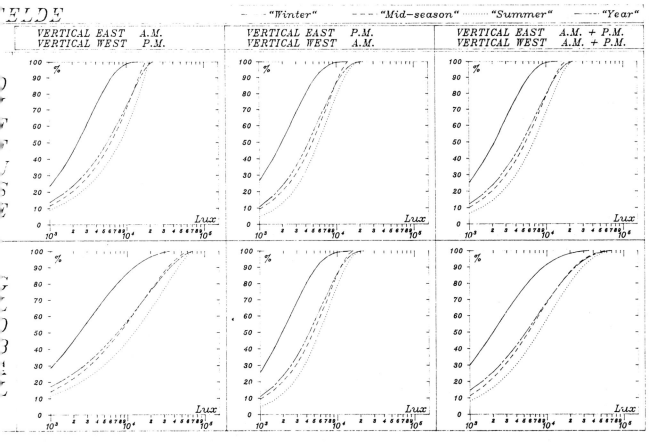

B.30 Daylighting in Architecture

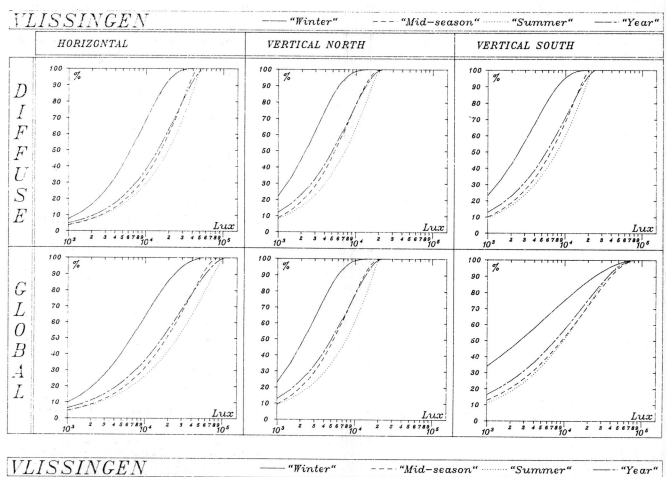

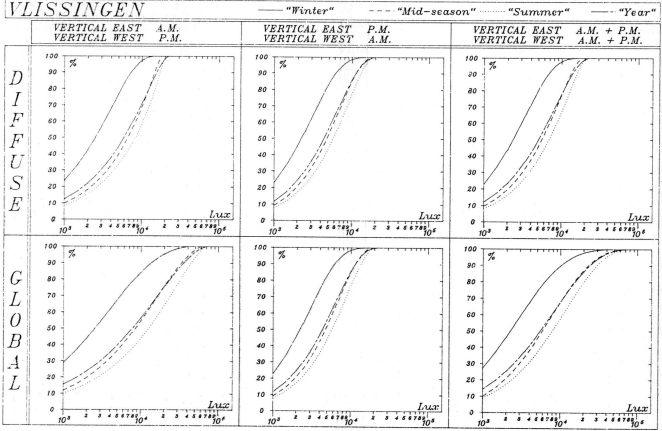

Daylight Availability B.31

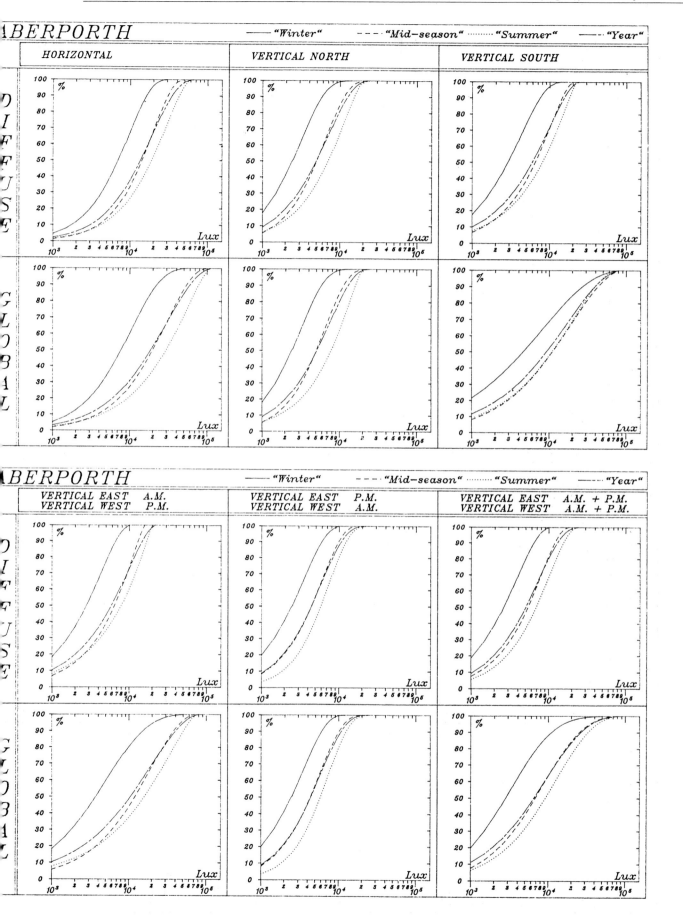

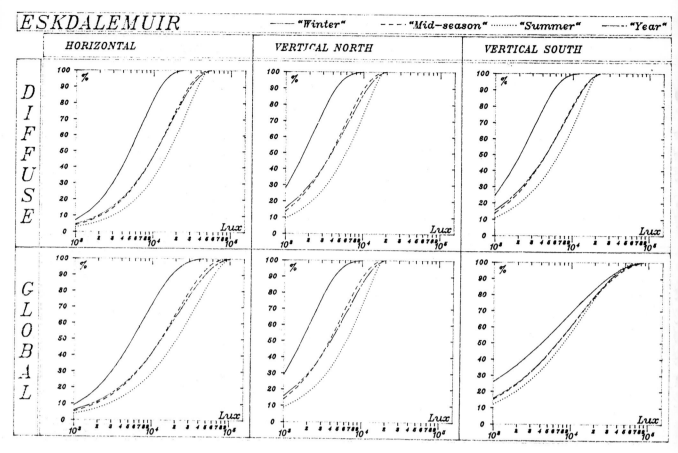
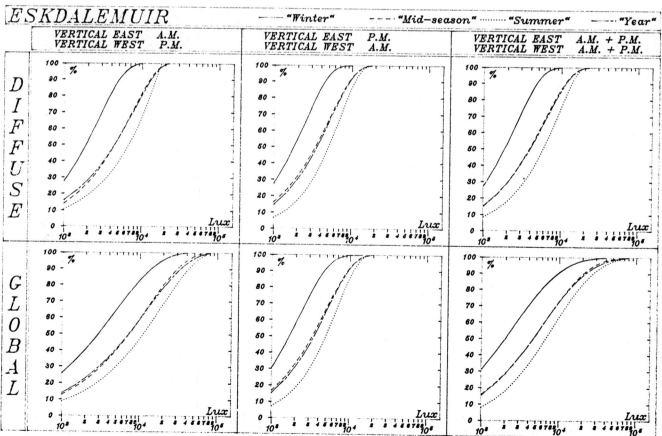

Daylight Availability **B.33**

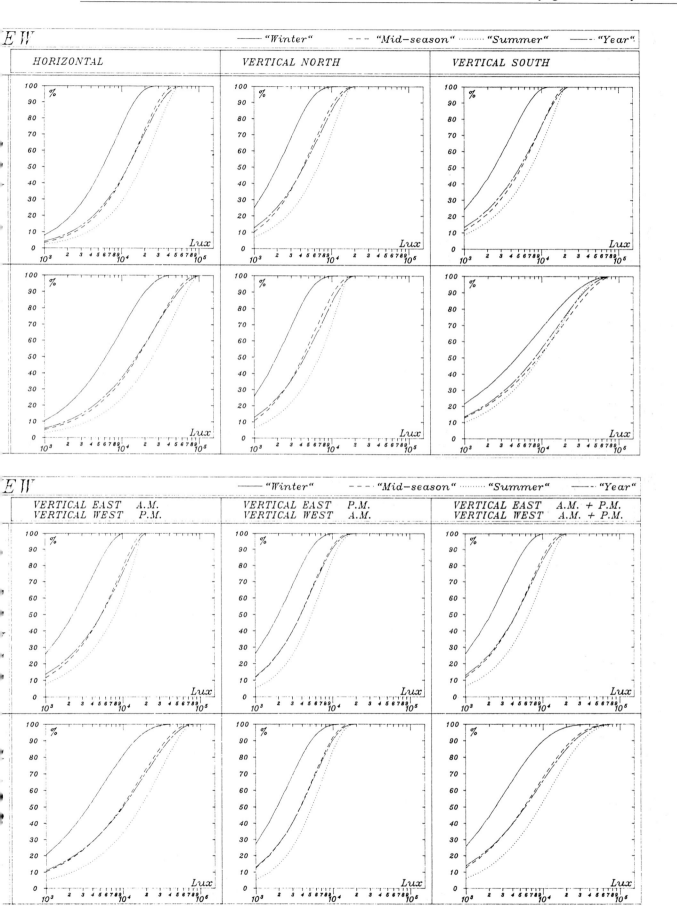

B.34 Daylighting in Architecture

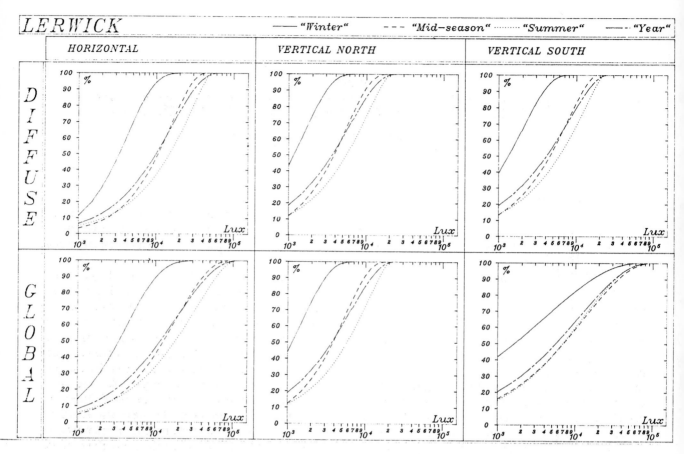
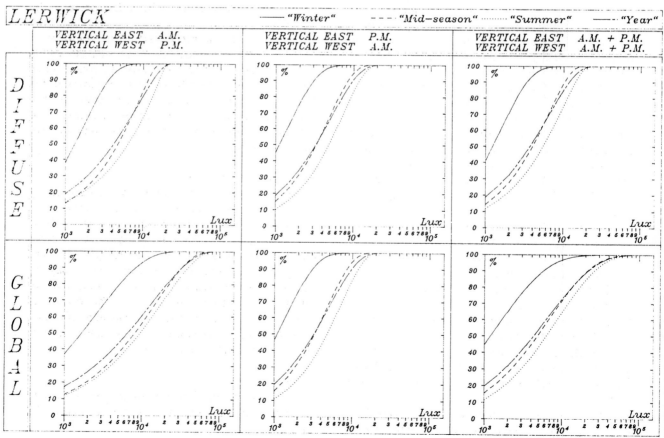

Appendix C
SURVEY OF LIGHT MEASURING INSTRUMENTS

LIGHT MEASURING INSTRUMENTS

Light measuring instruments are used for outdoor daylight availability measurements as well as for indoor illumination measurements. Photometry refers to the measurement of the visible radiation (light) with a sensor having a spectral responsivity curve equal to the average human eye.

Photometry is used to describe lighting conditions such as illumination of working areas, interior lighting, etc, where the human eye is the primary sensor. The spectral responsivity curve of the standard human eye at typical light levels is called the CIE Standard Observer Curve (photopic curve), and covers the waveband of 380-770nm. The human eye responds individually to light of different colours and has maximum sensitivity to yellow and green. In order to make accurate photometric measurements of various colours of light or from differing types of light sources, a photometric sensor's spectral responsivity curve must match the CIE photopic curve very closely.

For measuring the daylight inside a space two kinds of instruments are important:

Illuminance meter. The illuminance meter consists of a built-in or detachable photometer head (with cable link), a transducer and the display unit. Illuminance meters that are suitable for the measurement of the relevant quantities by means of attachable or built-in accessories are referred to as (1):
- instrument for the measurement of the illuminance (E meter)
- instrument for the measurement of the spherical illuminance (E_o meter)
- instrument for the measurement of the cylindrical illuminance (E_z meter)
- instrument for the measurement of the semi-cylindrical illuminance (E_{zh} meter):

Luminance meter. The luminance meter consists usually of a built-in or detachable photometer head (with cable link), and a display unit.

Both kinds of instruments use light-sensitive detectors which convert the incident light into an electrical quantity. The light sensitive detector is either made of selenium or silicon. The characterisation of photometers follows the recommendation of CIE (see Table I, (1)).

Errors

CIE Publication No. 69 (1) defines the sources of errors of light measuring instruments which the researcher must be aware of when conducting measurements.

Errors of the instruments are as follows - $V(\lambda)$ match, UV response, cosine response, E_o response, E_z response, E_{zh} response, linearity error, error of display unit, temperature coefficient, fatigue, modulated radiation, polarisation, range change, crest factor, lower frequency limit, upper frequency limit. The errors are largely independent of each other and are random in polarity and magnitude. Therefore they can be summed in quadrature for defining the total error of the instrument. The maximum errors for different classes of light measuring instruments are given in DIN 5032, Part 7 (2): there are given values for classes L, A, B and C, where L represents the best class.

When purchasing a radiation measuring instrument, it is necessary to insure that the spatial (cosine) and relative spectral response errors are as low as possible. These two errors depend upon the skill and the expertise of the designer and the manufacturer. Some manufacturers deliberately do not give specifications for these errors and the user can expect large errors.

User Application Errors

Besides these errors due to the manufacturing technology, there can be several user application errors of which the user should be aware. These typically include reflections or obstructions from clothing, buildings, trees, etc.; dust, flyspecks, bird droppings, etc.; shock causing permanent damage of optics within the sensor; rain, if the sensor is not completely waterproof; use of the incorrect calibration constant; incorrect interpolation of analog meters; using wrong meter function; failure to have sensors re-calibrated periodically.

Guidelines

For the use of light measuring equipment in scale models the following guidelines should be followed:

Illuminance meters.
- Use colour and cosine corrected photocells
- use instruments with remote detectors: instruments with built-in detectors are not

suitable for scale model measurements
- check the possibility of cable connections to the instrument longer than standard carefully with the manufacturer: the potential for errors in transferring the data increases with longer cables
- use instruments with a high sensitivity range from 1 to 150000 lux, if direct sunlight is being measured
- use instruments with an automatic range change, especially if sunlighting studies are to be carried out
- use photocells which are as small as possible for application in scale models.

Luminance meters use a reflex-type luminance meter which permits direct observation of the measured area.

The following pages present a survey of light measuring instruments available in Europe in order to facilitate the decision making process for purchasing an instrument. A few companies are missing in this list mainly due to the fact that there was no reply to the request to send technical material for the survey.

REFERENCES

(1) CIE. *Methods of characterising illuminance meters and luminance meters - Performance, characteristics and specifications.*, CIE publ. No. 69, 1987.

(2) DIN, *Lichtmessung, Klasseneinteilung von Beleuchtungsstärke und Leuchtdichtemeßgerärten*, DIN 5032, Teil 7, 1985.

TABLE I - Summary of parameters for individual characteristics for illuminance and luminance meters

Characteristics	Illuminance Meter Parameter Symbol (error)	Luminance Meter Parameter Symbol (error)
$V(\lambda)$ match	f_1'	f_1'
UV response	u	u
IR response	r	r
Directional response		$f_2(g)$
Effect from the surrounding field		$f_2(u)$
Cosine response	f_2	
E_o response	$f_{2,0}$	
E_z response	$f_{2,z}$	
E_{zh} response	$f_{2,zh}$	
Linearity error	f_3	f_3
Error of display unit	f_4	f_4
Temperature coefficient	α	α
Fatigue	f_5	f_5
Modulated radiation	f_7	f_7
Polarization	f_8	f_8
Range change	f_{11}	f_{11}
Errors of focus		f_{12}
Crest factor	c	c
Lower frequency limit	f_l	f_l
Upper frequency limit	f_u	f_u

TABLE II - Maximum errors for the individual characteristics of illuminance and luminance meters, Class L, (DIN 5032, Part 7) (2)

Characteristics	Illuminance Meter Parameter Symbol (error)		Luminance Meter Parameter Symbol (error)	
$V(\lambda)$ match	f_1'	1.5%	f_1'	2%
UV response	u	0.2%	u	0.2%
IR response	r	0.2%	r	0.2%
Directional response			$f_2(g)$	2%
Effect from the surrounding field			$f_2(u)$	1%
Cosine response	f_2	1.5%		
E_o response	$f_{2,0}$	10%		
E_z response	$f_{2,z}$	5%		
E_{zh} response	$f_{2,zh}$	5%		
Linearity error	f_3	0.2%	f_3	0.2%
Error of display unit	f_4	0.2%	f_4	0.2%
Temperature coefficient	α	0.1%/K	α	0.1%/K
Fatigue	f_5	0.1%	f_5	0.1%
Modulated radiation	f_7	0.1%	f_7	0.1%
Polarization	f_8		f_8	0.2%
Range change	f_{11}	0.1%	f_{11}	0.1%
Errors of focus			f_{12}	0.4%
Crest factor	c		c	
Lower frequency limit	f_l	40Hz	f_l	40Hz
Upper frequency limit	f_u	10^5Hz	f_u	10^5Hz

Survey of Light Measuring Instruments Available in Europe

MANUFACTURER	PHOTO	MODEL	SENSITIVITY RANGE	NUMBER OF RANGES	AUTOM. RANGE CHANGE	COSINE CORRECTIONS	SENSOR MATERIAL: SILICON	SENSOR MATERIAL: SELENIUM	SENSOR SIZE (mm): RECEIVER ø	SENSOR SIZE (mm): HEAD ø × H	DISPLAY: DIGITAL	DISPLAY: ANALOG	LUMINANCE	ILLUMINANCE	METRIC	ENGLISH	SIZE (mm) L × W × H	WEIGHT (KG): INSTRUMENT	WEIGHT (KG): SENSOR	PORTABLE	LENGTH OF CORD SENSOR TO READOUT	COST	DATE	COMMENTS RWTH AACHEN 5/1989
BRÜEL & KJAER DK-2850 Naerum Denmark Tel 45 2 800500 Fax 45 2 801405		Luminance and Contrast Meter 1100	0,01 cd/m² - 200 kcd/m²				•				•		•		•		200 × 139 × 132	1,7		•		DM 8.029	(5/89)	
Sales agent in Germany: Brüel & Kjaer Birkenweg 3-5 D-2085 Quickborn Tel 04106-4055 Fax 04106-69955		Luminance Meter 1101	0,1 cd/m² - 2.000 kcd/m²				•				•		•		•		198 × 83 × 133	2,3		•		DM 19.018	(5/89)	
Sales agents also in: Belgium, France Greece, Great Britain, Ireland, Italy, Luxembourg, Spain, Portugal, Netherlands		Photometer 1105	0,001 lx - 200 klx			•	•		8 mm		•			•	•		300 × 284 × 138	4,2	0,075	•	1,2 (50 m)	DM 17.190	(5/89)	
GEC Measurements St. Leonards Works Stafford ST 174 LX UK Tel 0785-3251 Fax 0785-212232		Minilux M	0 - 2.500 lx	4				•		83 × 15 mm		•		•	•		146 × 96 × 51	0,425	0,227	•	2 m			
Sales agent in Germany: H. Weutschreck Köttgen 11 D-56 Wuppertal 1 Tel 0202-702031		Minilux 2	0 - 10.000 lx	10		•		•		90 × 15 mm		•		•	•		146 × 96 × 68	0,440	0,220	•	2 m	£ 258	(10/85)	

C.4 Daylighting in Architecture

Survey of Light Measuring Instruments Available in Europe

MANUFACTURER	PHOTO	MODEL	SENSITIVITY RANGE	NUMBER OF RANGES	AUTOM. RANGE CHANGE	COSINE CORRECTIONS	SENSOR MATERIAL SELENIUM	SENSOR MATERIAL SILICON	SENSOR SIZE (MM) RECEIVER Ø	SENSOR SIZE (MM) HEAD Ø × H	DISPLAY ANALOG	DISPLAY DIGITAL	LUMINANCE	ILLUMINANCE	METRIC	ENGLISH	SIZE (MM, L × W × H)	WEIGHT (KG) INSTRUMENT	WEIGHT (KG) SENSOR	PORTABLE	LENGTH OF CORD SENSOR TO READOUT	COST	DATE	COMMENTS
GOSSEN gmbH Nägelsbachstraße 25 D-8520 Erlangen Tel 09131-827-1 Fax 09131-28895		Sixtolux	0 - 60.000 lx	2													64 × 94 × 28	0,085				DM 180	(5/89)	
		Panlux Electronic 2	0 - 20.000 fc or 0 - 200.000 lx	6					20 mm	32 × 105 × 29							79 × 110 × 35	0,35			1,5 m	DM 520	(5/89)	Luminance attachment DM 132
		Mavo Monitor	0,01 cd/m² - 19,99 kcd/m²	4					18,5 mm	32 × 105 × 95							86 × 153 × 25	0,65			1,5 m	DM 986	(5/89)	
		Mavolux digital	0,1 lx - 199,9 klx	4					20 mm	32 × 105 × 20							86 × 153 × 25	0,65			1,5 m	DM 986	(5/89)	Luminance attachment DM 132
HAGNER Box 2256 S-17102 Solna Schweden Tel 08-836150		Luxmeter EC 1 EC 1 ES EC 1 LS	1 - 200.000 lx 0,1 - 20.000 lx 10 - 2 Mlx	1													135 × 75 × 35	0,19				DM 632 DM 702 DM 868	(10/89)	also available: EC 1 UVA (ultra violet A) EC 1 IR (infrared)

RWTH AACHEN 5/1989

Survey of Light Measuring Instruments

SURVEY OF LIGHT MEASURING INSTRUMENTS AVAILABLE IN EUROPE

MANUFACTURER	PHOTO	MODEL	SENSITIVITY RANGE	NUMBER OF RANGES	AUTOM. RANGE CHANGE	COSINE CORRECTIONS	SENSOR MATERIAL SILICON	SENSOR MATERIAL SELENIUM	SENSOR SIZE (MM) RECEIVER ø	SENSOR SIZE (MM) HEAD ø X H	DISPLAY DIGITAL	DISPLAY ANALOG	LUMINANCE	ILLUMINANCE	METRIC	ENGLISH	SIZE (MM, L X W X H)	WEIGHT (KG) INSTRUMENT	WEIGHT (KG) SENSOR	PORTABLE	LENGTH OF CORD SENSOR TO READOUT	COST	DATE	COMMENTS	RWTH AACHEN 5/1989
Sales agent in Germany: Microscan GmbH Postfach 601705 D-2000 Hamburg 60 Tel 040-632003-0 Fax 040-6320349		Luxmeter E 2 E 2X (with remote detector)	0,01 - 199,90 lx	5		•	•		10 mm		•			•	•		180 x 100 x 25	0,45		•	2 m	E2 DM 2.060 E2X DM 2.190	(10/89)	Different detectors available SD 1 - SD11	
Sales agent in Italy: LSI spa Strada Prov. 161 I-20090 Settala Premenugo Mi Tel 02-95770541		Universal Photometer S 1 S 2	0 - 100.000 cd/m² 0 - 100.000 lx 0 - 100.000 cd/m² 0 - 100.000 lx	5 / 10		•	•		10 mm			•	• $1°$	•	•	•	240 x 110 x 90	1,6		•	2 m	S1 DM 5.555 S2 DM 5.952	(10/89)		
IL International Light 17 Graf Road Newburyport, MA 01950 Tel 508-465-5923 Fax 508-462-0759		IL 1351 Photometer System (IL 1350 + SCD 110 (fc) or SCD 110/CM433 (lx)	0,2 - 1999 fc 2 - 21.500 lx	4		•	•		7 mm	112 x 17 x 17mm	•		•	•	•		220 x 95 x 35	1,13		•	3 m	DM 6.503	(9/89)		
IL 1710 Research Photometer System (IL 1700 + SED 038/Y/W)			0,005 - 10⁶ lx		•	•	•			42 x 35 mm	•		•	•			220 x 240 x 90	2,3		•	3 m	DM 2.431	(9/89)	incl. RS 232 TTL Multiple Power Selection	
Agent in Germany: Starna Postfach 1206 D-6102 Pfungstadt Tel 06157-7953 Fax 06157-85564		A 415 Eight Channel Computer	Combination with 8 SED detectors + IL 1700																			DM 2.505	(9/89)		

C.6 Daylighting in Architecture

Survey of Light Measuring Instruments available in Europe

MANUFACTURER	PHOTO	MODEL	SENSITIVITY RANGE	NUMBER OF RANGES	AUTOM. RANGE CHANGE	COSINE CORRECTIONS	SENSOR MATERIAL: SILICON	SENSOR MATERIAL: SELENIUM	SENSOR SIZE (MM) RECEIVER ø	SENSOR SIZE (MM) HEAD ø × H	DISPLAY: DIGITAL	DISPLAY: ANALOG	LUMINANCE	ILLUMINANCE	METRIC	ENGLISH	SIZE (MM, L × W × H)	WEIGHT (KG) INSTRUMENT	WEIGHT (KG) SENSOR	PORTABLE	LENGTH OF CORD SENSOR TO READOUT	COST DATE	COMMENTS RWTH AACHEN 5/1989
	Type SCD	5 different Illuminance detectors	different sensitivity ranges			•	•			112 × 17 × 17 mm				•						•	3 m	SCD-Type DM 872 - 965 SED-Type DM 1650 (6/89)	also available: detectors for -irradiance -radiance -power
	Type SED	2 Luminance detectors	different sensitivity ranges				•		7 mm	42 × 35 mm			•	•						•	3 m	DM 1654 (6/89)	high gain detector
LI-COR Box 4425 Lincoln Nebraska 68504 Tel 402-467 3576 Fax 402-467 2819		LI-210SA Photometric Sensor	20 µA per 100 lx			•	•			24 × 25 mm				•					0,028	•	3 m	DM 1.035 (4/88)	
Agent in Germany: Walz Eichenring 10-14 D-8521 Effeltrich Tel 09133-871 Fax 09133-5395		LI 1000 Data Logger			•						•						210 × 114 × 69	1,6		•		DM 3.760 (4/88)	10 channels RS 232 32 K memory Battery/ power
Also agents in: France, Greece, Italy, Netherlands, Belgium, Spain, Great Britain, Ireland		LI 185B Photometer	0 - 30.000 lx	6								•					178 × 127 × 64	1,11	•	•		DM 2.245 (4/88)	additional: SS-3 Sensor selector (switchbox for up to 3 sensor input) DM 880

Survey of Light Measuring Instruments — C.7

MANUFACTURER	MODEL	SENSITIVITY RANGE	NUMBER OF RANGES	AUTOM. RANGE CHANGE	COSINE CORRECTIONS	SENSOR MATERIAL - SILICON	SENSOR MATERIAL - SELENIUM	SENSOR SIZE (mm) RECEIVER ø	SENSOR SIZE (mm) HEAD ø X H	DISPLAY - DIGITAL	DISPLAY - ANALOG	LUMINANCE	ILLUMINANCE	METRIC	ENGLISH	SIZE (mm, L x W x H)	WEIGHT (kg) INSTRUMENT	WEIGHT (kg) SENSOR	PORTABLE	LENGTH OF CORD SENSOR TO READOUT	COST	DATE	COMMENTS
LMT Lichtmeßtechnik GmbH Berlin Helmholtzstr. 9 D-1000 Berlin 10 Tel 030-3934028 Fax 030-3918001	Pocket-Lux	0,1 - 200.000 lx		•		•		10 mm		•			•	•		135 x 80 x 40	0,35	0,08		2 m	DM 1.650	(4/89)	optional: E_z and E_{sz} adapter, remote control
	Photometer Heads P 60 P 15 P 10					•		60, 30, 15, 10											•		DM 1.180 to 4.160	(4/89)	optional: Thermostatic stabilization watershielding
	System Photometer S 1000	10^{-5} - 500.000 lx		•	•	•		30 mm		•			•	•		485 x 300 x 133	4,0	0,7		5 - 100 m	DM 9.130 to 10.480	(4/89)	up to 100 readings/sec Option: IEEE 488
	Illuminance Meter B 250	0,01 - 200.000 lx optional 0,001 - 20.000 lx		•	•	•		30 mm		•		optional	•	•		165 x 115 x 55	1,5	0,7	•	3 m	DM 9.900	(4/89)	Explosion and fire damp proof Option: Luminance attachment
	Illuminance Meter B 360 E B 360 F,H B 360 J	0,01 lx - 600 klx 0,001 lx - 200 klx 0,1 mlx - 20 klx	6	•	•	•		15 or 30 mm		•			•	•		175 x 114 x 58	1,0	0,2	•	3 m	DM 3.750 to 5.000	(4/89)	optional: E_z and E_{sz} photometer heads

RWTH AACHEN 5/1989

C.8 Daylighting in Architecture

Survey of Light Measuring Instruments available in Europe

MANUFACTURER	PHOTO	MODEL	SENSITIVITY RANGE	NUMBER OF RANGES	AUTOM. RANGE CHANGE	COSINE CORRECTIONS	SENSOR MATERIAL - SILICON	SENSOR MATERIAL - SELENIUM	SENSOR SIZE (mm) RECEIVER ø	SENSOR SIZE (mm) HEAD ø x H	DISPLAY - DIGITAL	DISPLAY - ANALOG	LUMINANCE	ILLUMINANCE	METRIC	ENGLISH	SIZE (mm, L x W x H)	WEIGHT (kg) INSTRUMENT	WEIGHT (kg) SENSOR	PORTABLE	LENGTH OF CORD SENSOR TO READOUT	COST	DATE	COMMENTS	RWTH AACHEN 5/1989
LMT Lichtmeßtechnik GmbH Berlin		Illuminance Meter B 510	0,001 - 500.000 lx			•	•		30 mm		•			•	•		384 x 150 x 96	3,0	0,2	•	3 m	DM 6.800 to DM 7.450	(4/89)		
		Photometer head with amplifier				•	•		30 mm					•						•		DM 3.500	(4/89)	For outdoor installation Output voltage 0-10V or 0-10mA	
		Limit Control Illuminance Meter GB 1000	0,01 - 200.000 lx			•	•		30 mm		•			•	•		485 x 300 x 133				3 m	DM 8.150	(4/89)	optional: - IEEE 488 - digital Thermometer	
		Photocurrent Meter I 510	10^{-11}A - 10^{-2}A		•		•				•									•		DM 5.010	(4/89)		
		Photocurrent Meter I 1000	10^{-11}A - 10^{-2}A		•		•				•						485 x 300 x 133	4,0		•		DM 6.300	(4/89)	up to 100 readings/sec optional: - IEEE 488	

Survey of Light Measuring Instruments

Survey of Light Measuring Instruments Available in Europe

MANUFACTURER	PHOTO	MODEL	SENSITIVITY RANGE	NUMBER OF RANGES	AUTOM. RANGE CHANGE	COSINE CORRECTIONS	SENSOR MATERIAL SILICON	SENSOR MATERIAL SELENIUM	SENSOR SIZE (mm) RECEIVER ø	SENSOR SIZE (mm) HEAD ø × H	DISPLAY DIGITAL	DISPLAY ANALOG	LUMINANCE	ILLUMINANCE	METRIC	ENGLISH	SIZE (mm, L × W × H)	WEIGHT (KG) INSTRUMENT	WEIGHT (KG) SENSOR	PORTABLE	LENGTH OF CORD SENSOR TO READOUT	COST DATE	COMMENTS RWTH AACHEN 5/1989
LMT Lichtmeßtechnik GmbH Berlin		Luminance Meter L 1003 L 1009	$0.001-2\times10^6$ cd/m² $0.0001-2\times10^7$ cd/m²		•		•				•		• $3°/1°/20'$		•		240 × 160 × 110	6.0		•		DM 10,400 L 1003 DM 13,800 L 1009	Option: - IEEE 488 - glare lens
		Colorimeter Head CH 60 CHS 60					•													•			
		Colorimeter C 1200			•						•			•						•			up to 120 readings/sec
		C 2200																		•			
		C 3300																		•			
LSI Laboratori di Strumentazione industriale S.P. 161, km 8,5 I-20090 Settala Premenugo (Mi) Fax 2-95770594		Climalux N (Illuminance Meter)	0 - 100.000 lx		•		•			55 × 30 mm							87 × 47 × 150	0.30	0.15	•	1.5 m		2 different sensors available

C.10 Daylighting in Architecture

Survey of Light Measuring Instruments Available in Europe

MANUFACTURER	PHOTO	MODEL	SENSITIVITY RANGE	NUMBER OF RANGES	AUTOM. RANGE CHANGE	COSINE CORRECTIONS	SENSOR MATERIAL SILICON	SENSOR MATERIAL SELENIUM	SENSOR SIZE (mm) RECEIVER ∅	SENSOR SIZE (mm) HEAD ∅ x L	DISPLAY DIGITAL	DISPLAY ANALOG	LUMINANCE	ILLUMINANCE	METRIC	ENGLISH	SIZE (mm, L x W x H)	WEIGHT (kg) INSTRUMENT	WEIGHT (kg) SENSOR	PORTABLE	LENGTH OF CORD SENSOR TO READOUT	COST	DATE	COMMENTS
MEGGER Instruments Ltd. Archcliffe Road Dover, Kent CT 17 9EN Tel 0304-202620 Fax 0304-207342		Lightmeter LM 4	0 - 500 lx 0 - 2000 lx			•						•		•	•		105 x 80 x 45	0,250				DM 380	(6/88)	RWTH AACHEN 5/1989
MEGATRON 165 Marlborough Road London N19 4NE UK Tel 01-272 3739 Fax 01-272 5975		Lightmeter DA 12 DA 10 DA 10/L DA 7 DA 5	0,005 - 100.000 lx 0,05 - 100.000 lx 0,005 - 100.000 lx 0,5 - 25.000 lx 0,005 - 25 lx	5 - 12		•		•	45 mm	68 mm	•			•	•		295 x 135 x 75	1,3	0,075	•	2 m	£ 290 to 360	(6/88)	
		Digital Lightmeter DL 3	0,1 - 10.000 lx	3		•		•	40 mm		•			•	•		145 x 85 x 35	0,29		•		£ 185	(6/88)	
		Digital Lightmeter DL 4	0,1 - 50.000 lx	4		•		•	45 mm	68 mm	•			•	•		300 x 140 x 80	1,3		•	2 m	£ 305	(6/88)	
		Digital Lightmeter DL 5	0,001 - 100.000 lx	5		•		•	25 mm		•			•	•		300 x 145 x 70	1,3		•	2 m	£ 360	(6/88)	

Survey of Light Measuring Instruments Available in Europe

MANUFACTURER	PHOTO	MODEL	SENSITIVITY RANGE	NUMBER OF RANGES	AUTOM. RANGE CHANGE	COSINE CORRECTIONS	SENSOR MATERIAL SILICON	SENSOR MATERIAL SELENIUM	SENSOR SIZE (MM) RECEIVER Ø	SENSOR SIZE (MM) HEAD Ø × H	DISPLAY DIGITAL	DISPLAY ANALOG	LUMINANCE	ILLUMINANCE	METRIC	ENGLISH	SIZE (MM, L × W × H)	WEIGHT (KG) INSTRUMENT	WEIGHT (KG) SENSOR	PORTABLE	LENGTH OF CORD SENSOR TO READOUT	COST DATE	COMMENTS	RWTH AACHEN 5/1989
MEGATRON		Architectural Model Lightmeter AML 2	2 – 10.000 lx	3		•		•		21 mm		•		•	•		360 × 330 × 150	5.0	0,025	•	2 m	£ 1120 (6/88)	up to 12 cells simultaneously	
		Daylight Factor Meter DFM 2	2 – 100.000 lx 0,1 – 20 % DF 100 – 50.000 cd/m²	3		•		•	43 mm			•	•	•	•		360 × 330 × 150	5.5	0,065	•	5 × 2 m	£ 770 (6/88)	Daylight Factor Directional cell, luminance attachment	
		AML / DFM	5 – 100.000 lx 0,1 – 20 % DF 100 – 50.000 cd/m²	3		•		•	22 mm			•	•	•	•		360 × 330 × 150	5.1	0,025	•	2 m	£ 1490 (6/88)	Combination of AML & DFM	
METRAWATT GmbH Postfach 1333 D-8500 Nürnberg 50 Tel 0911-8602-1		MX 4	0 – 5.000 lx 10 – 500.000 lx (using a grey-filter)	4	•	•		•	38 mm	58 × 135		•	•	•	•	•	126 × 92 × 45	0,30	0,05	•	1,35 m	DM 463 (8/87)	Luminance attachment	
MINOLTA Meßtechnik Kurt-Fischer-Str. 50 D-2070 Ahrensberg Tel 04102-70-1 Fax 04102-40178		Illuminance Meter T-1 T-1H T-1M	0,01 – 99.900 lx 0,1 – 999.900 lx 0,001 – 99.900 lx		•	•	•		25 mm	72 × 55 × 30	•			•	•	•	170 × 72 × 33	0,22		•	1 – 10 m	T-1 DM 1.200 T-1H DM 1.400 T-1M DM 1.550 (2/89)	T-1M Mini sensor 16,5 Ø × 12, receiver 14 mm, lead length 1 m	

C.12 Daylighting in Architecture

Survey of Light Measuring Instruments available in Europe

MANUFACTURER	PHOTO	MODEL	SENSITIVITY RANGE	NUMBER OF RANGES	AUTOM. RANGE CHANGE	COSINE CORRECTIONS	SENSOR MATERIAL SILICON	SENSOR MATERIAL SELENIUM	SENSOR SIZE (mm) RECEIVER ø	SENSOR SIZE (mm) HEAD ø x H	DISPLAY DIGITAL	DISPLAY ANALOG	LUMINANCE	ILLUMINANCE	METRIC	ENGLISH	SIZE (mm, L x W x H)	WEIGHT (KG) INSTRUMENT	WEIGHT (KG) SENSOR	PORTABLE	LENGTH OF CORD SENSOR TO READOUT	COST DATE	COMMENTS	RWTH AACHEN 5/1989
MINOLTA		Luminance Meter LS-100 / LS-110	0,001-299.900 cd/m² / 0,01-999.900 cd/m²				•				•		• 1°, 1/3°		•	•	208 x 79 x 15	0,85		•		DM 3.900 (2/89)		
OPTRONIK GmbH Kurfürstenstraße 84 D-1000 Berlin 30 Tel 030-269191 Fax 030-2615152		Chroma-Meter CL-100 xy-1	5,1- 32.700 lx / 10 - 200.000 lx				•				•			•	•		170 x 72 x 33 / 195 x 12 x 39	0,29; 0,27		•		CL-100 DM 3.810 / XY-1 DM 2.730 (9/87)	remote probe various lab-chroma-meters available	
		Miniluxmeter Standard A B	0,01-199.900 lx	5	•	•	•		6 mm		•			•	•		150 x 80 x 30	0,25		•		DM 900 to 1.500 (7/87)	remote probe, shadowfree measurement	
		Digital-Luxmeter AS 501-A AS 501-B AS 501-C	1 mlx-199.900 lx / 0,1mlx-199.900 lx / 0,01mlx-199.900 lx	6,7,8	•	•	•		8 mm		•			•	•		325 x 240 x 100	3,0		•		A DM 4.105 B DM 4.785 C DM 6.240 (8/88)	amplifier switching for 2. photo meterhead IEC-bus available on request	
		Photometer Heads					•													•				

Survey of Light Measuring Instruments

MANUFACTURER	PHOTO	MODEL	SENSITIVITY RANGE	NUMBER OF RANGES	AUTOM. RANGE CHANGE	COSINE CORRECTIONS	SENSOR MATERIAL SILICON	SENSOR MATERIAL SELENIUM	SENSOR SIZE (mm) RECEIVER ø	SENSOR SIZE (mm) HEAD ø x H	DISPLAY DIGITAL	DISPLAY ANALOG	LUMINANCE	ILLUMINANCE	METRIC	ENGLISH	SIZE (mm, L x W x H)	WEIGHT (KG) INSTRUMENT	WEIGHT (KG) SENSOR	PORTABLE	LENGTH OF CORD SENSOR TO READOUT	COST DATE	COMMENTS	RWTH AACHEN 5/1989
PRC Krochmann GmbH Geneststraße 6 D-1000 Berlin 62 Tel 030-7517007/8 Fax 030-7510127		Photometer Heads				•	•		Stand. 8 mm Mini 6/4 mm				•	•			70 x 54, 34 x 20 Mini 14 x 12		0,75 0,11 0,01		3 m (10m)	DM 500 - 2.000 (8/88) Mini DM 1.000 (8/89)	optional: thermostated heads various luminance and illuminance adapters	
		Multiplexer 901															180 x 275 x 95	1,1				DM 5.150 (6/86)	up to 32 channels simultaneously possible (16 channels) 2.000	
		Luxmeter 106	$0,1 - 120.000$ lx	2	•	•	•		8 mm	34 x 20	•			•	•		103 x 25 x 262 180 x 80 x 37	0,21	0,11	•	3 m	DM 1.250 to 2.000 (4/87)	various adapters luminance attachment	
		Luxmeter 110	$10^{-4} - 200.000$ lx	5	•	•	•		8 mm	34 x 20	•			•	•		103 x 257 x 262	4,0	0,11	•	3 m	DM 4.095 (4/87)	various accessoires optional IEEE 488	
		Luxmeter 122	$10^{-5} - 200.000$ lx	6	•	•	•		8 mm	34 x 20	•	•		•	•		105 x 470 x 325	7,0	0,11	•	3 m	DM 7.410 (4/87)	various adapters optional IEEE 488	

C.14 Daylighting in Architecture

Survey of Light Measuring Instruments available in Europe

Manufacturer	Model	Sensitivity Range	Number of Ranges	Autom. Range Change	Cosine Corrections	Sensor Material: Silicon	Sensor Material: Selenium	Sensor Size (mm) Receiver ø	Sensor Size (mm) Head ø x H	Display: Digital	Display: Analog	Luminance	Illuminance	Metric	English	Size (mm, L x W x H)	Weight (kg) Instrument	Weight (kg) Sensor	Portable	Length of Cord Sensor to Readout	Cost / Date	Comments
PRC Krochmann	Photocurrent Meter 321	$10^{-14} - 10^{-3}$ A	7	•						•				•		478 x 155 x 365	7,0		•		DM 5.990 (4/86)	optional IEEE 488
	Irradiance Meter 202	$1 - 2000$ W/m²	1							•						280 x 320 x 105	4,0	0,83	•	5 m	DM 6.200 (4/86)	Radiometer accuracy: WMO class I; optional IEEE 488
	Calorimeter 420	$10^{-3} - 200.000$ lx								•			•			475 x 155 x 365	7,0	1,5		3 m	DM 35.000 (4/89)	Waterproofed thermostated calorimeter head
	ϱ_{diff}-Meter									•						110 x 260 x 270	5,0	0,3	•			measurement of the reflectance at hemispherical incidence
	Daylight Factor Meter D-Meter	$10^{-3} - 140.000$ lx 10^{-2} cd/m² - 140 cd/cm² 0,0 - 105 %	5	•	•	•		8 mm		•		•	•			105 x 470 x 325	10,5	2 x 0,11	•	3 m		optional IEEE 488 RS 232

RWTH AACHEN 5/1989

Survey of Light Measuring Instruments Available in Europe

MANUFACTURER	PHOTO	MODEL	SENSITIVITY RANGE	NUMBER OF RANGES	AUTOM. RANGE CHANGE	COSINE CORRECTIONS	SENSOR MATERIAL SILICON	SENSOR MATERIAL SELENIUM	SENSOR SIZE (MM) RECEIVER ∅	SENSOR SIZE (MM) HEAD ∅ × H	DISPLAY DIGITAL	DISPLAY ANALOG	LUMINANCE	ILLUMINANCE	METRIC	ENGLISH	SIZE (MM, L × W × H)	WEIGHT (KG) INSTRUMENT	WEIGHT (KG) SENSOR	PORTABLE	LENGTH OF CORD SENSOR TO READOUT	COST	DATE	COMMENTS	RWTH AACHEN 5/1989
PRC Krochmann		Daylight Meter Head 910 GV			•	•	•		8 mm		•	•		•	•		230 × 200 × 220	8.5			10 m	DM 14,095	(4/89)	4 vertical 1 horizontal ill. values thermostated shading ring ∅ 600 mm width 60mm	
		Sky Scanner	0 – 20.000 cd/m²		•		•						• 10°		•							DM 57,000	(4/89)	151 measured directions thermostated runtime/scan 45 sec, IEEE 488 incl. Data Acquis. Sys.	
SOLEX international 95 Main Street Broughton Astley Leicestershire LE9 6R⌀ Tel 0455-283486 Fax 0455-283912		Digital Luxmeter SL-100	0 – 50.000 lx	3	•										•		108 × 72 × 23	0.16		•		59 to 65	(10/88)		
TEKTRONIX Inc. P.O. Box 500 Beaverton Oregon 97005, USA Tel 503-644-0161 Telex 36-0485		Digital Radiometer/ Photometer J 16									•				•	•	123 × 63 × 203	1.5		•		DM 5,734	(10/85)		
Agent in Germany: Tektronix GmbH Sedanstraße 13-17 D-5000 Köln 1 Tel 0221-7722-1 Fax 0221-3798-500		Illuminance Probes J 6501 J 6511	0,01 – 19.900 lx			• ○	•													•	8 m	J6511 DM 2.605 J6501 DM 2.696	(10/85)		

C.16 Daylighting in Architecture

SURVEY OF LIGHT MEASURING INSTRUMENTS AVAILABLE IN EUROPE								
MANUFACTURER	TEKTRONIX							
PHOTO								
MODEL	Irradiance Probes J 6502 J 6512	Luminance Probes J 6503 J 6523						
SENSITIVITY RANGE	0,001 - 1.999 microwatts/cm²	1 cd/m² - 1.999 kcd/m² ; 1 cd/m² - 199,9 kcd/m²						
NUMBER OF RANGES								
AUTOM. RANGE CHANGE								
COSINE CORRECTIONS								
SENSOR MATERIAL — SILICON		•						
SENSOR MATERIAL — SELENIUM								
SENSOR SIZE (MM) — RECEIVER ø								
SENSOR SIZE (MM) — HEAD ø X H								
DISPLAY — DIGITAL								
DISPLAY — ANALOG								
LUMINANCE		• 8°, 1°						
ILLUMINANCE								
METRIC								
ENGLISH								
SIZE (MM, L X W X H)								
WEIGHT (KG) — INSTRUMENT								
WEIGHT (KG) — SENSOR								
PORTABLE	•	•						
LENGTH OF CORD SENSOR TO READOUT	2 m							
COST / DATE	J6502 DM 2.696 J6512 DM 2.627 (10/85)	J6503 DM 2.696 J6523 DM 7.630 (10/85)						
COMMENTS	RWTH AACHEN 5/1989							

Appendix D
GUIDE TO SCALE MODELS

GUIDE TO SCALE MODELS

The Scale of the Models

Physical models can be built and studied at all stages of the design process. According to the stage of design different types of models are appropriate depending on the nature of question being asked. Generally there are three types of scale models to define the performance of a design:
- massing models for examining solar access for purposes of site planning, building location and orientation
- models for studying the performance characteristics of the building, including daylight penetration and distribution, illuminance levels, glare and contrast
- models for investigating individual apertures, including glazing, shading devices, light directing elements, new materials and other characteristics of the aperture.

Appropriate scales are typically -
- massing models 1:500
- building performance models 1:100 and 1:50
- models for individual apertures 1:20, 1:10 or if necessary 1:1.

Depending on the scale of the models the accuracy has to be adequate for the information wanted. Massing models can be built very crudely if the information wanted is more general. Full scale mock-ups may be used to examine the effectiveness and workability of control elements and new materials and therefore have to be built with a high degree of accuracy.

Besides the stage of the design process, the scale of the model depends on the kind of visual observation (eye or camera) used for studying qualitative and quantitative aspects. Using a camera the centre of the viewport has to be at the height of the eye-level in the scale model, normally in the scale height of 1.5 to 1.7m. Depending on the diameter of the lens of the camera (normally 50 mm) a scale greater than 1:20 is required.

Another aspect which influences the scale of the model is the final size of the model and the conditions where the studies are being carried out: under real sky conditions or under artificial sky conditions. If the studies are made under an artificial sky the size of the model is limited according the size of the artificial sky. Testing under real sky conditions does not normally impose constraints upon the size of the model.

Materials and Construction of the Model

- The geometry of the space and the building has to be duplicated very carefully in a smaller scale as well as the reflective and the transmissive characteristics of the building materials. If the characteristics of the materials are not known they may be estimated by using an illuminance meter.
- Large symmetrical buildings can be built as half-models and a mirror used to provide an image in place of the remaining space. This technique however is applicable only for overcast sky conditions and can create substantial error when direct sunlight enters the space.
- Furniture and other obstructions should be modelled with correct geometry and reflectance characteristics.
- Depending on the information needed, the model has to be changeable (interior reflectances, window sizes, geometry of the space, slope of the ceiling, etc), movable and wind resistant (for testing outside).
- All edges have to be sealed very carefully to avoid unwanted light penetration.

For the construction of the scale model almost any common architectural modelling material may be used. All materials should either match the color, reflectivity and texture of the surfaces they represent or, if not, they can be painted to match.

An important factor in deciding what materials to use is whether the photographic analysis is to be in color or in black and white. Color analysis requires close matches of both color and reflectivity. For black and white photography only materials with the required reflectivity are used.

The transmittance of nominally opaque materials (eg) foamboard should be checked before use by holding them up to bright sun. If transmitted light is visible, the outside of the model must be painted black. A further check is to measure illuminance in the model when all apertures are blocked - any light will then be due to the transmission of "opaque" parts, or through cracks due to poor construction. This can be of considerable significance in spaces with low daylight factors.

D.2 Daylighting in Architecture

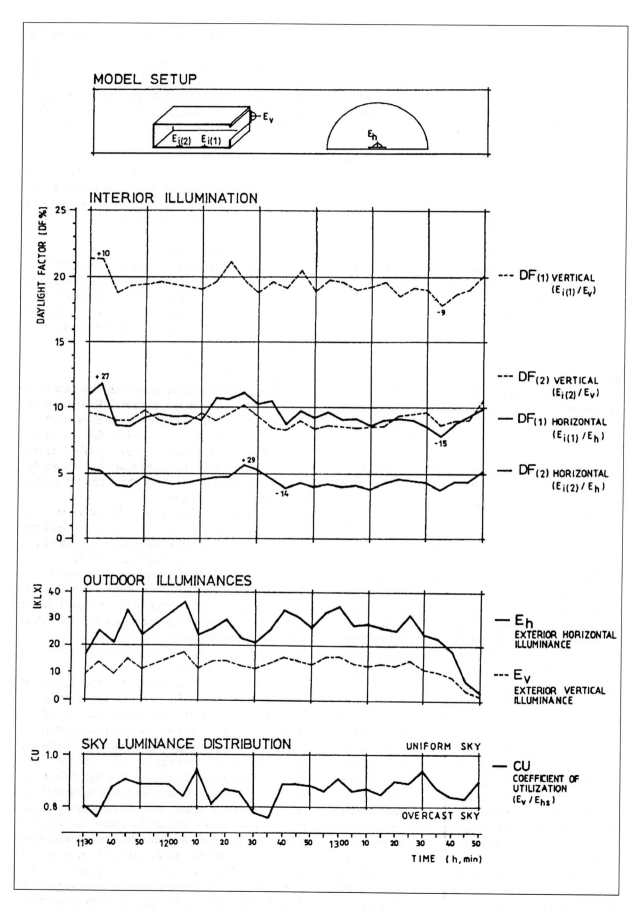

Figure 1 - *Effect on the Daylight Factor in the space for an overcast day related to changing luminance distribution of the sky, studied with the use of a scale model (1).*

For unglazed models, a transmission factor must be applied to the experimental result, to allow for the attenuation of the glass. If the principal source of daylight enters the aperture at angles of incidence greater than about $60°$, the glazing material must be included in the model to establish the proportion of daylight reflected off the glazing. If the glazing is tinted or semi reflective, or has particular geometric properties (eg prismatic) then the actual material should be used.

If the actual reflectivity of a material used is not known, its value can be determined approximately by measuring the ratio of reflected light to incident light using a luxmeter. The material has to be illuminated with a diffuse source of light. The transmission of diffuse light can be measured in a similar way. Accuracy of this method is given as +/- 10% (2,3,4).

The table or baseboard must also be of realistic reflectivity to model accurately ground reflected light. The internally reflected component is particularly sensitive to this.

Real Sky Conditions

Daylighting studies. The testing should take place under completely overcast sky conditions on a relatively unobstructed site. Under overcast sky conditions time of day is not critical as this sky condition is commonly regarded as a good sky configuration providing an even luminance distribution. However the luminance distribution is permanently changing and therefore has to be monitored carefully if quantitative analysis is required.

Actual measurements, however, indicate a rapid change in luminance distribution from uniform to CIE overcast sky. The impact of different luminance distributions on the illumination in the model is considerable and in order to receive comparable results the sky has to be checked before measuring the model. Figure 1 shows the measured daylight factors for two positions in the room, and for both horizontal and vertical external illuminance. The values of daylight factors vary during the experimental period in response to the variation in the sky luminance distribution by up to 29%. Thus for real skies it is recommended that a large number of sensors be set up on a grid, and illuminances measured simultaneously. This generates a huge amount of data, requiring a computer to control the measurements, and store and process the data.

Sunlighting Studies. Studies of massing models for shadows and shadow patterns are very easy to carry out by using a sundial and tilting the model according the desired time of the year and day. Models for studies of penetration and distribution of sunlight in buildings should not be tilted as the different portions of the sky seen by the model affect the interior illuminance. Tilting the model +/- $10°$ may be allowed, as the error of a higher or smaller impact of the ground reflectance can be accepted for visual observation.

For quantitative analysis, illuminance levels have to be determined. For this reason rotating the model along its vertical axis to simulate different sun angles provides the designer with more realistic data as the sky seen from the window is that of the real space. To find the appropriate illuminance for different times of the year, daylight availability data tables can be used by relating the global horizontal illuminance from the table to the measured global horizontal illuminance. The interior illuminance will be adapted by the ratio defined earlier (2).

Artificial Sky Conditions

Daylighting studies. The advantage of artificial skies is the stable luminous environment, but it is more expensive and only a few establishments are equipped with an artificial sky (see List of Artificial Skies, Appendix H). In general there are two types of skies:
- hemispherical skies
- mirror box skies.

Most of the artificial skies can simulate only overcast sky conditions, either CIE or uniform luminance distribution. The use of an artificial sky is easier than a real sky, but restricts the size of the scale model. Also, color photographs taken do not reproduce accurately the colors used in the model setup unless good colour rendering lamps are used in the artificial sky luminaire.

Sunlighting studies. To test the use of sunlight in the building an artificial sun (very expensive) can also be used. The artificial sun can be used to assess the visual effects of sunlighting by direct observation. Also the measured results received with an artificial sun define only the impact of the sun on the space not taking into account the impact of the sky light. For these testing procedures real sky (clear sky) is more appropriate.

Analysis of the Testing

Visual observation. There are many aspects of daylighting which do not led themselves to quantitative measurement. Such issues as glare, contrast and visual comfort can be studied only by direct visual observation. For qualitive analysis the space should in general be more realistic (eg furniture, carpeting, curtains, texture and reflectivity of surfaces). Also models for qualitative analysis require viewports at eye level corresponding to the predominant views within the space. Usually for studies on glare from windows the predominant view should be from the wall perpendicular to the window. Critical to any qualitative evaluation are the points of view, observation ports or viewing paths as they allow for subjective appraisal of the appearance of the space. As the qualitative analysis is based on subjective criteria of the observer several viewports should be provided for comparing the same illumination from different points of view. A viewpoint can be a model opening such as a door, or a removable observation section that can later be hidden behind a movable wall or wall painting. Care should be taken that these openings do not allow unwanted light penetration.

Photographic documentation. Often visual analysis needs to be documented for later reference. Photography provides a permanent record of daylighting conditions inside the space (real or scale model). This technique provides an evaluation method for observation of quality of light and a comparison to other design options. The limitations of this method are based on the sensitivity of the film as the human eye is much more sensitive than the most sensitive film. In order to find the illumination most appropriate to the visual observation it is recommended to take several exposures of each design option and select the one which is closest.

For the purpose of further comparisons each photographic documentation should include all information which is special for this design option: sky condition, window azimuth, solar altitude and a brief description of the design option.

Quantitative analysis. The photometric evaluation can be done with one light meter with one remote probe which has to be moved within the space to different locations and which is also used to measure the exterior illuminance. Due to the steady conditions of an artificial sky, serial measurements can be made using one movable sensor. Photometers which incorporate the sensor and the display in the same enclosure are not suited for model studies due to their large size and the difficulty of reading the illuminance values without obstructing the illumination.

Evaluating the Results

Data presentation. Daylighting model studies usually generate a very large amount of quantitative data; therefore it is desirable to present these data graphically to facilitate the designer's understanding of both the quantity and distribution of light as well as to compare the performance of various design alternatives. Two graphic formats are most widely used for this purpose:
- the iso-lux contour plan
- the sections daylight factor graph.

(The daylight dactor is the ratio of the indoor illuminance on the work plane at a specific location to the simultaneously value on the outdoor horizontal plane for an unobstructed sky).

The iso-lux contour plan. Daylight Factor data from physical model studies can be presented graphically in the form of contours of equal Daylight Factors plotted over a building floor plan. This method allows assessment of illuminance distribution throughout the room and indicates zones of different amount of daylight available. Zones with a Daylight Factor greater than 6% are considered to be bright, and those with a Daylight Factor of less than 2% are considered to be poorly daylit. The primary value of the iso-lux contour plans is their indication of light distribution, which is the quantitative factor most representative of visually perceived illumination quality in an architectural environment.

Sectional Daylight Factor graph. The Daylight Factor values are presented graphically overlayed on a building section cut through the fenestration. The daylight factor curve for each fenestration reveals much design information above and beyond the individual illuminances used for the plot. The average illuminance is represented by the total area under the curve and provides a basis for comparing total illumination on the workplane.

A more important criterion is the shape of the curve. Flatter curves indicate a more uniform distribution of light, but may also be a symptom of potential contrast glare in the occupant's field of view. Steep curves near the window indicate potential veiling reflections. An alternative design with the use of a lightshelf produces a flatter curve.

The slope of the curve at any point is a measure rate of illuminance change. Due to the eye's adaptive capability, this gradient is a better measure of perceived brightness changes than are absolute illuminance values. The changes in distances between curves as well as slope differences provide important information on the impact of the different fenestration options onthe illuminance distribution of the space.

Results of sunlighting studies with scale models are presented similarly to the daylight factor values either as:
- sunlight factor curves plotted over a section of the building as well.
- or sunlight factor curves for specific positions in the space related to different solar altitudes.

Sunlight factors indicate the contribution of sunlight to the illumination of the space. A high Sunlight Factor for low sun angles indicates a high contribution during winter time, whereas a high sunlight factor for high sun angles indicates a high contribution during summer time which is not wanted from the thermal point of view.

Combination with Electric Lighting

In models it is difficult to integrate artificial lighting and natural daylighting. It is especially difficult to get any accurate quantitative measurements. However qualitative effects can be modelled and visualised, as the main interest is the optical appearance of a space.

As there are no scaled luminaires the effect of the electric lighting has to be modelled. Therefore any translucent modelling material is appropriate as almost any light source. Interesting effects can be simulated using scaled light bulbs from a hobby shop. More advanced techniques involve the use of fibre optics to conduct light from an external artificial source to scale model luminaires.

Only the full scale mock-up provides the opportunity to test day and night situations with real fixtures and their luminance distribution; qualitative data from these 1:1 mock-ups can be considered as real.

REFERENCES

(1) Willbold-Lohr, G., *European Concerted Action on Daylighting,* Intermed. Rep. No.3, March 1988.
(2) Moore, F., *Concepts and Practice of Architectural Daylighting,* Van Nostrand Reinhold, New York, 1985.
(3) Robbins, C. L., *Daylighting Design and Analysis,* Van Nostrand Reinhold, New York, 1986.
(4) Schiler, M.(ed.), *Simulating Daylight with Architectural Models,* Univ. of Southern California, Los Angeles.

Appendix E
SURVEY OF CONTROL SYSTEMS

TECHNICAL SPECIFICATIONS FOR INTERIOR LIGHTING SYSTEM CONTROL			
MANUFACTURER	*Siemens, Munich (D)*		
MODEL	*Altomat*		
TYPE OF CONTROL	*Continuous light regulation*		
MAIN AREA OF USE	*Offices, schools, etc., of various sizes*		
SENSORS	MODEL	*LDR (photoelectric cell)*	
	MEASURE	*Room illuminance*	
	NOTES		
CONTROL SYSTEM	TYPE	*Electric/electrochemical*	
	POWER	*Single and 3-phase, from 2kVA to 3 x 8kVA*	
	PROGRAMMABLE	ILLUMINANCE LEVELS	*Yes*
	Yes	TIME SWITCH	
		OTHER	
	CONTROL WITH RESPECT TO DAYLIGHT		*Yes*
	CONTROL WITH DELAY		*Yes*
	NOTES		
LIGHTING UNITS	TYPE	*Incandescent and fluorescent lamps*	
	ACTUATORS	*Electronic ballast is needed on fluorescent lamps*	
	CONTROL RANGE	*25-100% (fluorescent), 0-100% (incandescent)*	
	NOTES	*fluorescent tubes require change of ballast, lamps, etc.*	
NOTES	*Suitable for new installation systems*		

TECHNICAL SPECIFICATIONS FOR INTERIOR LIGHTING SYSTEM CONTROL			
MANUFACTURER	*Siemens, Munich (D)*		
MODEL	*Altoswitch*		
TYPE OF CONTROL	*Step (on / off)*		
MAIN AREA OF USE	*From small office to large building interior*		
SENSORS	MODEL	*LDR (photoelectric cell)*	
	MEASURE	*Room illuminance*	
	NOTES		
CONTROL SYSTEM	TYPE		
	POWER	*5A / 220V, 1100VA*	
	PROGRAMMABLE	ILLUMINANCE LEVELS	*Yes*
	Manual value input	TIME SWITCH	
		OTHER	
	CONTROL WITH RESPECT TO DAYLIGHT		*Yes*
	CONTROL WITH DELAY		*Yes*
	NOTES	*Not much modification to existing system required*	
LIGHTING UNITS	TYPE	*Any*	
	ACTUATORS	*None*	
	CONTROL RANGE		
	NOTES		
NOTES	*Only allows lamps to be switched on or off*		
	Does not require modification to existing lighting units		

TECHNICAL SPECIFICATIONS FOR INTERIOR LIGHTING SYSTEM CONTROL			
MANUFACTURER	*Philips, Eindhoven (NL)*		
MODEL	*Integrated Function System (IFS)*		
TYPE OF CONTROL	*Step (on / off)*		
MAIN AREA OF USE	*Average size or large rooms*		
SENSORS	MODEL	*LMC (light measuring cell)*	
	MEASURE	*Room illuminance*	
	NOTES	*Other sensor types can be used (e.g. people sensors)*	
CONTROL SYSTEM	TYPE	*Microprocesor controlled / electronic*	
	POWER	*Any*	
	PROGRAMMABLE	ILLUMINANCE LEVELS	*Yes*
	Manual value input	TIME SWITCH	*Yes*
		OTHER	
	CONTROL WITH RESPECT TO DAYLIGHT		*Yes*
	CONTROL WITH DELAY		*Yes*
	NOTES	*External photoelectric cell can be added*	
LIGHTING UNITS	TYPE	*Any*	
	ACTUATORS	*None*	
	CONTROL RANGE	*0 / 100% (on / off)*	
	NOTES		
NOTES	*This system can be used for other on / off control situations: Publicity lighting; external lighting; etc.*		

TECHNICAL SPECIFICATIONS FOR INTERIOR LIGHTING SYSTEM CONTROL			
MANUFACTURER	*Philips, Eindhoven (NL)*		
MODEL	*HF Dimming (LRA 110 amplifier)*		
TYPE OF CONTROL	*Continuous (automatic)*		
MAIN AREA OF USE	*Even for small or individual offices*		
SENSORS	MODEL	*LRF 101 (photoelectric cell)*	
	MEASURE	*Room illuminance*	
	NOTES	*Other sensor types can be used (e.g. people sensors)*	
CONTROL SYSTEM	TYPE	*Electronic (signal amplifier)*	
	POWER	*500 ballasts*	
	PROGRAMMABLE	ILLUMINANCE LEVELS	*Manual or potentiometer*
	No	TIME SWITCH	
		OTHER	
	CONTROL WITH RESPECT TO DAYLIGHT		*Yes*
	CONTROL WITH DELAY		*Yes*
	NOTES	*LRA 110 control system*	
LIGHTING UNITS	TYPE	*HF neon tubes rated at 32W and 50W*	
	ACTUATORS	*Single or twin tube "dimming" electronic ballasting*	
	CONTROL RANGE	*25-100% (continuously), switched off below 25%*	
	NOTES		
NOTES	*No programming possible. External control devices can be added (e.g. timers, people detectors, etc.). All equipment is designed specifically for the system.*		

TECHNICAL SPECIFICATIONS FOR INTERIOR LIGHTING SYSTEM CONTROL		
MANUFACTURER	*Multipoint Control Co., Washington (USA)*	
MODEL	*Mark VI POL, Mark VIII*	
TYPE OF CONTROL	*Step regulation (on / off)*	
MAIN AREA OF USE	*Interior lighting regulation*	
SENSORS	MODEL	*Photodiode*
	MEASURE	*Room illuminance*
	NOTES	*A photocell with 0-500 fc sensitivity (Mark VIII)*
CONTROL SYSTEM	TYPE	*Electromechanical*
	POWER	*Any*
	PROGRAMMABLE	ILLUMINANCE LEVELS
	Fixed regulation with	TIME SWITCH
	on/off potentiometer	OTHER
	CONTROL WITH RESPECT TO DAYLIGHT	*Yes*
	CONTROL WITH DELAY	
	NOTES	
LIGHTING UNITS	TYPE	*Any*
	ACTUATORS	*None*
	CONTROL RANGE	
	NOTES	
NOTES	*The systems differ in their types of output.*	
	The Mark VII requires a pulsed output.	

TECHNICAL SPECIFICATIONS FOR INTERIOR LIGHTING SYSTEM CONTROL		
MANUFACTURER	*Conservolite Inc., Pennsylvania (USA)*	
MODEL	*Eclipse*	
TYPE OF CONTROL	*Continuous (normal and automatic)*	
MAIN AREA OF USE	*Small rooms and individual lights*	
SENSORS	MODEL	*Optical sensor (not specified)*
	MEASURE	*Room illuminance*
	NOTES	
CONTROL SYSTEM	TYPE	*Electronic*
	POWER	*From 96 to 384 W single phase 120/277 VA*
	PROGRAMMABLE	ILLUMINANCE LEVELS *Yes*
	Yes	TIME SWITCH
		OTHER
	CONTROL WITH RESPECT TO DAYLIGHT	*Yes*
	CONTROL WITH DELAY	
	NOTES	*Automatic (optical sensor) and normal regulator*
LIGHTING UNITS	TYPE	*Rapid start, slimline fluorescent tubes*
	ACTUATORS	
	CONTROL RANGE	*30-70% continuously*
	NOTES	*Installed on existing units, without special apparatus*
NOTES		

TECHNICAL SPECIFICATIONS FOR INTERIOR LIGHTING SYSTEM CONTROL			
MANUFACTURER	XO Industries Inc., California (USA)		
MODEL	Automatic Tuning Controls		
TYPE OF CONTROL	Continuous		
MAIN AREA OF USE	Large and small rooms or offices		
SENSORS	MODEL	Not specified	
	MEASURE	Room illuminance	
	NOTES		
CONTROL SYSTEM	TYPE		
	POWER	1 - 100 lighting units (power not specified)	
	PROGRAMMABLE	ILLUMINANCE LEVELS	
	Not specified	TIME SWITCH	
		OTHER	
	CONTROL WITH RESPECT TO DAYLIGHT	Yes	
	CONTROL WITH DELAY		
	NOTES		
LIGHTING UNITS	TYPE	Fluorescent tubes	
	ACTUATORS	Electronic ballast (XO Electronic Tuning Ballast)	
	CONTROL RANGE	35-100% continuously	
	NOTES		
NOTES			

TECHNICAL SPECIFICATIONS FOR INTERIOR LIGHTING SYSTEM CONTROL			
MANUFACTURER	Lutron Electronics Co. Inc., Pennsylvania (USA)		
MODEL	Paesar		
TYPE OF CONTROL	Continuous		
MAIN AREA OF USE	Large and average sized rooms		
SENSORS	MODEL	Photometric head (not specified)	
	MEASURE	Room illuminance	
	NOTES		
CONTROL SYSTEM	TYPE		
	POWER	90 fluorescent tubes (not specified)	
	PROGRAMMABLE	ILLUMINANCE LEVELS	Yes
	Yes	TIME SWITCH	
		OTHER	
	CONTROL WITH RESPECT TO DAYLIGHT	Yes	
	CONTROL WITH DELAY	Yes	
	NOTES		
LIGHTING UNITS	TYPE	Fluorescent tubes / discharge lamps	
	ACTUATORS	None needed	
	CONTROL RANGE	50-100% continuously	
	NOTES		
NOTES			

TECHNICAL SPECIFICATIONS FOR INTERIOR LIGHTING SYSTEM CONTROL			
MANUFACTURER	*General Electric Company, Rhode Island (USA)*		
MODEL	*Smart Remote Control*		
TYPE OF CONTROL	*Step (on / off model)*		
MAIN AREA OF USE	*Average to large shops, factories and public buildings*		
SENSORS	MODEL	*Photorelay (not specified)*	
	MEASURE	*Room illuminance*	
	NOTES		
CONTROL SYSTEM	TYPE	*Electromechanical (auxiliary switching relay)*	
	POWER	*Modular: each relay can control 20A at 277 VA*	
	PROGRAMMABLE	ILLUMINANCE LEVELS	*Yes*
	Yes	TIME SWITCH	*Yes*
		OTHER	
	CONTROL WITH RESPECT TO DAYLIGHT		*Yes*
	CONTROL WITH DELAY		
	NOTES		
LIGHTING UNITS	TYPE	*Any*	
	ACTUATORS	*None*	
	CONTROL RANGE	*0 / 100%, on / off*	
	NOTES		
NOTES			

TECHNICAL SPECIFICATIONS FOR INTERIOR LIGHTING SYSTEM CONTROL			
MANUFACTURER	*General Electric Company, Rhode Island (USA)*		
MODEL	*Programmable Lighting Control*		
TYPE OF CONTROL	*Step (on / off model)*		
MAIN AREA OF USE	*Large office complexes, factories and connected offices*		
SENSORS	MODEL	*Photosensor (not specified)*	
	MEASURE	*Room illuminance*	
	NOTES		
CONTROL SYSTEM	TYPE	*Computer run electronic relay equipment*	
	POWER	*Modular: each relay can control 20A at 277 VA*	
	PROGRAMMABLE	ILLUMINANCE LEVELS	*Yes*
	Yes	TIME SWITCH	*Yes*
		OTHER	
	CONTROL WITH RESPECT TO DAYLIGHT		*Yes*
	CONTROL WITH DELAY		*Yes*
	NOTES		
LIGHTING UNITS	TYPE	*Any*	
	ACTUATORS	*None*	
	CONTROL RANGE	*0 / 100%, on / off*	
	NOTES		
NOTES			

TECHNICAL SPECIFICATIONS FOR INTERIOR LIGHTING SYSTEM CONTROL			
MANUFACTURER	*ETEQ Ltd., Corby (UK)*		
MODEL	*Power-miser Model Alpha and Beta*		
TYPE OF CONTROL	*Continuous*		
MAIN AREA OF USE	*Interior lighting control*		
SENSORS	MODEL	*Not specified or always necessary*	
	MEASURE	*Room illuminance*	
	NOTES		
CONTROL SYSTEM	TYPE	*PCB control systems*	
	POWER	*2, 4, 6 - 30 kW (single phase)*	
	PROGRAMMABLE	ILLUMINANCE LEVELS	*Yes*
	Yes	TIME SWITCH	
		OTHER	
	CONTROL WITH RESPECT TO DAYLIGHT		*Yes*
	CONTROL WITH DELAY		
	NOTES	*Single control channel*	
LIGHTING UNITS	TYPE	*Neon tubes or discharge tubes*	
	ACTUATORS		
	CONTROL RANGE		
	NOTES		
NOTES	*The power to the tube is adjusted so that it is lower than the nominal value. Lights are on during occupancy and when light levels are below the set level.*		

TECHNICAL SPECIFICATIONS FOR INTERIOR LIGHTING SYSTEM CONTROL			
MANUFACTURER	*ETEQ Ltd., Corby (UK)*		
MODEL	*Stopwatch*		
TYPE OF CONTROL	*On / off*		
MAIN AREA OF USE	*Interior lighting control*		
SENSORS	MODEL	*Not specified*	
	MEASURE	*Room illuminance*	
	NOTES		
CONTROL SYSTEM	TYPE	*Electronic*	
	POWER	*All (not system based; output contacts provided)*	
	PROGRAMMABLE	ILLUMINANCE LEVELS	*Yes*
	Yes	TIME SWITCH	*Yes*
		OTHER	
	CONTROL WITH RESPECT TO DAYLIGHT		
	CONTROL WITH DELAY		*Yes*
	NOTES	*Sophisticated multi-channel versions available*	
LIGHTING UNITS	TYPE	*All*	
	ACTUATORS	*None*	
	CONTROL RANGE	*0 / 100%, on / off*	
	NOTES		
NOTES			

SPECIFIC DEVICES: FACADE SCREENING EQUIPMENT			
MANUFACTURER	*Merlo Solar Screen Brise Soleil, Paris (F)*		
MODEL	*Brise Soleil Type A*		
MAIN AREA OF USE	*Solar screen suitable for buildings with very tall windows*		
TYPE OF SCREEN	TYPE	*Vertical louvers*	
	MATERIAL	*Galvanised steel*	
	MOVEABLE	*Yes*	
	RETRACTABLE	*No*	
MOVEMENT SYSTEM	NOTES	*Rotation around central pivot*	
	DRIVE SYSTEM	*Electric motor*	
	CONTROL SYSTEM	MANUAL	*Yes*
		CENTRAL	*Yes*
		AUTOMATIC	*Yes*
CONTROL SENSORS	SOLAR POWER PACK	*Yes*	
	ANEMOMETER	*No*	
	OTHER		
	NOTES		
NOTES	*The electronic control unit receives input from a roof-mounted sensor and adjusts the screen acoording to the sun's position*		

SPECIFIC DEVICES: FACADE SCREENING EQUIPMENT			
MANUFACTURER	*Merlo Solar Screen Brise Soleil, Paris (F)*		
MODEL	*Brise Soleil Type G*		
MAIN AREA OF USE	*Solar screen suitable for medium height window facades*		
TYPE OF SCREEN	TYPE	*Vertical louvers*	
	MATERIAL	*Aluminium*	
	MOVEABLE	*Yes*	
	RETRACTABLE	*No*	
MOVEMENT SYSTEM	NOTES	*Rotation around central pivot*	
	DRIVE SYSTEM	*Electric motor*	
	CONTROL SYSTEM	MANUAL	*Yes*
		CENTRAL	*Yes*
		AUTOMATIC	*Yes*
CONTROL SENSORS	SOLAR POWER PACK	*Yes*	
	ANEMOMETER	*No*	
	OTHER		
	NOTES		
NOTES	*The electronic control unit receives input from a roof-mounted sensor and adjusts the screen acoording to the sun's position*		

SPECIFIC DEVICES: FACADE SCREENING EQUIPMENT				
MANUFACTURER	*Griesser Italiana SPA, Como (I) / Griesser SA, Aadorf (CH)*			
MODEL	*Solomatic 82*			
MAIN AREA OF USE	*Venetian blinds to screen windows*			
TYPE OF SCREEN	TYPE	*Movable vertical strip venetian blinds*		
	MATERIAL	*Aluminium*		
	MOVEABLE	*Yes*		
	RETRACTABLE	*Yes*		
MOVEMENT SYSTEM	NOTES			
	DRIVE SYSTEM	*Electric motor*		
	CONTROL SYSTEM	MANUAL	*Yes*	
		CENTRAL	*Yes*	
		AUTOMATIC	*Yes*	
CONTROL SENSORS	SOLAR POWER PACK	*Yes*		
	ANEMOMETER	*Yes*		
	OTHER			
	NOTES	*Timer for cumulative control can be added*		
NOTES	*Blinds are raised, lowered or tilted according to solar radiation on facade.*			
	Manual and cumulative controls possible (e.g. closing at night; during rain).			

SPECIFIC DEVICES: FACADE SCREENING EQUIPMENT				
MANUFACTURER	*Griesser Italiana SPA, Como (I) / Griesser SA, Aadorf (CH)*			
MODEL	*Metalunic*			
MAIN AREA OF USE	*Venetian blinds for public offices or large glazed facades*			
TYPE OF SCREEN	TYPE	*Movable vertical strip venetian blinds*		
	MATERIAL	*Aluminium*		
	MOVEABLE	*Yes*		
	RETRACTABLE	*Yes*		
MOVEMENT SYSTEM	NOTES	*Automatic stop when lowered (optional)*		
	DRIVE SYSTEM	*Electric motor*		
	CONTROL SYSTEM	MANUAL	*Yes*	
		CENTRAL	*Yes*	
		AUTOMATIC	*Yes*	
CONTROL SENSORS	SOLAR POWER PACK	*Yes*		
	ANEMOMETER	*Yes*		
	OTHER			
	NOTES	*Timer for cumulative control can be added*		
NOTES	*Blinds are raised, lowered or tilted according to solar radiation on facade.*			
	Manual and cumulative controls possible (e.g. closing at night; during rain).			

SPECIFIC DEVICES: PEOPLE DETECTORS		
MANUFACTURER	*Stafa Control Systems, Stafa (CH)*	
MODEL	*Series SCS-Dimo Model FR-B1*	
MAIN AREA OF USE	*Turning air conditioning plant on or off*	
TECHNICAL DATA	MEASURING DEVICE	*Infraredsensor*
	POWER SUPPLY	*12-17V DC*
	RESPONSE DELAY	*3 minutes*
	CUT OFF DELAY	*10 minutes*
	INSTALLMENT	*Wall*

SPECIFIC DEVICES: PEOPLE DETECTORS		
MANUFACTURER	*Novitas Inc., Santa Monica (USA)*	
MODEL	*LIGHT-O-MATIC*	
MAIN AREA OF USE	*Indivual rooms; switches off light when unoccupied*	
TECHNICAL DATA	MEASURING DEVICE	*Not specified*
	POWER SUPPLY	*120/277 VA*
	RESPONSE DELAY	*Lights are switched on manually*
	CUT OFF DELAY	*Programmable from 30 seconds to 12 minutes*
	INSTALLMENT	*Wall - it replaces a normal light switch*

Appendix F
REVIEW OF DESIGN TOOLS

EQUATIONS

BRS Interreflection Formula
   Authors: R.G. Hopkinson, J. Longmore and P. Petherbridge (GB)
   Type: Analytical - Internally reflected component
   Applicability: Vertical windows. CIE standard overcast sky luminance distribution.
   Output: Internally reflected component at mid-depth and back of side lit rooms. Allows for light reflected from ground, transmission losses, influences of structural members and from external obstructions, internally reflected light.
   Discussion: This method is known also as "split-flux principle". It divides the flux entering the room through the window into two parts: the first enters the room directly from the sky or from the external obstructions above the horizon, the second enters the room directly from the ground. The two parts of flux are considered to be modified by the average reflectance of the surfaces of the rooms as seen from the sky and from the ground.
   References: 1, 2, 3

Fin
   Author: J. Navarro Casas (Spain)
   Type: Analytical - Direct component
   Applicability: Vertical and horizontal windows. Standard overcast sky luminance distribution.
   Output: Sky factor and sky component at single point.
   Discussion: The method works as a normal split flux one, but different components (SC, ERC) are obtained through the integration of an illuminance vector for different window shapes whether horizontal or vertical. The output of this model is a formula where one can find E_x, E_y or E_z for each point of the room: the illuminance at a definite point over the three spatial planes.
   Reference: 4

Higbie's Formula
   Author: H.H. Higbie (USA)
   Type: Analytical - Direct component
   Applicability: Windows in a vertical plane. Uniform sky luminance distribution.
   Output: The value of direct illumination on a reference point.
   Discussion: This is the first exact and perfectly general solution to give the intensity of light at any point of the room from a window of any dimension. This formula is the basic formula for successive simple and complex daylight analysis methods.
   Reference: 5

Higbie - Levin Formula
   Authors: H.H. Higbie and A. Levin (USA)
   Type: Analytical - Direct component
   Applicability: Vertical and sloping windows. Uniform sky luminance distribution.
   Output: The value of direct illumination on a reference point.
   Discussion: This formula is based on Higbie's Formula but allows sloped apertures to be considered.
   Reference: 6

Kittler and Ondrejicka's Formula
   Authors: R. Kittler and S. Ondrejicka (Czechoslovakia)
   Type: Analytical - Direct component
   Applicability: Any rectangular window aperture with arbitrary slope, with or without glass. Uniform and CIE standard overcast sky luminance distribution when ground is snow - covered.
   Output: Sky factor and sky component at single points.
   Discussion: The exact relations are given by mathematical formulae using angular as well as dimensional parameters. Partial results can be obtained from simple tables or graphs and, for practical purposes, fairly approximate protractors have been developed for 45° 60°, 75° and 90° tilted apertures.
   References: 7, 8, 9

Pugno's Method
   Author: G.A. Pugno (Italy)
   Type: Analytical - Direct component
   Applicability: Rooflights. Uniform and CIE standard overcast sky luminance distribution.
   Output: Illumination value at single points. Allowances: Glazing losses.
   Discussion: It proposes an analytical formula for the determination of illumination value to a reference point. A simplified graphical method to determine the sky factor due to a vertical aperture is provided.

Two graphs, for the rapid determination of sky factor with uniform and CIE standard overcast sky luminance distribution, are developed.
Reference: 10

Spencer's and Stakutis' Integral Equation
Authors: D.E. Spencer and V.J. Stakutis (USA)
Type: Analytical - Single stage
Applicability: Vertical windows. CIE Standard Overcast sky luminance distribution.
Output: Daylight factor at single points
Discussion: This method applies the integral-equation method (as developed by Buckley, Moon and Spencer) to daylighting problems. It considers a non-uniform sky (as studied by Moon and Spencer) and the effects of the interreflection component. The general interreflection problem is formulated for very long rooms with windows on one wall. A complete solution is then developed for the special case of a perfectly diffusing window on one wall. These results are applicable to rooms lighted by diffusing glass blocks.
Reference: 11

SINGLE STAGE METHODS

Average Daylight Factor (Italian Regulation)
Author: Ministero dei Lavori Pubblici (Italy)
Type: Analytical - Single stage
Applicability: Any type of fenestration. Uniform sky luminance distribution.
Output: Average daylight factor in a room Allows for light reflected from ground, transmission losses, influence of structural members, internally reflected light.
Discussion: The proposed formula allows one to consider the direct component entering the room and the internally reflected component after the first reflection from the walls in a single stage. It takes into account: glazing area of the window, average value of illumination on the glazing area (expressed as window-factor), transmission losses, average reflectance of internal surfaces, whole area of room's surfaces (walls, ceiling and floor).
Reference: 12

CEBS
Author: Commonwealth Experimental Building Station of Sydney (Australia)
Type: Graphical - Single stage
Applicability: Vertical windows, sawtooth rooflights. CIE standard overcast sky luminance distribution.
Output: Indoor daylight level at single points for vertical windows. Average daylight level over working planes for sawtooth roof buildings.
Allowances: Skylight reflected from ground, transmission losses, influence of structural members, internally reflected light.
Discussion: A very easy method that gives approximate results which may be used to evaluate the design of the room and window dimensions. It shows the influence of geographical latitude on daylight design in Australia. This method, together with the DLNS Method, was used by CIE for setting-up the simple method proposed in the International Recommendations of Publication CIE No. 16, 1970.
Reference: 13

CIE Simple Method
Authors: Department of Labour and National Service of Melbourne, and Commonwealth Experimental Building Station of Sydney (Australia)
Type: Graphical - Single stage
Applicability: Vertical windows and rooflights. CIE standard overcast sky luminance distribution.
Output: The minimum value of the daylight factor in a room of known depth. This value is calculated for rooms that have windows in one wall at a point 0.6 m away from the rear wall, for rooms with windows in two opposite walls at a point near the central axis, for rooms with rooflights, as the average level of daylight available over the whole working area under these rooflights.
The method takes into account light coming directly from the sky, light reflected from surfaces inside and outside the building, the effects of exterior obstructions, light transmission properties of glazing, and the probable effect of dirt accumulation on the glazing.
Discussion: This method was selected by CIE as an example of a simple method of daylight prediction and was proposed in the Daylight International Recommendations of 1970. It is designed for simple and easy estimation of the adequacy of daylight in uncomplicated rooms, under average conditions. The accuracy to be expected is not high, but it is still adequate enough for the design requirements.
References: 13, 14, 15

Daylight Design Diagrams
Author: Department of Labour and National Service of Melbourne (Australia)
Type: Graphical - Single stage
Applicability: Vertical windows, skylight, sawtooth and monitor roofs. CIE standard overcast sky luminance distribution.
Output: Daylight factor at a single point in a room with vertical windows in one or two opposite walls. Daylight factor averaged over working plane for skylights, sawtooth and monitor roofs.
The diagrams allow for direct sky light, externally and internally reflected light, influence of structural members, transmission losses. Effects of sky glare control are only given for few special cases.

Discussion: The method provides a series of diagrams that relate the daylight factor with the room depth and with the dimensions of the windows. The accuracy is good if the conditions do not differ too much from those assumed in the development of the diagrams. This method, together with CEBS methods above, was used by the CIE in setting-up the Simple Method proposed in the International Recommendation of Publication CIE No.16, 1970.
Reference: 14

Daylight Factor Meter
   Author: N. Baker (GB)
   Type: Tool - Single stage
   Applicability: Any type of windows. Standard overcast sky luminance distribution.
   Output: A general evaluation of the illuminance value on a internal surface or partition of a scale model.
   Discussion: The principle of this manual tool is to allow the operator to compare simultaneously indoor and outdoor illuminances, on the basis of the comparison of the brightness of two surfaces. The apertures related to external illuminance at the top of the light tower can be continuously reduced or enlarged. The operator varies the aperture of the meter until the brightness of two surfaces observed in the meter are identical. At this point, a scale in the aperture displays the daylight factor.
   Reference: 16

LUMEN METHODS

Frühling's Formula
   Author: H.G. Frühling (Germany)
   Type: Analytical - Single stage
   Applicability: Vertical windows. Some reference to other types of fenestration. Uniform sky luminance distribution.
   Output: Average daylight factor on a horizontal reference plane as a function of the ratio between glazing and floor areas. Allows rooms of different dimensions, different window sizes and different interior surface reflectances to be taken into account. Frühling published tables to evaluate the effects of external obstruction.
   Discussion: The method is very simple but it only gives a rough evaluation of the effects, due to the changes of internal surface reflectances. Nevertheless, Frühling's method was the first to apply to daylight calculations the procedure commonly used in artificial lighting calculation that include a utilisation coefficient. Therefore, using this method, it is possible to recognise daylight calculation, based on a flux-transfer method known as "Lumen Method".
   Reference: 17

Modified Lumen Method
   Authors: J.W. Griffith, W.J.Jr. Arner and E.W. Conover (USA)
   Type: Analytical - Single stage
   Applicability: Vertical windows and rooflights. Uniform, overcast and clear sky luminance distribution without sun.
   Output: Illumination on the work plane at three points of the room. The three reference points are located on a work plane 0.76 m. above the floor. The three points are centred and located in a row perpendicular to the window wall. One point is located 1.5 m from the fenestration, a point at the centre of the room, and a point 1.5 m from the interior wall opposite the fenestration.
   The method takes into account three different models of sky, ground brightness, transmission of glass, effects of dirt, and interior surface reflectances.
   Discussion: The method is a modified version of the Biesele-Arner-Conover formula known as the "Lumen Method". It is a calculation procedure that has been adopted by the American IES as part of their Recommended Practice of Daylighting.
   The illumination values are obtained from an analytical formula in which the effects of different types of sky, glass, interior surface reflectances, room dimensions are expressed by illumination coefficients. These coefficients may be derived graphically from a series of nomograms. The use of three points allows the evaluation of the distribution of light levels in the room. The method is relatively simple and quick to use.
   References: 18, 19

SERI Lumen Input Method
   Authors: K.C. Hunter and C.L. Robbins (USA)
   Type: Analytical - Single stage
   Applicability: Vertical windows. CIE standard, clear or overcast, clear sky distribution without sun on windows. Clear sky with sun, using a shading device to prevent direct sun from entering the room.
   Output: Illuminance at three reference points in a room located on a work plane 0.76 m high. The three points are centred and located in a row perpendicular to the window wall, at 1.5 m from the glazing, mid-point at the centre of the fenestration, and at 1.5 m from the interior wall.
   The method allows three different models of sky, ground brightness, transmission of glass, effects of dirt and interior surface reflectances. For a clear sky with sun, the method takes into account the position of the sun with respect to the fenestration, intensity of sunlight, date and time location.
   Discussion: This method is based on a method developed by Griffith-Arner-Conover above, but it reduces the number of coefficients required for calculation from four to two. New coefficient charts for vertical windows are developed.
   Reference: 20

TABLES

BRS Tables
 Authors: R.G. Hopkinson, J. Longmore, A.M. Graham, Murray, G.A. (GB)
 Type: Tables - Direct component
 Applicability: Vertical windows. Uniform and CIE standard overcast sky luminance distribution.
 Output: Sky factor, sky component and externally reflected component at singles points. Allows for rectangular windows unglazed or glazed, effects of simple obstructions.
 Discussion: The tables were derived from summated values of sky components obtained graphically from a large-scale Waldram Diagram. The input to the tables was the ratio of H/D and W/D (the height (H) and the width (W) of the window and the distance (D) of the reference point from the window). The use of the tables is very simple but it is difficult to study complex cases with this method. The accuracy is good if the influence of structural members is not relevant. The method is very good for initial design stages
 References: 3, 21, 22

Daylight Factor Tables
 Authors: H. Nakamura and M. Oki (Japan)
 Type: Tables - Direct component
 Applicability: Vertical windows. Intermediate sky (defined as sky condition of all types of sky except the clear and the overcast sky).
 Output: Sky factor, sky component. Allows for rectangular windows, effects of external simple obstructions.
 Discussion: The method deals with a complex problem in a simple way by using two tables of data. The first table gives the direct daylight factor per configuration factor of window: the data input is the altitude and azimuth angles of sky element. The second table gives the direct daylight factor on window surfaces: the input is the average height of obstructions and the azimuth of the window.
 Reference: 23

Fenestra Method
 Author: E.W. Conover (USA)
 Type: Tables - Single stage
 Applicability: Vertical windows and clerestorey glazing. Uniform sky luminance distribution.
 Output: Levels of illumination at three single points allowing assessment of externally reflected sun light, externally reflected skylight from the ground, transmission losses, rough influence of structural members, internally reflected light.
 Discussion: The method is based on the flux-transfer-method. It is a collection of simple tables valid for room lengths or widths not exceeding 10 m and ceiling heights between 2.6 and 4 m.
 Reference: 24

NPL Graded Sky Factor Tables
 Authors: T.S. Smith and E.D. Brown (GB)
 Type: Tables - Direct component
 Applicability: Vertical windows. Uniform sky luminance distribution.
 Output: Sky factor, allowing for the effects of dirt, effects of transmission losses, effects of simple obstruction, influence of structural members.
 Discussion: The method provides depth and half breadth of daylight penetration, and daylight area for a given sky factor, for windows of different sizes and with different degrees of obstruction. This method has a particular merit where contours of sky factors are required. It can be used very efficiently at the initial design stages.
 References: 25, 26

Rivero's Tables
 Author: R. Rivero (Uruguay)
 Type: Tables - Direct component
 Applicability: Vertical windows and horizontal skylights. Uniform and CIE standard overcast sky.
 Output: Sky factor and sky component at single points.
 Discussion: The method consist of 12 tables from which either sky factors or sky components can be read off directly as a function of window size, and their distances from a window. The calculations are developed with Higbie's formula above subdividing small elements.
 Reference: 27

NOMOGRAMS

BRS Interreflection Nomograms
 Author: R. G. Hopkinson, J. Longmore and P. Petherbridge (GB)
 Type: Nomogram - Internally reflected component
 Applicability: Vertical windows. Horizontal, vertical, 30° and 60° tilted rooflights. CIE standard overcast sky luminance distribution.
 Output: Internally reflected component at mid-depth (nomogram I), at back of sidelit rooms (nomogram II) and averaged over the working plane (nomogram III). The method takes account of light reflected from ground, transmission losses, influence of structural members, internally reflected light.
 Discussion: This method provides a relatively simple technique. It includes all the relevant parameters and gives values of sufficient accuracy. Nomograms I and II are fit for vertical windows, nomogram III for horizontal, vertical and tilted rooflights.
 References: 1, 2, 3, 22, 28

Bull's Nomogram
Author: H. S. Bull (USA)
Type: Nomogram - Direct component
Applicability: Vertical windows. Uniform sky luminance distribution.
Output: Illumination value at single points.
Discussion: This nomogram is based on Higbie's formula above for vertical windows. The input data are the width and the height of the window, and the distance of the reference point from the window.
Reference: 29

CSTB Abacus
Authors: R.Dogniaux, J.Leroy, P.Chauvel, J. Dourgnon and D. Fleury (Belgium, France)
Type: Nomogram - Direct component
Applicability: Vertical windows of rectangular shape. Uniform, CIE standard overcast, clear sky luminance distribution.
Output: Sky factor and sky component at single points taking into account influences of orientation and of structural members, transmission losses.
Discussion: This method is derived from the CSTB (*Centre Scientifique et Technique du Bâtiment*). "Diagram" below over which it has advantages and limitations.
Reference: 30

CSTB Diagram
Authors: J.Dourgnon, F.Fleury and P.Chauvel (France)
Type: Nomogram - Direct component
Applicability: Vertical and horizontal windows of rectangular shape. Uniform and CIE standard overcast sky luminance distribution.
Output: Sky factor and sky component at single points taking into account transmission losses and influence of structural members.
Discussion: This method has been developed in successive stages at the CSTB (*Centre Scientifique et Technique du Bâtiment*). It is very simple: it consists of a graph in which the direct component is determined with the graphical tracing of the azimuthal and of the elevation angles under which the reference point sees the window. The accuracy is acceptable for windows having small width but not for those with a great width.
References: 31, 32

Daylight Design Nomograms
Authors: Burt, Hill, Kosar and Rittelmann (USA)
Type: Nomogram - Single stage for commercial buildings
Applicability: Vertical windows and skylights. Clear sky conditions.
Output: Average daylight level, lighting energy consumption, estimated energy costs. Takes into account weather data, shape of building, required lighting level and lighting systems.
Discussion: This method allows appropriate decisions about daylighting to be made from the initial stage of the design process. The energy nomograms are an energy design tool which calculate the annual energy consumption of commercial buildings including lighting, heating, cooling, domestic hot water, fans, pumps and miscellaneous items. The calculation procedure is made with the simple use of a graphical process in a nomograms set.
Reference: 33

Fin-Graphico
Author: J. Navarro Casas (Spain)
Type: Graphical - Direct component and internally reflected component
Applicability: Vertical and horizontal rectangular windows. CIE standard overcast sky luminance distribution.
Output: Sky factor and sky component by means of isolux contours taking into account illuminances on a horizontal and a vertical plane.
Discussion: A complete set of ISO-DF contours for most representative window sizes is available through computer graphics.
The effect of rows of window can be determined as the ISO-DF set for each window is additive and thus new overlays for window rows or special arrangments should be obtained by the designer
Reference: 4

Kittler's Nomogram
Author: R. Kittler (Czechoslovakia)
Type: Nomogram - Single stage
Applicability: Vertical windows. CIE standard overcast sky luminance distribution.
Output: Sky component and daylight factors at single points. Allows for skylight reflected from ground, transmission losses, influence of structural members, internally reflected light.
Discussion: The nomogram allows one to obtain the total daylight factor in a very simple way. It can be used to determine either the daylight factor given by the window dimensions, or vice versa.
Reference: 34

Kittler's Semi-Logarithmetric Chart
Author: R. Kittler (Czechoslovakia)
Type: Graphical - Internally reflected component
Applicability: Vertical windows. CIE standard overcast sky luminance distribution.
Output: Internally reflected component at any point of the horizontal working plane in side-lit rooms. Takes account of light reflected from the ground, transmission losses, structural members.
Discussion: This method provides a relatively simple graphical technique of determining the IRC
Reference: 35

SERI Clear Sky IRC Nomogram
   Authors: C.L. Robbins and K.C. Hunter (USA)
   Type: Nomogram - Internally reflected component
   Applicability: Vertical windows. Clear sky luminance distribution.
   Output: Averaged internally reflected component. The nomogram takes orientation and transmission losses into account.
   Discussion: The clear sky nomogram was purposely designed so that it could be used following the same general procedures used with the overcast sky nomograms. The major difference is that the SERI clear sky nomogram establishes an initial internal reflectance component (iIRC) rather than the actual IRC. The IRC nomogram is used for all clear sky situations, regardless of aperture orientation. A series of orientation functions is used to establish IRC based on the nomogram value, the solar altitude, the aperture azimuth with respect to the sun, and the aperture area/room area ratio.
   References: 36, 37

Window-Factor Chart
   Author: M. Vio (Italy)
   Type: Nomogram - Direct component
   Applicability: Vertical windows with rectangular shape. Uniform sky luminance distribution.
   Output: Sky factor on a medium point (barycentre) of the window's glazing area taking into account effects of the shape of the window and of the structural members.
   Discussion: The nomogram provides the sky factor on the medium point of the window taking into account the effects of the window shape and the thickness of the wall. The Italian daylighting regulation requires this value for the application of its daylight evaluation by means of a single-stage method.
   Reference: 38

PROTRACTORS

BRE Daylight Protractor
   Author: A. F. Dufton (GB)
   Type: Tool - Direct component
   Applicability: Vertical, horizontal and tilted windows. Uniform sky luminance distribution
   Output: Sky factor, sky component at single points. Takes into account influence of structural members, effects of external obstructions.
   Discussion: The method is based on the daylight protractor concept. It consists of five series of pairs (primary protractor or daylight scale, and an auxiliary protractor or correction factor scale) of protractors, each pair corresponding to a particular slope of glazing and, in one case, to unglazed openings. The analytical formulae on which the method is based is the projected solid angle principle. The accuracy of the responses is good but it is difficult to use for irregular shape windows. These protractors have become the standard manual daylighting design tool and are still used throughout the world to design day lighting in buildings.
   References: 3, 22, 70

BRS Daylight Calculator
   Authors: R.G. Hopkinson, P. Petherbridge, and J. Longmore (GB)
   Type: Tool - Internally reflected component
   Applicability: Roof lights. CIE standard overcast distribution.
   Output: Internally reflected component. Allowing for external obstructions and average internal reflectance.
   Discussion: The tool consists of a circular calculator with five particular scales.
   References: 3

BRS Daylight Factor Calculator (Daylight Factor Slide Rule)
   Authors: R.G. Hopkinson and J. Longmore (GB)
   Type: Tool - Single stage
   Applicability: Vertical windows. CIE standard overcast sky luminance distribution.
   Output: Sky component and total daylight factor at a penetration of 1.6, 3.3, 5 m depth on the centre-line of a side lit room. light reflected from the ground, internally reflected light, effect of sky glare control is taken into account.
   Discussion: The slide rule gives daylight factors in one step on the working plane. It takes into account only rooms with a height between 2.6 and 4 m and any combination of internal surface reflectances. This tool is based on model measurements in rectangular rooms with a single side window. This method can be used at the initial design stage.
   References: 2, 39

BRS Daylight Protractor (Series 2)
   Author: J. Longmore (GB)
   Type: Tool - Direct component
   Applicability: Vertical, horizontal and tilted windows. Uniform and CIE standard overcast sky luminance distribution.
   Output: Sky factor, sky component at single points. Allowances for the influence of structural members, effects of various types of glass and effects of external simple obstructions.
   Discussion: The method is based on the daylight protractor component developed with BRS

Protractors by Dufton (see above). In this method the primary and auxiliary scale were combined into a single circular protractor.

Two sets of five protractors were published: one for uniform and one for CIE standard overcast luminance distribution. Each protractor of a set was developed for a particular slope. The analytical formulae on which the method is based is the projected solid angle principle. The accuracy of the responses is good but it is difficult to use for irregular shape windows.

References: 37, 40

LRL (Lighting Research Laboratory)
   Author: B.F. Jones (USA)
   Type: Tool - Direct component
   Applicability: Vertical, horizontal and tilted fenestration. Any sky luminance without or with the sun.
   Output: Sky factor, sky component and externally reflected factor. Allows for effects of external obstruction and structural members.
   Discussion: The original contribution of the method is a protractor, known as the "lumen protractor", that measures the amount of light arriving at a given point from a source. This measure is given as a factor (ratio between the illumination at the point and the luminance of the source) and for this reason the method allows one to consider different values of luminance either for the sky or for external obstructions.
   Only two simple equations are needed in addition to the protractor, and all the calculations may be done by hand. The protractor is useful for any slope. The accuracy can be as great as desired.
   References: 41

SERI Daylight Protractor
   Authors: C. Robbins, K.C. Hunter and N. Carlisle (USA)
   Type: Tool - Direct component
   Applicability: Vertical windows. Clear sky and standard overcast sky luminance distribution without sun.
   Output: Sky factor, sky component and externally reflected component. Allows for effects of glass, dirt, frame, and simple external obstruction and structural members.
   Discussion: This protractor is based, on the concept of the BRS protractors and on a first version for clear sky proposed by Bryan (43) in 1982 but uses the flux-transfer-method known as Lumen Method. Only one protractor, having circular shape, is necessary to analyse all vertical apertures under clear sky conditions regardless of aperture orientation, time of day and solar position. It is designed to be used with a plan and section drawing, of any scale. The use of only one protractor simplifies the lecture proceeding. Simple formulae to calculate the correct factor for glazing, dirt and frame effects are provided. To take into account different exteriors, illuminance data for two types of sky (clear and overcast) are provided: SERI illuminance data available on a hourly basis for the 21st of each month for 219 cities in the USA and for eight orientations

References: 37, 42

Sky Component Protractors for Clear Sky
   Authors: H. J. Bryan and D. B. Carlberg (USA)
   Type: Tool - Direct component
   Applicability: Vertical windows. CIE standard overcast and clear sky luminance distribution.
   Output: Sky factor, sky component and externally reflected component. Takes into account effects of glazing and effect of simple external obstruction.
   Discussion: In 1982 Bryan developed a clear sky protractor based upon the CIE clear sky. It was identical in format to BRS Series 2 protractors.(see above). However, a separate protractor was needed for each aperture orientation and differing sun position for a total set of 41 protractors. In 1983 Bryan proposed a modification to his original protractors in which the set was reduced to nine protractors combining the five orientations. This modified series is different in format from the BRS protractors. The method is based on the Daylight Factor Method. Its high accuracy was validated against the Quicklite I daylight program.
   References: 37, 43

Turner-Szymanowsky Protractor
   Author: W. Turner-Szymanowsky (USA)
   Type: Tool - Direct component
   Applicability: Vertical windows and rooflights. Uniform sky luminance distribution.
   Output: Value of illumination on single points, taking into account external obstructions and transmission losses.
   Discussion: This is perhaps the first method that uses the protractor concept for daylight calculation. It is based on the Higbie-Levin Formula (see above). The author provides a particular nomogram over which a transparent protractor with a transparent arm must be superposed. Correction coefficients for transmission losses in window glass are given. Comparison of measured and predicted values of illumination are presented. The accuracy is very good.
   References: 44

DOT DIAGRAMS

Dot Diagram (Pepper Pot Diagram)
Author: Pilkington Brothers (GB)
Type: Graphical tool - Direct component
Applicability: Vertical window. CIE standard overcast sky luminance distribution.
Output: Sky factor at single point taking into account effects of glazing and simple external obstructions.
Discussion: This method is known as the "Pepper Pot Diagram". It consists of a dotted chart that represents a plane projection of the sky vault. The dots are distributed as a function of the luminance distribution. The vertical window must be projected on an imaginary screen placed at a fixed distance from the reference point. The projection of the window must be superposed over the dotted chart and from the number of dots contained in the projection of the window the sky factor upon the reference point is determined.
References: 45, 49

Full-Field Camera Technique
Author: J. Longmore (GB)
Type: Tool - Direct component
Applicability: any shape of fenestration. Uniform sky luminance distribution.
Output: Sky factor, sky component and externally reflected component at single points. Allows for influence of structural members and of external and internal obstructions.
Discussion: This method is an alternative to the methods that use the stereographic projections. The particular use of the equidistant projection in the present instance is in the interpretation of photographs obtained with a R. Hill full-field camera.
References: 46, 47

Perspective Projection Chart
Author: I. Swarbrick (GB)
Type: Graphical Tool - Direct component
Applicability: Vertical windows and skylights. Uniform sky.
Output: Sky factor at single point allowing for influence of structural members and of obstructions.
Discussion: A perspective view of the window and obstructions is prepared either by a graphical method on one of a number of Swarbricks's Perspective Projection Charts, or by photography (Swarbricks's Photo-Theodolite). The sky factor is found by superimposing a "Daylight Factor Grille". The method has been used mainly for legal purposes but may be adapted to provide sky components incorporating allowances for the CIE sky and for transmission losses. The limited angle encompassed by each chart is sometimes a disadvantage.
References: 48

Pleijel's Diagram
Author: G. Pleijel (Sweden)
Type: Graphical tool - Direct component
Applicability: Vertical, horizontal and tilted windows. Standard overcast sky luminance distribution.
Output: Sky factor, sky component and externally reflected component on a single reference point. The method takes into account daylight illumination on a horizontal and on a vertical plane and effects of complex external obstructions.
Discussion: Pleijel's Diagrams are part of a more comprehensive system that permits the calculation, not only of daylight, but also of sun penetration and solar radiation from sun and sky. The method uses a "screen figure" on which one projects the visible sky about the reference point. This screen is placed over the diagram. The number of dots on the diagram which lie within the boundary of the patch of visible sky as seen through the window are counted and related to the total number of dots on the complete diagram. The method is a little complex but has a good level of accuracy.
References: 49

WALDRAM DIAGRAMS

Farrell's Method
Author: R. Farrell (USA)
Type: Graphical - Direct component
Applicability: Any type of fenestration. Clear sky and uniform sky luminance distribution.
Output: Sky factor, and externally reflected factor on a reference point. Allows for orientation of windows, time, glazing losses, effects of external obstruction and of structural members. The method allows one also to take account of very complex landscapes
Discussion: This method is both based on the Waldram Diagram, corrected for glazing losses, and on the method of Pleijel. Onto this Diagram is superposed a grid which divides the half vault of the sky in patches corresponding to equal solid angles. The value of illumination for each patch is determined with the Kittler's Chart (see above) approved by CIE Technical Committee TC-4.2. This value is expressed with a suitable number of dots. Each dot corresponds to a sky factor equal to 0.1. The successive proceedings are the same as those of the Waldram Method. The sky factor on the reference point is given by the sum of the dots seen through the window. The grids of dots proposed are calculated for different solar altitude angles. For careful calculations it is necessary to have many grids and for very large windows the calculation of the total number of dots may be tedious. Apart from these factors, the method may be very interesting:
References: 50

Graphical Method for the Direct Factor

Author: M. Vio (Italy)

Type: Graphical - Direct component

Applicability: Vertical and horizontal windows. Uniform and CIE overcast sky luminance distribution.

Output: Sky factor, sky component and externally reflected component at single points. The method takes into account the influence of structural members and of complex external obstructions, but only parallel and/or perpendicular to the plane of the window. It allows one to consider horizontal obstruction above the window. The reference point may be inside or outside the room.

Discussion: The method is derived from the basic formula for the calculation of direct daylight on a reference point, from the Waldram Diagram and from the IUAV method below.

The method provides graphs that consist of a series of droop-lines from which, entering directly with the plan of analysing configuration and with the ratio H/D (between the height of the window and the distance of the reference point), the illumination factor is determined. The method is very simple and quick to use and its accuracy also as good as the Waldram Diagram.

References: 51, 52

IUAV

Author: G. Rossi and M. Vio (Italy)

Type: Graphical - Direct component

Applicability: Vertical and horizontal windows. Uniform sky luminance distribution.

Output: Sky factor, sky component and externally reflected component at single points. Takes into account influence of structural members and of complex external obstructions, but only those parallel and/or perpendicular to the plane of the window. It allows one also to consider horizontal obstruction above the window.

Discussion: The method is derived from the basic formula for the calculation of direct daylight on a reference point and from the Waldram Diagram. It provides a graph that consists of a series of droop-lines from which, entering with the values of azimuthal and elevation angles under which the reference point views the window, the illumination factor is determined. The method is very simple and quick and its accuracy is as good as the Waldram Diagram.

References: 53

IUAV Solar and Daylight Charts

Author: A. Fanchiotti, G. Rossi and M.Vio (Italy)

Type: Graphical tool - Daylight and energy performance

Applicability: Vertical windows. Uniform sky luminance distribution.

Output: Sky Factor, winter and summer solar radiation factors at single points. Allows for effects of external obstructions, structural members, shading devices, location, time, and orientation.

Reference: 54

Simplified Daylighting Design Methodology

Author: H.J. Bryan (USA)

Type: graphical and analytical - Direct component

Applicability: Vertical windows. Clear Sky without sun.

Output: Sky factor on a reference point. The methodology indicates a way to compute the external and the internal reflected components with split-flux-method. It takes into account transmission losses, effects of complex external obstructions and of structural members.

Discussion: This method is based on a Waldram Diagram corrected for transmission losses. On this base the method superposes a grid which divides the half vault of the sky in patches corresponding to equal solid angles. The value of illumination for each patch is determined with the Kittler's Chart (see above) approved by CIE Technical Committee TC-4.2. The successive proceeding is the same as that in the Waldram Method. The sky factor on the reference point is the sum of the values of illumination given by each patch of sky seen through the window. A simplification of the proceeding, for vertical rectangular windows, is proposed in a graph which gives the sky component as a function of the ratios W/D and H/D where H and W are respectively the height and the width of the window, and D is the distance from the window of the reference point. The grid proposed is calculated for a solar altitude of $40°$ and for a window having its azimuth from the sun of $+90°$. For different situations, different grids must be developed. For this reason the method seems tedious.

References: 55

Waldram Diagram

Author: P.J. Waldram (GB)

Type: Graphical - Direct component

Applicability: Any type and shape of window and rooflight in planes tilted at any angle to the reference plane. Uniform and CIE sky luminance distribution.

Output: Sky factor, sky component and externally reflected component at single point taking into account transmission losses; influence of structural members and complex external obstructions; influence of the complex. room shape.

Discussion: The method is based on a graphical resolution of the integral formula for the calculation of the sky component based on the projected solid angle principle. The original diagram is an "equal area" diagram based on an orthographic projection of the hemispherical sky vault. Other diagrams have

been developed subsequently in which the constant proportionality between areas and corresponding direct daylight values have been sacrificed in favour of other advantages. Direct component can be calculated with great precision. The method is particularly useful when external obstructions have complex shapes.
References: 3, 22, 56, 57, 58

METHODS OF URBAN ANALYSIS

Building Daylight Analysis
   Author: M. Vio (Italy)
   Type: Parametric analysis - Daylight and energy performance
   Applicability: Uniform sky luminance distribution.
   Output: For four building typologies the analysis provides sky component at any point of the façades, average sky factor on the façades as function of dimensional parameters of the building, and energetic evaluations.
   Discussion: Daylight characteristics for four different building typologies are analysed. The analysis leads to a better understanding of design criteria (building height and length) which influence illuminance and energy characteristics. The considered parameters are valid for European towns.
   References: 59

Daylight and Energy Diagrams
   Author: G. Rossi and M. Vio (Italy)
   Type: Graphical tool - Daylight and energy performances
   Applicability: Uniform sky luminance distribution.
   Output: Sky factor, winter and summer solar radiation factors at single points. Takes into account external obstructions, location, time, orientation, structural members and shading devices.
   Discussion: A simple methodology for the control of daylight and energetic performance at urban scale is presented. Graphical proceedings are based on the Waldram Diagram, on the Solar Chart and on the Solar Radiation Diagram. The performances corresponding to the analysed urban configuration are synthetically expressed by three parameters: sky factor, winter solar radiation factor and summer solar radiation factor on a reference point. The analysis is made with the superimposition of three suitable graphs. The same proceeding may be used to analyse the performances of vertical windows.
   References: 60

Daylight and Urban Form
   Author: J.A. Lazerwitz (USA)
   Type: Parametric analysis - Daylight performance
   Applicability: Overcast and Clear Sky without sun.
   Output: Daylight characteristics of four urban typologies allowing for surface reflectivity, street width and orientation.
   Discussion: Daylight characteristics for four different building typologies were compared through the SUPERLITE computer simulation program. The analysis has led to a better understanding of zoning criteria (building height and bulk, exterior materials and upper level setbacks) which influence The parameters were valid for US towns: the analysed building heights ranged from 25 to 55 storeys.

Permissible Height Indicators
   Author: R.G. Hopkinson (GB), C.L. Robbins, (USA)
   Type: Graphical tool
Applicability: Uniform, overcast and clear sky.
   Output: Height of the urban obstruction to obtain a certain sky factor at the reference point.
   Discussion: This tool is used to ensure that all or part of the building has views of enough sky dome to ensure sufficient daylighting can reach the façades.
   References: 3, 37

Sky-Dome Projection
   Author: H. J. Bryan, and S. Stuebing
   Type: Photographic tool - daylight performances
   Applicability: Any.
   Output: Daylight performance.
   Discussion: This work presents some practical examples of the use of full-field camera technique or sky-dome projection applied to the urban tissues.
   References: 61

GLARE CONTROL

BCD Evaluation
   Author: M. Luckiesh - S.H. Guth
   Type: Analytical - Glare control
   Output: Index of visual comfort. Takes into account brightness and angular size of the source, position of the source in the visual field, average brightness of visual field.
   Discussion: The method consists of empirical equations based on careful laboratory studies. Tables and graphs were presented to take into account the position of the source in the visual field. Methods of treating multiple sources were studied and empirical rules given to determine an equivalent single source. The formula is difficult to use and does not predict glare accurately for larger solid angles.
   References: 62

BRS Glare Constant Nomogram
   Author: R.G. Hopkinson and J. Longmore (GB)
   Type: Nomogram - Glare control
   Applicability: Uniform sky luminance distribution.
   Output: Glare constant useful for determining the glare index. Brightness and angular size of the source, position of the source in the visual field and average brightness of visual field are taken into account.
   Discussion: The use of the nomogram gives the Glare Constant for the window under analysis. If there are a number of windows, the Glare Constant for each must be derived from the nomogram and summed.
   References: 63

BRS Glare Formula
   Author: P. Petherbridge and R.G. Hopkinson (GB)
   Type: Analytical - Glare control
   Applicability: Vertical and horizontal windows.
   Output: Index of subjective sensation of glare discomfort. Allowances: brightness and angular size of the source, position of the source in the visual field, average brightness of visual field.
   Discussion: The subjective sensation of glare discomfort is related to the luminance of sky, apparent size of the window, position in the visual field, and adaptation conditions in the room. The method provided empirical formulas to evaluate a Glare Index value for any configuration.
   References: 3, 64

BRS Glare Index Tables
   Author: R. G. Hopkinson and J. Longmore (GB)
   Type: Tables - Glare control
   Applicability: Vertical and horizontal windows. Uniform sky luminance distribution.
   Output: Index of subjective sensation of glare discomfort.
Brightness and angular size of the source, position of the source in the visual field, average brightness of visual field are taken into account.
Discussion: The tables give the Glare Index directly or with only one or two auxiliary steps. Limiting values of glare index recommended by the IES for interiors lit by symmetrical arrays of lighting fittings are provided. Conversion terms to add to Glare Index allowing for skies of average luminance higher than 500 ft-L are suggested.
   References: 65

Cornell Formula
   Author: Cornell University and BRS (USA and GB)
   Type: Analytical - Glare control
   Output: Glare Constant useful for determining Glare Index. The formula takes into account brightness and angular size of the source, position of the source in the visual field, average brightness of visual field.
   Discussion: The formula has been studied jointly by Cornell University and the Building Research Establishment. The equation can be applied to any light source.
   References: 66

Daylighting Glare Index
   Author: R.G. Hopkinson (GB)
   Type: Analytical - Glare control
   Applicability: Overcast and clear sky luminance distribution.
   Output: Glare Constant useful to determine the Glare Index. Takes account of brightness and angular size of the source, position of the source in the visual field, average brightness of visual field, luminance at the plane of the aperture.
   Discussion: An equation derived from the original equation of glare constant is developed for the daylighting due to overcast and clear sky conditions. Graphs for the rapid evaluation of solid-angles for vertical and horizontal windows are provided.
   References: 67, 68

Solid Angle Diagram with Position Correction
   Author: P. Petherbridge and J. Longmore (GB)
   Type: Graphical - Glare control
   Applicability: Any type of fenestration. Uniform sky luminance distribution.
   Output: Modification to reduce glare effects of sources located off the line of sight to evaluate the glare constant. Taking into account position and size of the glare source, effects of internal and external structural members and obstructions.
   Discussion: The glare source is plotted on this diagram exactly as a window on the Waldram Diagram. The diagram allows one to know the solid angle that is then used in the methods for evaluating the glare constant and the glare index.
   References: 69

REFERENCES

(1) Hopkinson, R.G., Longmore, J., Petherbridge, P., "An Empirical Formula for the Computation of Indirect Component of Daylight Factor", *Trans. IES*, vol. 19, No. 7, p. 201-219, 1954.

(2) Hopkinson, R.G., *Architectural Physics: Lighting*, HMSO, London, 1963.

(3) Hopkinson, R.G., Longmore, J., Petherbridge, P., *Daylighting*, Heinemann, London, 1966.

(4) Navarro Casas, J., *Sobre iluminacion natural en arquitectura*, Publicationes de la Universidad de Sevilla, 1983.

(5) Higbie, H.H., "Prediction of Daylight from Vertical Windows", *Annual Convention of the Illuminating Engineering Society,* Briacliff Manor, New York, 27-30 October 1924.

(6) Higbie, H.H., Levin, A., "Prediction of Daylight from Sloping Windows", *Trans. IES,* vol. 21, p. 273-373, 1926.

(7) Kittler, R., Ondrejicha, S., "Exact Determination of the Daylight (Sky Component) from Rectangular Sloping Window Apertures with a CIE Overcast Sky", *Monograph of the IES* No. 6, London, 1963.

(8) *Compte Rendu CIE Quinzième Session,* Vienne, June 1963, Publ. CIE No. 11 B, 1964.

(9) Kittler, R., Ondrejicha, S., "A More Exact Calculation Method of Natural Daylight" (in russian), Izvestya Akad. Stroy., 55 No.1, 1963.

(10) Pugno, G.A., *"Illuminazione naturale con copertura a sheds", La Luce,* No. 1, Anno XX.

(11) Spencer, D.E., Stakutis, V.J., "The Integral-Equation Solution of the Daylighting Problem", *Journal of the Franklin Institute,* vol. 252, p. 1507-1512, 1951.

(12) *Circolare Ministero dei Lavori Pubblici* No. 3151, Rome, 22 May 1967.

(13) *Note of the Science of Building,* No. 56: "Daylighting of Building", Comm. Exper. Building Station, Sydney, 1962.

(14) *Industrial Data Sheet* L2, "Natural Lighting of Buildings: Daylight Design Diagrams", Dep. of Labour and Nat. Service, Melbourne, 1963.

(15) *Daylight: International Recommendations for the Calculation of Natural Daylight.* Publication CIE No. 16, 1970.

(16) Baker, N., "Low Cost Instrument for Measuring Daylight Factors", *Proc 2nd Eur. Conf. on Architecture,* Paris Kluwer Acad. Publ., Dordrecht, 1989.

(17) Frühling, H.G., *"Tagesbeleuchtung von Innenräumen, ihre Messung und ihre Berechnung nach der Wirkungsgradmethode",* Doctoral Thesis, Technical University, Berlin, 1928.

(18) Griffith, J.W., Arner, W.J., Conover, E.W., "A Modified Lumen Method of Daylight Design", *Illuminating Engineering,* vol. 50, p.103, 1955.

(19) IES, "Recommended Practice of Daylighting", *Illuminating Engineering,* vol 57, p 517, 1962.

(20) Hunter, K.C., Robbins, C.L., *"Simplification of the Lumen Input Method for Predicting Daylighting in Buildings"*, SERI/TR-254-184, Golden, Colorado, 1984.

(21) Hopkinson, R.G., Longmore, J., Murray, "Simplified Daylight Tables", Nat. Buil. Stud. *Special Report,* No. 26, HMSO, London, 1958.

(22) *British Standard Code of Practice*, CP 3, Ch.1, Part 1, "Daylighting", London, 1964.

(23) Nakamura, H., *et al.* "Calculation of Daylight Factor Dominated by Intermediate Sky", *Proc. 1983 International Daylighting Conf.*, Phoenix, Arizona, p. 57-59, 1983.

(24) Conover, E.W., *The Fenestra Method for Predicting Daylight in Class Rooms and Offices*, Fenestra Inc, Detroit, 1957.

(25) Smith, T.S., Brown, E.D., *The Natural Lighting of Houses and Flats with Graded Daylight Factor Tables*, DSIR National Physical Laboratory, HMSO, London, 1944.

(26) *British Standards Code of Practice*, CP 3: Ch.1, British Standard Institution, London, 1949.

(27) Rivero, R., *"Illuminacion Natural - Calculo del factor de dia directo para ventanas sin vidrios y con vidrios para cielos uniformes y no uniformes"*, Instituto de la Construccion de Edificios, Dpto. de Acondicionamento, Facultad de Arquitectura, Montevideo, Uruguay, 1958; English translation: BRS Library Communication No. 860, London, 1958.

(28) Robbins, C.L., Hunter, K.C., "Reflected Daylight", *Arch. Journal,* vol. 120, No. 3107, 1954.

(29) Bull, H.S., "A Nomogram to Facilitate Daylight Calculation", *Trans. IES,* vol. 23, 1928.

(30) Dogniaux, R., Leroy, J., Chauvel, P., Dourgnon, J., Fleury, D., *"Abaques pour la détermination de la composante directe du facteur du jour par ciel uniforme, ciel Moon et Spencer et ciel serein"*, *Cah Cen. Sci. Tech. Bâtim.,* Paris, 1963.

(31) Dourgnon, J., Fleury, F., *"Graphiques pour la détermination rapide de la composante directe du facteur de lumi<138>re du jour dans le cas d'un ciel suivant la loi de Moon et Spencer",* Cah. Cent. Sci. Bâtim., No. 33, Cahier 271, 1958.

(32) Dourgnon, J., Chauvel, P., *"Détermination de la composante directe du facteur de lumi<138>re du jour tenant compte du taux de transmission des vitrages"* Cah. Cent. Sci. Bâtim., No. 45, Cahier 358, 1960.

(33) Burt, Hill, Kosar, Rittelmann, "Energy Nomographs as Design Tool for Daylighting", Energy and Building, No. 6, 1984.

(34) Kittler, R., *"Prispevek praktickemu narhovaniu okien bocne osvetlovanych miestonosti"*, Stavebnicky Casopid, SAV 7 No. 5, 1959, "A Contribution to the Practical Design of Windows in Side-lit Rooms", English translation: BRS Library Communication No. 986, July 1960.

(35) Kittler, R., "Development of Methods for the Calculation of Natural Illumination in Buildings, Taking into account Reflected Light", *Svetotekhnika* No. 2 p. 188, March 1956 (in Russian); English translation: BRS Library Communication No. 790, March 1957.

(36) Robbins, C.L., Hunter, K.C., *"A Method for Determining Intereflected Daylight in Clear Climates"*, SERI/TR-254-184, Golden, Colorado, 1984.

(37) Robbins, C.L., Daylighting, Van Nostrand Reinhold. Company, New York, 1986.

(38) Vio, M., *"Illuminazione naturale nell'architettura tradizionale veneziana"*, *Recuperare*, No. 29, 1987.

(39) Hopkinson, R.G., Longmore, J., "A Study of the Interreflection of Daylight Using Model Room and Artificial Skies", *National Building Studies, Research Paper*, No. 24, HMSO, London, 1954.

(40) Longmore, J., *Daylight Protractors*, HMSO, London, 1968.

(41) Jones, B.F., *Very Simple Hand Calculations Daylighting*, Lighting Research Laboratory, PO Box 6193, Orange, California, 1983.

(42) Robbins, C.L., Hunter, K.C., and Carlisle, N., *"SERI Daylight Protractor for Clear and Overcast Sky"* SERI/TR-253-2588, Golden, Colorado, 1984.

(43) Bryan, H.J., Carlberg, D.B., "Development of Protractors for Calculating the Effects of Daylight from Clear Skies", *Journal of IES*, April 1985.

(44) Turner-Szymanowsky, W., "A Rapid Method for Predicting the Distribution of Daylighting in Buildings" *Engineering Research Bulletin*, No. 17, 1931.

(45) Pilkington Brothers, *Windows and Environment*, London, 1969.

(46) Longmore, J., "The Full-Field Camera Photogrammetric Technique in Lighting Research", *The Photographic Journal*, No. 5, 1964.

(47) Hill, R., "A Lens for Whole Sky Photographs", *Quart. J.R. Met. Soc.*, vol. 50, No. 227, 1924.

(48) Swarbrick, J., *Easements of Light*, vol. II B.T. Batsford Ltd, London, 1933.

(49) Pleijel, G., "The Computation of Natural Radiation in Architecture and Town Planning", *Bulletin* No. 25, *Statens Namnd for Biggnadsforskining*, Stockholm, 1954.

(50) Farrell, R., "Calculating Direct Illumination from the Sky Conditions", *IES Conference*, San Francisco, California, 1975.

(51) Vio, M., "Graphical Method for Direct Factor of Daylight Evaluation", *Proc. Passive Solar Architecture International Conference*, Bled, Yugoslavia, 1988.

(52) Vio, M., "Graphical Method of Computing Sky Factor for Uniform and CIE Overcast Sky Luminance Distribution", *Proc. VI Lux European Congress*, Budapest, 3-5 October 1989.

(53) Rossi, G., Vio, M., *"Valutazione dell'illuminamento naturale negli ambienti chiusi"*, *Condizionamento dell'aria, riscaldamento, refrigerazione*, December 1985.

(54) Fanchiotti, A., Rossi, G., Vio. M., "A Simple Method for Evaluating the Effect of Urban and Building Configurations on Energy and Daylighting Performances", *Proc. 4th. International Conference PLEA* Venice, Italy, 1985.

(55) Bryan, H.J., *A Simplified Daylighting Design Methodology for Clear Sky*, Pub. of Dep. of Arch., Univ. of California, Berkeley, California, 1979.

(56) Waldram, P.J., *A Measuring Diagram for Daylight Illumination*, B.T. Batsford Ltd, London, 1950.

(57) Walsh, J.W.T., "Ascertaining the Sky Factor" Arch. Journal, vol. 129, No. 3342, 1959.

(58) Walsh, J.W.T., *The Science of Daylight*, Macdonald, London, 1961.

(59) Vio, M., *"Per una valutazione sintetica delle prestazioni iluminotecniche ed energetiche di alcune tipologie edilizie"*, *Energie Alternative HTE*, No. 51, 1988.

(60) Rossi, G., Vio, M., *"Metodi di valutazione energetici ed illuminotecnici del tessuto urbano"*, *Energie Alternative HTE*, No. 44, 1986.

(61) Bryan, H.J., Stuebing, S., "Daylight: the Third Dimension of the City", *Sun World*, vol. 11 No. 2. 1987.

(62) Luckiesh, M., Guth, S.K., "Brightness in Visual Field at Bordeline Between Comfort and Discomfort, *Trans. IES*, 44, 1949.

(63) Hopkinson, R.G., Longmore, J., "Nomogram for Glare Index Calculations", *Light and Lighting*, No. 55, 1962.

(64) Hopkinson, R.G., "Evaluation of Glare", *Illuminating Engineering*, No. 52 1957

(65) Hopkinson, R.G., Longmore, J., "Tables for Glare Index in Daylit Interiors", *Nat. Build. Studies, Special Report*, HMSO, 1966.

(66) Chauvel, P., et al., *Glare from Windows: Current Views of the Problem*, Building Research Establishment, HMSO, 1980.

(67) Hopkinson, R.G., "Glare from Windows", *Journal of Construction Research and Development*, Nos. 2 and 3, 1970-1971.

(68) Hopkinson, R.G., "Glare from Daylighting in Buildings", Applied Ergonomics, No. 3, 1972.

(69) Petherbridge, P., Longmore, J., "Solid Angles Applied to Visual Comfort Problems", *Light and Lighting*, No. 47, 1954.

(70) Dufton, A.F., "Protractors for the Computation of Daylight Factors", DSIR Building Research Tech. Paper n. 28, HMSO, London, 1946.

Appendix G
REVIEW OF COMPUTER CODES

MICROCOMPUTERS

BELYS
Developed by: Lysteknisk Laboratorium, Lyngby, Denmark
Contact: Peder Obro
Hardware: HP 1000
Software: Operating System RTE-A
Modelling capabilities:
- uniform sky luminance
- fixed position of the sun for each calculation
- any obstructions
- any room shape
- diffuse or specular room surface
- any fenestration
- clear or diffusing glass

Output features:
- numerical
- illuminance distribution on surfaces
- no energy analysis
- no graphic output included, may be attached

CONTROLITE
Developed by: Lawrence Berkeley Laboratory, California, USA
Contact: Francis Rubinstein
Hardware: IBM PC XT/AT or true compatible; 256K
Software: FORTRAN (free)
Modelling capabilities:
- uniform, overcast and clear sky, and ground reflection
- no obstructions
- rectangular room shape
- diffusing room surface
- glazing in vertical plane
- clear or reflective glass

Output features:
- horizontal illuminance
- no graphic output

DALITE
Developed by: National Bureau of Standards, Maryland, USA
Contact: Gary Gillette
Description: The daylighting program is a system of FORTRAN subroutines designed for inclusion into larger building energy simulation programs. Once incorporated, these subroutines will allow the existing program to account for the energy trade-offs associated with natural illumination. The daylighting model comprises three separate routines: the first generates hourly sky luminances, sky illuminances and direct sun illuminance taking solar radiation and sun position as input. The second predicts interior daylight illumination at various points within a room due to any number of windows, skylights or clerestories. The last routine adjusts the electric lighting load in response to the available daylight. It is a dynamic model designed to study how conditions change with time.

Intended users: Those studying the energy performance of daylighting
Hardware: Generic
Software: Can be incorporated into any hardware or software system with ANSI 77 FORTRAN building energy program; program source code listing available
Models and algorithms: Radiant flux model for predicting luminances within space; sky luminance model uses a phasing technique between clear and overcast driven by the ratio of clear to diffuse sky.

Modelling capabilities:
- sky luminances, illuminance at horizontal, direct solar illuminance
- exterior horizontal or vertical obstructions, parallel obstructions
- rectangular room shape
- fenestration: windows, skylights, clerestories, roof monitors

Output features:
- numerical
- illuminance: as desired from within external energy program
- energy analysis: designed to work within other energy analysis programs

Reference: 1

DAYLIT
Contact: Murray Milne
Description: DAYLIT is a user-friendly design tool for architects at the very beginning of a project. It uses the IES method and RP-21 to calculate and display daylight illumination levels. Electric lighting levels are then calculated for up to three zones, and each may have a different control strategy. Heating and cooling loads in the space are calculated.

Various HVAC systems can be specified with a variety of control strategies. Electricity and gas consumption is then computed and actual hour-by-hour costs are displayed, based on locally utility rate. All this is plotted out in 3-D for each hour of each month. DAYLIT has a variety of powerful options for combining and comparing designs.

Intended users: Architects, lighting designers and students.
Hardware: IBM PC 256K, with maths co-processor
Software: FORTRAN compiled into EKE files
Models and algorithms: IES Lumen Method of Side lighting (sometimes called LOF method)
Modelling capabilities:
- uniform, overcast and clear sky, direct sunlight, ground reflectance
- rectangular room shape
- diffusing room surface
- glazing in the vertical plane
- clear or reflective glass
- simple shading devices (opaque overhang or fin)

Output features:
- numerical
- horizontal illuminance
- annual, monthly or hourly energy analysis
- graphic output: room sections, 2-D plots, 3-D plots, annual figures of daylight, total lighting use in kWHr, BTU, dollars, etc

References: 2, 3

DAYLITE
Developed by: Solarsoft, California, USA
Contact: Christine Ashton
Hardware: IBM PC compatible, APPLE MACINTOSH
Software: PASCAL
Modelling capabilities:
- overcast and clear sky, direct sunlight
- rectangular room shape
- diffusing room surface
- vertical, sloping or roof glazing
- clear or reflective glass
- simple shading devices (opaque overhang or fin)

Output features:
- numerical
- annual, monthly or hourly energy analysis
- 2-D and 3-D isolumen contours

ECAP
Developed by: Tennessee Valley Authority, Tennessee, USA
Contact: Jeff Jansen
Hardware: IBM PC and compatibles, 256 K, one disk drive
Software: Written in BASIC Interpreter
Modelling capabilities:
- standard overcast sky only
- simple overhangs only
- rectangular room shape
- any room surface finish
- rectangular windows in the vertical plane only
- any glazing type
- simple shading devices (overhang or fin)

Output features:
- numerical
- energy savings
- graphic output

FENESTRA
Developed by: School of Architecture, Victoria University, Wellington, New Zealand
Contact: Kit Cuttle
Description: A design procedure that includes evaluation of daylighting performance for determining shape, size and position of windows
Hardware: PC under MS-DOS or PC-DOS, 256K
Software: BASIC, 18K for storage, 12K for each data file
Modelling capabilities:
- uniform sky, with direct sunlight, ground reflectance
- diffusing obstructions with complex horizon
- rectangular room shape
- diffusing room surface
- vertical fenestration
- clear or reduced transmittance glazing
- simple shading device (opaque overhang)

Output features:
- numerical
- mean room surface illuminance
- perspective of window wall and horizon visible through windows from inside the room

References: 4, 5

LUMEN MICRO
Developed by: Lighting Technologies Inc, Colorado, USA
Contact: Thomas P. Swanson
Hardware: IBM PC XT/AT or compatible, 640K, EPSON dot-matrix printer.
Modelling capabilities:
- uniform, overcast and clear sky, direct sunlight, ground reflectance
- diffuse or specular obstructions
- rectangular room shape
- diffuse or specular room surface
- vertical or roof glazing
- clear or reflective glass

Output features:
- numerical
- horizontal and vertical illuminance
- luminance and exitance of room surface, RVP, VCP, ESI
- iso-contour shading, full colour perspective graphic images of modelled room

Reference: 6

MICRO-DOE
Developed by: Acrosoft International Inc, Colorado, USA
Contact: Gene Tsai
Description: Micro-DOE2 is an enhanced micro-computer version of DOE2.1C which performs energy use analysis for residential and commercial buildings. This program also includes analysis on daylighting and sunspace.
Intended users: Engineers and architects
Hardware: IBM PC XT / AT and compatibles
Models and algorithms: Hour-by-hour daylight illuminance and glare calculation.
Modelling capabilities:
- overcast and clear sky, direct sunlight, partly cloudy sky, ground reflectance
- diffusing obstructions
- rectangular room shape
- diffusing room surface
- vertical, sloping or roof glazing
- clear, reflective or diffusing glass
- simple shading devices (opaque overhang or fin)

Output features:
- numerical
- horizontal illuminance
- annual, monthly and hourly energy analysis
- no graphic output

References: 7, 8

MICROLITE
Developed by: Designer Software Exchange, Massachusetts Institute of Technology, Massachusetts, USA
Contact: Harvey J. Bryan
Hardware: IBM PC AT or compatible, 128K
Software: The program is written in C and has been compiled
Modelling capabilities:
- uniform, overcast and clear sky, direct sunlight, ground reflectance
- rectangular room shape
- diffusing room surface
- vertical fenestration only
- clear glass

Output features:
- numerical
- horizontal illuminance
- the user can utilise the AutoCAD graphic editor to create room configuration; isolumen graphics, axonometrics

References: 9, 10, 11, 12

QUICKLITE
Developed by: Lawrence Berkeley Laboratory, California, USA
Contact: Robert Clear
Hardware: TRS 80, TI 59
Software: BASIC, FORTRAN

Modelling capabilities:
- uniform, overcast and clear sky, ground reflectance
- diffusing obstructions
- rectangular room shape
- diffusing room surface
- vertical fenestration
- clear or reflective glass

Output features:
- numerical
- horizontal illuminance
- graphic output

References: 13, 14

SERILUX
Developed by: Ecotech Design Ltd, Sheffield, UK
Contact: Cedric Green
Hardware: IBM PC and compatibles; 640 K
Software: Used in conjunction with Scribe and SERI-RES
Modelling capabilities:
- daylight factors, ground reflectance, CIE
- horizontal or vertical obstructions
- any room shape
- any room surface
- any fenestration shape
- any glazing type

Output features:
- numerical output for use in SERI-RES thermal modelling system and daylight grid values
- daylight grid of rooms

References: 15, 16

SHADE
Developed by: Tennessee Valley Authority, Tennessee, USA
Contact: Bruce Reeves
Hardware: IBM PC and compatibles (64 K), one disk drive
Software: written in TURBO PASCAL
Modelling capabilities:
- hourly direct and diffuse sunlight for eight orientations for the 21st of each month
- five obstructions at any tilt and orientation
- vertical fenestration
- clear or reflective glass
- simple shading devices (overhang or fin)

Output features:
- numerical
- annual, monthly and hourly energy analysis
- graphic output

G.4 Daylighting in Architecture

MINI AND MAINFRAME COMPUTERS

CITYLIGHT
Developed by: J.J. Kim
Hardware: DEC or other mainframe
Software: FORTRAN
Modelling capabilities:
- uniform, overcast and clear sky, direct sunlight, ground reflection
- specular or diffusing obstructions

Output features:
- numerical
- illuminance on a tilted or horizontal plane
- graphic output

DOE
Developed by: Lawrence Berkeley Laboratory, California, USA
Contact: Fred Winkelmann
Hardware: CDC, IBM, DEC, ELXSI, CRAY, SUN
Software: Free from LBL
Modelling capabilities:
- overcast and clear sky, direct sunlight, partly cloudy sky, ground reflectance
- diffusing obstructions
- rectangular room shape
- diffusing room surfaces
- vertical, sloping or roof glazing
- clear, reflective or diffusing glass
- simple shading devices (opaque overhang or fin)

Output features:
- numerical
- horizontal illuminance
- annual, monthly and hourly energy analysis
- graphic output

References: 7, 8

GENELUX
Developed by: Laboratoire Sciences de l'Habitat, Ecole Nationale des Travaux Publics de l'Etat (LASH/ENTPE), Vaulx-en-Velin, France
Contact: Marc Fontoynont
Description: Computes illuminance on all surfaces inside a building for given outdoor conditions (clear and overcast sky, sun). Building configurations can be complex (up to 100,000 polygons) and various photometries can be simulated (diffuse, specular, spread or complex). Graphic output includes luminance and brightness computation.

It was developed to offer a powerful design tool in the design of daylighting systems. For this reason the priority was to simulate complex shapes and realistic photometries. The user friendly approach relies heavily on graphics. Graphic output in colours increases the facility of the analysis.

Intended users: Design firms and schools of architecture

Hardware: Apollo workstation with colour graphics
Software: PASCAL, FORTRAN, C; Apollo's GMR.3D (PHIGS) graphic subroutines; DIALOG's interface manager.
Models and algorithms: "Photons Generation" (similar to ray-tracing, but from source to surfaces). Monte Carlo method for diffuse and spread reflections and transmissions. Photon generation used also to simulate multiple reflections by the exterior environment. Photons are tracked with a octree algorithm.
Modelling capabilities:
- clear, overcast and uniform sky, sun
- any horizontal, vertical or tilted obstruction, specular or diffuse
- room shape: up to 100,000 polygons
- any fenestration from library of photometric data
- any glazing type from library of photometric data

Output features:
- numerical
- illuminance for each surface: MAX, MIN, AVERAGE and for each element
- energy analysis determines heat fluxes through aperture
- graphic output in 256 simultaneous colours, smoothing, multiple windows

Reference: 17

LUZDIA
Developed by: Laboratorio Nacional de Engenharia Civil (LNEC), Lisbon, Portugal
Contact: Licinio Carvalho
Description: The program permits detailed characterisation of daylighting conditions for a given room. The program is used only by LNEC
Hardware: VAX 8700
Software: FORTRAN
Models and algorithms: Illuminance resulting from external sources: double integral with spherical coordinates over the aperture. Internal reflected illuminances: sum of the contributions of many point sources, representing the internal surfaces.
Modelling capabilities:
- uniform, overcast and clear sky (Kittler and Gusev), direct sunlight, sky with luminance values defined by an array, reflected light from ground and buildings
- diffusing obstructions
- rectangular room shape
- diffusing room surface
- vertical fenestration (window or skylight)
- clear or reflective glass
- any opaque shading device

Output features:
- numerical
- total illuminance on an arbitrary plane; internal reflected and external illuminance (sky and sun);

vector-scalar illumination (magnitude and direction)
- luminance
- graphic output
- isolux contours

Reference: 18

NATUREL

Developed by: Centre Scientifique et Technique du Bâtiment (CSTB), Nantes, France

Contact: Michel Perraudeau

Description: Calculation of the interior illuminances (or luminances) for simple room shape, and visualisation of interior luminous environment.

Hardware: VAX 750 and LEXIDATA graphic processor

Software: FORTRAN and EGOS library

Models and algorithms: Radiosity method

Modelling capabilities:
- uniform, overcast and clear sky, direct sunlight, ground reflectance
- diffusing obstructions
- rectangular room shape
- diffusing room surface
- vertical or roof glazing
- clear or reflective glass

Output features:
- numerical
- illuminance on horizontal or any arbitrary plane
- luminance
- annual, monthly and hourly energy analysis
- isolux contours and visualisation of the interior of the room

RADIANCE

Developed by: Lawrence Berkeley Laboratory, California, USA

Description: RADIANCE is a collection of programs for the graphical simulation and analysis of lighting. It uses the technique of backwards ray-tracing to calculate scene illuminance and has the ability to account accurately for both diffuse and specular inter-reflection in complex spaces. Artificial light sources can be modelled as surfaces that have non-Lambertian distributions.

Intended users: Architects, engineers and researchers

Hardware: Workstations

Software: Free (copyright)

Models and algorithms Backward ray-tracing. The contribution of indirect diffuse light to the total illumination is calculated at various points in the scene and interpolated between these. The spacing of calculation points responds to the rate of change of irradiance. RADIANCE allows user-directed preprocessing. A first pass simulation determines the output distribution of window or opening, which is then used in the subsequent rendering. A hierarchical instancing of objects allows scenes with many millions of surface primitives to be modelled efficiently.

Modelling capabilities:
- complex room geometry
- diffusing or specular surfaces

Output features:
- numerical
- 3-D full colour perspectives

SUPERLITE

Developed by: Lawrence Berkeley Laboratory, California, USA

Contact: Michael Wilde

Description: SUPERLITE is a daylighting analysis program designed to predict illuminance in buildings. The program can compute workplane illumination, as well as room surface luminances and relative daylight factors. The program allows the modelling of any complex building geometry, but all complex surfaces must be broken into a set of flat elements. SUPERLITE also allows the modelling of exterior shading surfaces, such as fins and the effect of nearby buildings.

Intended users: Architects, engineers, lighting designers, researchers and educators

Hardware: CDC, 7600 DEC (370K), IBM 3033 with MTS operating system (mainframe computers)

Software: FORTRAN (program listings are available upon request).

Models and algorithms: The algorithms are a combination of Monte Carlo method and finite integration method for sky component calculation, and recursive iteration method for reflected component calculation.

Modelling capabilities:
- uniform and overcast sky, CIE clear sky, with and without direct sunlight, ground reflection
- maximum of two outdoor enclosures with up to 30 diffusing surfaces
- complex room geometry, up to 30 diffusing surfaces
- diffusing room surface
- vertical fenestration, sloped glazing and skylight
- clear or diffusing glass
- simple overhangs or fins (opaque and transparent)

Output features:
- numerical
- illuminance on workplane and interior surfaces
- luminance of interior surfaces
- graphic plotting file generated, no graphic program publicly available

References: 19, 20, 21, 22, 23

REFERENCES

(1) Gillette, G., Kusuda, T., *Journal of IES*, vol. 12, No. 2. pp. 78-85, January 1983.

(2) Milne, M.,Liggett, R., Campbell, C., Lin, L. and Tsay, Y., *Proc.1983 International Daylighting Conference*, Phoenix, Arizona, February 16-18, 1983.

(3) Ander, G., Milne, M. and Schiler, M., *Proc. 1986 National Daylighting Conference*, Long Beach, California, November 4-7, 1986.

(4) Leslie, S. F., and Trethovan, H. A., New Zealand *Energy Research and Development Committee*, Report No. 21, University of Auckland.

(5) Cuttle, K., and Baird, G., *Proc. 1986 International Daylighting Conference*, Long Beach, California, November 4-7, 1986.

(6) Di Laura, D.L., *Journal of IES,* vol.7, No.1, p. 2-14, October 1978.

(7) Winkelmann, F. and Selkowitz, S., *Energy and Buildings*, No. 8, 271, 1986.

(8) Winkelmann, F., *LBL Report*, Lawrence Berkeley Laboratory, California, May 1983.

(9) Bryan, H.J. and Clear, R.D., *Journal of IES*, vol. 10, No. 4, July 1981.

(10) Bryan, H.J. and Krinkel, D., *Proc. 7th National Passive Solar Conference*, Knoxville, Tennessee, August 30 - September 2, 1982.

(11) Bryan, H. and Krinkel, D., *Proc. Int. Daylight. Conf.*, Phoenix, Arizona, February 16-18, 1983.

(12) Bryan, H. and Fergle, R., *Proc. 1986 International Daylighting Conference*, Long Beach, California, November 4-7, 1986.

(13) Bryan, H. J. and Clear, R. D., *Journal of IES*, July 1981.

(14) Bryan, H. J., Clear, R. D., Rosen, J. and Selkowitz, S. *Solar Age*, August 1981.

(15) Haves, P., *Proc. UK-ISES Conference on Design Methods for Passive Solar Buildings*, London, October 1983.

(16) Green, C. and Haves, P., *Proc. PLEA Conference*, Pechs, Hungary, September 1987.

(17) Fontoynont, M., *ISES Conference*, Hamburg 1987.

(18) Carvalho, L. C., *Calculo automatico da iluminacao natural*, LNEC, 1984.

(19) Modest, M.F., *Energy and Buildings*, No.. 5, p. 69-79, 1982.

(20) Selkovitz, S., Kim, J.J., Navvab, M. and Winkelmann, F., *Proc. 7th National Passive Solar Conference*, Knoxville, Tennessee, August 30 - September 1, 1982.

(21) Modest, M.F., *Journal of IES,* July 1983.

(22) Windows and Daylighting Group, "SUPERLITE 1.0: Program Description and Summary", *LBL Report* DA-205, Lawrence Berkeley Laboratory, California, July 1985.

(23) Windows and Daylighting Group, *LBL Report* No. 19320, Berkeley, California, January 1985.

Appendix H
ARTIFICIAL SKIES

Table I - List of artificial skies in European countries

Organisation/Address	Contact	Type of Sky	Additional Equipment	Publication
DENMARK				
Copenhagen School of Architecture Royal Academy of Art Institute of Building Science Lighting Laboratory COPENHAGEN	Mr. Christensen	Mirror CIE Overcast 4.5m x 4.5m	Artificial Sun Photometric Equipment	Lighting Research & Technology, Vol. 8, No. 4, 1976
FRANCE				
CSTB Nantes Division Eclairage et Colorimetrie 11, Rue Henri Picherit 44300 NANTES	Mme. P. Chauvel 40-594255	Mirror CIE Overcast 4.5m x 4.5m	Photometric Equipment	
GERMANY				
RWTH Aachen Fakultät für Architektur Lehrstuhl für Baukonstruktion II Schinkelstraße 1 5100 AACHEN	Mrs. G. Willbold-Lohr 0241-803894	Mirror CIE Overcast 1.7m x 1.7m	Photometric Equipment	
TU Berlin Fakultät für Architektur Institut für Ausbau- und Innen- raumplanung Straße des 17. Juni 152 1000 BERLIN 12	Prof. H. Schreck 030-31423668	Dome CIE Overcast Artificial Sun	Photometric Equipment	
Fachhochschule Lippe, Abtlg. Detmold Fachbereich Architektur und Innenarchitektur 4930 DETMOLD	Prof. V. Schultz 05231-47009	Mirror CIE Overcast 3m x 3m	Photometric Equipment	db 1/1987: V. Schultz: Funktionen eines Licht- labors
Fachhochschule Köln Fachbereich Architektur Betzdorferstraße 2 5000 Köln 21	Mr. Rosenkranz 0221-82752812	Dome CIE Overcast 4.5 diameter	Artificial Sun Photometric Equipment	
ERCO Technisches Zentrum Postfach 2460 5880 LÜDENSCHEID	Mr. Hoffmann 02351-5510	Mirror CIE Overcast 5m x 5m	Photometric Equipment Mock-up room	Bauwelt No. 45, 25.11.88 Technisches Zentrum der ERCO Lüdenscheid Artificial Sun

H.2 Daylighting in Architecture

Organisation/Address	Contact	Type of Sky	Additional Equipment	Publication
TU München Fakultät für Architektur 8000 MÜNCHEN	Prof. F. Kurrent	Dome CIE Overcast	Artificial Sun	AIT 3/89 Licht 3-4/89
Fraunhofer-Institut für Bauphysik Nobelstraße 12 7000 STUTTGART	Mr. M. Szermann	Dome luminance distribution variable	Artificial Sun Photometric Equipment	

GREAT BRITAIN [1]

Organisation/Address	Contact	Type of Sky	Additional Equipment	Publication
School of Architecture and Building Engineering Claverton Down BATH, Avon BA2 7AY	Mr. M.A. Wilkinson 0225-826826	Mirror CIE Overcast 4m x 4m	Heliodon Photometric Equipment	
Bartlett School of Architecture Gower Street, Euston LONDON, N W 1	Mr. D. Loe 01-3877050	Mirror CIE Overcast 3m x 3m	Artificial Sun Photometric Equipment Heliodon	
Birmingham School of Architecture Birmingham Polytechnic Perry Barr BIRMINGHAM B42 2SU	Miss L.J. Heap 021-3365106	Mirror CIE Overcast 2.4m x 2.4m	Heliodon Illuminance Meters Daylight Factor Meters	
Department of Construction & Environmental Health, Bristol Polytechnic Ashley Down Road BRISTOL BS7 9BU	Mr. P.J. Innocent 0272-241241 ext. 2105	Mirror CIE Overcast 1.2m x 1.2m	Heliodon Photometers	
Brunel University Department of Mechanical Engineering UXBRIDGE	Mr. J. Smith 0895-74000 ext. 2880	Mirror CIE Overcast 2m x 2m	Heliodon Photometric instruments	
Building Research Station Garston WATFORD WD2 7JR Herts	Dr. P.J. Littlefair 0923-664874	Mirror CIE Overcast 4.5m x 4.5m	Heliodon Mock up room Photometric Equipment	
Research in Building Polytechnic of Central London 35 Marylebone Road LONDON NW1 5LS	Mr. C. Hancock 01-4865811	Mirror CIE Overcast 4m x 4m	Photometric instruments	
Essex Institute of Higher Education School of the Built Environment Victoria Road South CHELMSFORD CM1 1LL	Mr. I. Frame 0245-493131 ext. 256	Mirror CIE Overcast 3m x 3m	Heliodon Computer Program	
Heriot Watt University Department of Building RICCARTON Edinburgh EH14 4AS	Mr. R.S. Webb 031-4495111	Mirror CIE Overcast 1.7m x 1.7m	Heliodon	
Humberside College of Higher Education Hull School of Architecture Strand Close HULL HU2 9BT	Mr. J. Lynes 0482-25938	Fresnel CIE Overcast table top	Heliodon Photometric instruments	

Artificial Skies H.3

Organisation/Address	Contact	Type of Sky	Additional Equipment	Publication
School of the Built Environment Leicester Polytechnic P.O. Box 143	Dr. K. Lomas 0533-551551 ext. 2512	Mirror Uniform 2.4m x 2.4m	Heliodon Solarscope Photometric instruments	
School of the Built Environment Liverpool Polytechnic 98 Mount Pleasant LIVERPOOL L3 5UZ	Mr. N.S. Sturrock 051-2073581 ext. 2708/9	Mirror CIE Overcast 1.5m x 1.5m	Heliodon Photometric instruments	
The Martin Centre Department of Architecture University of Cambridge CAMBRIDGE CB2 2EB	N V Baker 0223 332981	Mirror CIE Overcast 2.4m x 2.4m	Heliodon Photometric instruments	
School of Architecture and Building Engineering University of Liverpool P.O. Box 147 LIVERPOOL L69 3BX	Dr. D.J. Carter 051-7942000	Mirror CIE Overcast 2.5m x 2.5m	Photometric Equipments Motorised Heliodon	
School of Architecture University of Mancester Oxford Road MANCHESTER M20 OGP	Mr. W. Burt 061-2756934/5	Mirror CIE Overcast 2m x 2m	Solarscope	
Faculty of Art & Design Manchester Polytechnic Loxford Tower Lower Chatham Street MANCHESTER M15 6HA	Dr. G. McKennan 061-2286171 ext. 2876	Mirror Unknown 1.2m x 1.2m	Photometric Equipment	
Building Science Section School of Architecture University of Newcastle-upon-Tyne NEWCASTLE NE1 7RU	Prof. T.J. Wiltshire 091-2226007	Mirror CIE Overcast 3m x 3m	Heliodon Photometric instruments	
Department of Architecture University of Nottingham University Park NOTTINGHAM NG2 7RD	Dr. P. Tregenza 0602-484848	Dome CIE Overcast 3.3m diameter	Heliodon Photometric instruments	
Low Energy Architecture Research Unit Department of Architecture & Interior Design, Polytechnic of North London Holloway Road LONDON N7 8DB	Mr. M. Wilson 01-6072789 ext. 2178	Dome 3.6m diameter Mirror CIE Overcast 2.4m x 2.4m	Heliodon Photometric instruments	
Department of Building Science University of Sheffield SHEFFIELD S10 2TN	Dr. S. Sharples 0742-768555 ext. 4713	Mirror CIE Overcast 2m x 2m	Heliodon Photometric instruments	
Department of Construction and Management Polytechnic of Southbank Wandsworth Road LONDON SW8 2JZ	B Worsfold	Mirror CIE Overcast 2.4m x 2.4m	Photometric instruments	
Institute of Environmental Technology Polytechnic of South Bank Borough Road LONDON SE1 0AA	Dr. J. Frost 01-9288989	Mirror CIE Overcast 1.6m x 1.6m	Photometric instruments	

H.4 Daylighting in Architecture

Organisation/Address	Contact	Type of Sky	Additional Equipment	Publication
Department of Building University of Ulster Jordanstown BELFAST	Dr. G.L. McCullghen 0232-365131	Mirror CIE Overcast 4.5m x 4.5m	Photometric Equipment	
Department of Building Engineering UMIST P.O. Box 88 MANCHESTER M60 1QD	Dr. K.M. Letherman 061-2363311	Mirror 1m x 1m	Heliodon	
Welsh School of Architecture UWCC P.O. Box 25 CARDIFF CF1 3XE	Mr. C.M. Parry 0222-874000 ext. 5961	Mirror CIE Overcast 1.5m x 1.5m	Heliodon Photometric Equipment	

NETHERLANDS

Organisation/Address	Contact	Type of Sky	Additional Equipment	Publication
Technische Universiteit Delft Faculty of Architecture and Town Planning, Berlageweg 1 2628 DELFT	Prof. A.J. Hansen Mrs. C. van Santen	Mirror CIE Overcast 3.8m x 3.8m	Photometric Sun Simulator	BOUW no. 14/15, 5/19-7-1980 T. Haartsen: Daglichtkamer een uitstekend hulphiddel bij het ontwerpen ELECTRICAL DESIGN Oct 1986 A.J. Hansen: Will the sun shine in?

[1] Information on Artificial Skies in the UK provided by the Building Research Station, Garston, (Dr. Paul Littlefair)

Table II - List of special artificial skies in other countries

Organisation/Address	Contact	Type of Sky	Additional Equipment	Publication
AUSTRIA				
Lichtplanung Christian Bartenbach Rinnerstraße 14 A-6071 ALDRANZ/Innsbruck	Mr. C. Bartenbach (43) 5222-615910 distribution	Dome luminance Equipment variable 9m diameter	Artificial Sun Photometric	
CZECHOSLOVAKIA				
Institute of Construction and Architecture Dept. of Lighting Slovak Academy of Sciences Dubravska cesta 9 842 20 BRATISLAVA	Dr. R. Kittler 427-3782892	Dome CIE Overcast Uniform Clear Skies 8m diameter	Sun Simulator Photometric Equipment	Lighting Research & Technology, Vol.6, No.4, 1974, p. 227 ff
USA				
Lawrence Berkeley Laboratory Window and Daylighting Group BERKELEY, CA 94720	Mr. S. Selkowitz 415-486-4000	Dome Uniform CIE Overcast Clear Skies	Sun Simulator Photometric Equipment	Proc. of the 6. Passive Solar Conf., Portland, Oregon, 1981
USSR				
NIISF Gosstroy of the USSR Pushinskaya Str. 26 MOSCOW 103828	Dr. A. Spiridonov	Dome Uniform CIE Overcast Clear Skies 21m diameter	Sun Simulator Photometric Equipment	

INDEX

Aalto, A., 11.8
absorptance, GL.24
absorption, GL.24
active division, 5.20
actuators
    electric lighting, 7.5, 10.7, E.2-7
    facade, 7.7, 10.7, E.8-10
adaptation, GL.16
aerogels, 4.16
age and vision, 2.4-5
airports, 11.10
arcades, 1.5
architects
    Aalto, A., 11.8
    Ahrends, Burton and Koralek, 1.18
    Albini, F., 1.8
    Arup Associates, 1.16
    Barbeyer, P., 1.12
    Cullinan, E., 1.11
    Foster, Sir Norman, 11.10
    Galí, Quintana and Solanas, 11.12
    Gasson, B., 1.8
    Hertzberger, H., 1.16
    Jourda and Perraudin, 11.11
    Kahn, L. I., 11.15
    Krohn & Hartvig Rasmussen, 11.9
    Le Corbusier, 1.1
    Los and Pulitzer, 11.13
    Scarpa, C., 1.8
    SETOM, 1.17
    Soane, Sir John, 1.6, 1.12-13
    Stirling and Wilford, 1.8
    Tombazis, A. N., 1.18
    Unwin, R., 1.10
    Wright, F. L., 1.6
architectural design
    generative analysis, 11.1
    method, 11.2
    non-intuitive design, 11.2
    pattern language, 11.2
    precedents, 11.1
    process, 11.1, 11.3
    proposed method, 11.2
    shape grammar, 11.2
    traditional methods, 11.1
    typological grammar, 11.1-2
art galleries, 1.6, 1.8, 1.13
    Conservation Movement, 1.7
    museum fatigue, 1.8
    "strong box" aesthetic, 1.8
    "temple" and "store", 1.8
artificial light, 1.1, 6.1 - 8
    colour, 6.5, 6.8, 6.10, 6.11
    controls, 7.1, 7.12, E.1-7
        energy savings, 7.12, 10.10-12
    cost per lumen, 1.3
    for plants, 1.15
    gas lighting, 1.5
    strategies, 6.1
artificial skies, D.3, H.1
assembly halls, 11.15
atria, 1.15-17, 5.11, 5.34-49
awning, 5.20

BRE daylight factor protractors, F.6
BRS
    daylight calculator, F.6
    daylight protractors, 1.14, F.6
baffle, 5.23
balcony window, 5.13
ballast, types of, 6.5
BELYS program, G.1
blue sky luminance distribution, 3.5, 3.6
brise-soleil, 5.25
Bull Nomogram, F.5

Candela, GL.15
case study buildings, 11.3-4, 11.8-15
chapels, 1.1, 1.2
chromatic and achromatic information, 2.3
    CIE 1974 index, GL.18
churches, 1.2
CIE standard sources, GL.19
CITYLIGHT program, G.4
clerestory, 1.7, 5.14
climate-adapted building (selective), 11.4
climate-rejecting building (exclusive), 1.8, 11.4
cloud, 2.9-10
    amount, GL.32
    ratio, GL.32
cognitive process, 2.1
colleges, 1.3
colour
    artificial light, 6.5, 6.8, 6.10, 6.11
    discrimination, 2.3

Munsell system, 2.6
perception, 2.6
pigment, 2.6
rendering, 2.7, 2.15, GL.18
   index, 2.7
shift after reflection, 4.8
shift after transmission, 4.8
temperature, 2.6, 4.6, GL.20
colourimetric systems
   CIE 1931 standard, GL.19
   CIE 1964 supplementary standard, GL.19
comfort
   luminous, 2.1, 2.5, 2.14
      complaints about, 2.5
      visual display units, 2.5
   thermal, 1.10, 2.1
   visual, 2.1, 2.14
      requirements, 2.12-18
components
   control elements, 5.3-6
   experimental analysis, 5.33
   fields of application, 5.26
   for schools, 5.27, 5.29, 5.30, 5.31-32
   for small buildings, 5.30-32
   offices, 5.28-30
conservatories, 1.5
contrast, GL.17
   detection, 2.4
   sensitivity, 2.14
control elements, 5.3-6, 5.19-26
   design recommendations, 5.30-32
   flexible screens, 5.3-5
   rigid screens, 5.3-5
   separator screens, 5.3-5
   solar filters, 5.3, 5.6
   solar obstructors, 5.3, 5.6
control gear for lamps, 6.11
CONTROLITE program, G.1
Cornell formula, 2.17, F.11
courtyard, 5.10
CSTB
   abacus (nomogram), F.5
   diagram, F.5
curtain, 5.21
curtain wall glazing, 5.14

D

Daylight, GL.31
   and artificial light, 6.1, 7.1-3
   availability, B.1
   coefficient, 8.1, 10.2
   data, 3.1-13
      luminance distribution
         blue sky, 3.5, 3.6
         intermediate sky, 3.5
         overcast sky, 3.1, 3.4, 3.5
   factor, 1.10, 1.13-15, 8.1, 9.3, 10.2-3, D.2-4
      dynamic, D.2
      externally reflected component, 1.14, GL.33

      internally reflected component, 1.14, GL.34
      meter, F.3
      sky component, 1.14, GL.33
      tables, F.4,
   Glare Index, 2.17
   glare perception scale, 2.18
   locus, GL.20
   overcast skies, 2.9-10
   protractors, 1.14, F.6
DAYLIT program, G.1-2
DAYLITE program, G.2
densitometer, GL.27
design checklist, 5.64
design choices, 2.10-11
design tools, 1.12, 1.12, 1.17, 9.1, F.1
   characteristics, 9.4
   comparison, 9.8-9
   dot diagrams, 9.7, F.8
   equations, 9.7, F.1
   evaluation in use, 9.6
   glare control, 9.8, F.10-11
   LT Method, 1.17, 10.19-21
   lumen methods, 9.7, F.3
   nomograms, 9.7, F.4-6
   protractors, 9.7, F.6-7
   single stage, 9.7, F.2-3
   tables, 9.7, F.4
   urban analysis, 9.7-8, F.10
   validation, 9.8-9
   Waldram diagram, 1.14, 9.6-7, F.9
diffraction, 5.33
   holographic optical elements, 5.60
diffusion, GL.21
   factor, GL.21
   indicatrix, GL.25
DIM programs, 8.12
dimming, 7.2, 10.11
dispersion, GL.25,
   angle, GL.25
DOE program, G.4
dome, 5.17
dot diagrams, 9.7, F.8

E

ECAP program, G.2
educational buildings
   case study, 11.11
   schools, 1.9-12, 5.26, 5.27, 5.29-32
   see also schools
electric lighting, 6.1-18
electrochromic devices, 4.16
energy
   analyses
      ESP, 10.1-12
      HEATLUX, 10.13-18
      LT Method, 1.17, 10.19-21
   conservation, 1.12
   consumption
      control strategies, 7.1-3, 10.10-12

offices, 1.17
primary, 1.17
simplified calculation method, 10.19
management, 7.7-14
systems
continuous control, 7.1, 7.6
mixed control, 7.6
step regulation, 7.5
environmental issues
CO_2, 1.17
global, 1.12, 1.18
pollution, 1.2, 1.7, 1.9
ESP program, 10.1-12
equations, 9.7, F.1
European Concerted Action, 9.9
European research and development, 1.16
externally reflected component, 1.14, GL.33
eye
accommodation, 2.3
adaptation, 2.3
neural components, 2.3
non-visual aspects, 2.5-6
optic components, 2.3

Farrell Method, F.8
FENESTRA program, G.2
fibre optics, 4.17
fin, 5.23
fire, 1.2
internal pollution, 1.2
radiant environment, 1.2
fluorescent concentrators, 4.17
Foster Associates, 11.10
Frühling's Formula, F.3

Gallery, 5.9
Galleries
see arcades or art galleries
GENELUX program, G.4
glare, 2.15-18, GL.17
by reflection, 2.15, GL.17
constant, 2.16
control, 6.13-14, 9.8, F.10-11
criteria, 2.17
direct, 2.15, GL.17
discomfort, 2.16, GL.17
from windows, 1.10, 2.17-18
Index (GI), 2.16, 10.7
daylight (DGI), 2.17
limiting values of, 2.13, 2.17
glass, 1.1
cylindrical, 1.4
float, 1.4
for horticulture, 1.4
leaded panes, 1.3
plate, 1.4
stained, 1.2

tinted, 1.15
glazing index, GL.33
glossmeter, GL.27
goniophotometer, GL.26
greenhouse, 5.10

Health, 2.5-6
HEATLUX program, 10.13-18
Higbie's Formula, F.1
Higbie - Levin Formula, F.1
holographic films, 4.15-16, 5.19
advantages, 5.63
materials, 5.60-61
scale model tests, 5.61-62
holographic optical elements, 5.60-63
hospitals, 1.18

Illuminance, 1.15, GL.14
global, GL.31
in art galleries, 1.8
meter, GL.26
recommended levels, 2.13
service, GL.14
uniformity ratio of, GL.30
vector, GL.29
illuminant, GL.19
reference, GL.18
illumination
skylight, 8.2-3
sunlight, 8.2
Integrated Environment Design (IED), 1.11
integrated sphere approximation, 8.14
integrating, GL.26
internally reflected component, 1.14, GL.34
interreflection, 8.8-9, GL.30
formula, F.1
nomograms, F.4
irradiance, GL.13
cylindrical, GL.14
spherical, GL.14
irradiation, GL.13
iso-lux contours, D.4

Kahn, L. I., 11.15
Kittler's
and Ondrejicka Formula, F.1
nomogram, F.5
semi-logarithmetric chart, F.5

Laboratories, 11.13
Lam, W. M. C., 11.2, 11.14
Lambert's Law, GL.22
Lambertian surface, GL.22
lamps, 6.1
choosing, 6.1

compact fluorescent, 6.6-8
high pressure discharge, 6.7, 6.9-10
metal halide, 6.9-10
sodium, 6.9-11
incandescent, 6.1-2
low voltage tungsten halogen, 6.3-4
mains voltage tungsten halogen, 6.3
reflector, 6.2-3
tubular fluorescent, 6.4-6
lantern rooflight, 1.7, 5.17
Le Corbusier, 1.1
Libby Owens Ford Method, 10.2
libraries, 11.8
case study building, 11.12
light
borrowed, 1.9
damage, 1.7
loss factor (maintenance factor), GL.30
measuring instruments, C.1
errors, C.1
switching, 7.1-3
probability, 10.6, 10.12
well, 1.13
light-duct, 5.11
lighting
control systems, 7.1, E.1-7
energy savings from, 7.12, 10.10-12
ideal linear dimming, 10.11
occupant switching, 10.12
people detectors, E.10
two increment step-down switching, 10.11
diffused, GL.29
direct, GL.28
directional, GL.29
indirect, GL.28
lightness, GL.16
lightshelf, 1.11, 1.17-18, 5.22, 5.50-56
Linke turbidity factor, GL.31
louvre, 5.24
LT Method, 1.17, 10.19-21
Lui-Jordan procedure, 10.15
lumen, GL.15
method, 8.14, 9.5, 9.7, 10.2, F.3
LUMEN MICRO program, G.2
luminaires, 6.11
choosing, 6.1
classification, 6.12-13
electrical functions, 6.14
electrical safety, 6.15
glare control, 6.13-14
mechanical functions, 6.11-13
mounting systems, 6.16
electrical, 6.16-17
mechanical, 6.16
optical functions, 6.13
luminance, GL.13
acceptable levels, 2.15
coefficient, GL.24
factor, GL.24

from energetic data, 3.9
ground, 2.11
meter, C.1-17, GL.26
ratios, 2.14
sky distribution, 3.4-6
luminosity, GL.16
luminous efficacy, 1.13, 1.15, GL.15
daylight, 3.2-3
overcast sky, 10.4
sunlight, 3.4, 10.4, 10.16
luminous flux, GL.12
luminous intensity, GL.12
distribution curve, 4.1
lux, GL.15
LUZDIA program, G.4

Maintenance
factor (light loss factor), GL.30
of glazing, 5.64
materials
classification, 4.2-15
complex prismatic, 4.3-4, 4.12, 4.15
diffuse, 4.3-4, 4.10, 4.14
diffuse and specular, 4.3-4, 4.11, 4.14
experimental set-up for classification, 4.6
scatter narrow, 4.3-4, 4.11, 4.14
scatter wide, 4.3-4, 4.10
selection 4.7
specular, 4.3-4, 4.13
measuring instruments, C.1
medium
opaque, GL.25
translucent, 4.10-12, GL.25
transparent, GL.25
membrane, 5.18
MICRO-DOE program, G.3
MICROLITE program, G.3
monitor roof, 5.15
Monte Carlo Method, 8.9
morphological box, 11.4-5
museums
see art galleries

Naturel program, G.5
NATLIT algorithm, 10.5
NBSLD code, 10.13
nebulosity index, 10.4, GL.32
nomograms, 9.7, F.4-6
north-light roof, 5.15

Occupant interaction, 10.6
offices, 1.6, 1.16, 5.26, 5.28-33
case studies, 11.9, 11.14
illumination conditions, 5.29
typologies, 5.28
optical

division, 5.19
radiation spectrum, 2.6
overcast skies, 2.9-10
    CIE standard, 2.7, 3.1, 8.1, 8.3, GL.32
    luminance distribution, 3.1, 3.4, 3.5
    luminous efficacy, 10.4
overhang, 5.21

P
Page, J. K., 2.18
passive zone, 10.20
photometer, GL.26
    see goniophotometer
photometry, 4.1, GL.26
photosensor control, 10.5-6
physical scale models, 9.1-3, D.1-5
Planckian locus, GL.20
Pleijel's diagram, 1.14, F.8
pollution, 1.2, 1.7, 1.9
    CO_2, 1.17
    global, 1.12, 1.18
    internal pollution, 1.2
porch, 5.9
position factor, 10.7-8
prismatic devices, 4.15
    advantages and disadvantages, 5.58
    division, 5.20
    Luxfer prisms, 5.57-8
    sidelighting, 5.59
    systems, 5.57-59
probability switching function, 10.6
PSALI, GL.28
Pugno's method, F.1

Q
QUICKLITE program, G.3

R
Radiance, GL.13
    coefficient, GL.23
    factor, GL.23
RADIANCE program, G.5
radiant flux, GL.12
radiation spectrum, 2.6
ray tracing
    backward, 8.11
    forward, 8.10-11
ray tracking, 8.11
reflectance, GL.22
    diffuse, GL.23
    Fresnel, 8.6
    regular, GL.23
reflection, GL.21
    bi-directional, 8.4
    diffuse, GL.22
    isotropic diffuse, 8.5
    mixed, 8.5-6, GL.22
    operator, 8.8
    specular, 8.5, GL.21

reflectometer, GL.26
Rivero's tables, F.4
room index, GL.29

S
Scale models, 9.1-3
    artificial sky, D.3
    as design tools, 9.1
    dynamic daylighting, D.2
    materials, D.1
    photographic documentation, D.4
    scaling, 9.1-3, D.1
    shadow patterns, D.3
    sky conditions, D.3
    sunlight, D.3
    testing
        photographic, D.4
        photometric, D.4
        visual, D.3-4
Scarpa, C., 1.8
schools, 1.9-12, 5.26, 5.27, 5.29-32
    classrooms, 1.9, 5.27
        height-to-depth ratio, 1.10
    daylighting components, 5.29
    illumination conditions, 5.29
    refurbishment, 1.11
    Robson, E.R., 1.9-10
    rules of thumb, 1.9
    system-built, 1.10
    typologies, 5.27
    UK Education Act 1870, 1.9
seasonal affective disorder (SAD), 1.18, 2.6
sensors
    and simulation, 10.5-6
    lighting, 7.4-5
    people detectors, 7.5
    temperature and humidity, 7.5
SERI daylight protractor, F.7
SERILUX program, G.3
SHADE program, G.3
shading
    control systems, E.8-9
    devices, 1.11
shadows and sky type, 2.9
shadows, shape from, 2.14-15
shelter, 1.1
shutter, 5.25
sick building syndrome (SBS), 1.18
sill, 5.22
skies, artificial, H.1
sky component, 1.14, GL.33
sky types
    CIE standard clear, 2.8, 3.2, GL.32
    CIE standard overcast, 2.7, 3.1, 8.1, 8.3, GL.32
    Moon and Spencer overcast, 3.1, 8.3
    real, 2.8, 3.2,
        classification, 3.4
        Gillette and Treado, 3.2
        Littlefair, 3.2

　　　　Nakamura and Oki, 3.2
　　　　Pierpoint, 3.2
　　　　Tregenza, 3.2, 8.1
　　　　Winkelman and Selkowitz, 3.2
　　types, 3.4
　　　　probability of occurrence, 3.9, A.1
　　uniform, 2.8, 3.1
skylight, 5.16
Soane, Sir John, 1.6, 1.12-13
solar filters, 5.3, 5.6
　　solar radiation, GL.30
　　　　diffuse, GL.30
　　　　direct, GL.30
　　　　global, GL.31
　　　　reflected, GL.31
solar obstructors, 5.3, 5.6
spectral luminous efficiency, GL.11
Spencer's and Stakutis' Integral Equation, F.2
split-flux method, 1.15, 8.13, 10.16, F.1
starters, 7.6-7
sun-duct, 5.12
sunlight, 8.2
　　factor, GL.33
　　　　externally reflected component, GL.33
　　　　internally reflected component, GL.34
sunshine duration, GL.32
　　relative, GL.32
　　relative daily, GL.32
SUPERLITE program, 10.5, G.5

Thermochromic devices, 4.17
thermochromic films, 5.19
translucent surface, 5.18
transmission, 8.4, GL.21
　　bi-directional, 8.6
　　diffuse, GL.22
　　direct, 8.7, GL.21
　　isotropic diffuse, 8.7
　　mixed, 8.7, GL.22
transmittance, GL.23
　　diffuse, GL.23,
　　multi-layer fenestration, 8.7
　　regular, GL.23,
turbidity factor, GL.31

Turner-Szymanowsky protractor, F.7

Uniform sky, 2.8, 3.1

Veiling reflections, 2.15, GL.17
ventilation, 1.2
verandah, 1.10
Vermeer, J. 1.3
view factors, 8.9-10
vision and age, 2.4-5
visual
　　acuity, 2.4
　　　　and illumination, 2.4-5
　　comfort, 1.9, 2.1
　　display unit (VDU), 2.5, 5.50
　　field, 2.4
　　noise, 2.18
　　perception, 2.1
　　testing, D.3-4
Vitruvius, 1.13

Waldram diagram, 1.14, 9.6-7, F.9
window, 1.2, 5.7, 5.12
　　controls, 5.8
　　fenestration, 5.7
　　oriel, 1.3
　　orientation, 5.8
　　position, 5.8
　　shape, 5.7-8
　　shutters, 1.2
　　size, 5.7
　　tax, 1.4
　　type, 5.7
　　unglazed, 1.2
　　ventilation, 1.2
working plane, GL.29
Wright, F. L., 1.6

Z-factor, 9.5
Zwicky, F. and Wilson, A. G., 11.2, 11.3